FIELD MARGINS: INTEGRATING AGRICULTURE AND CONSERVATION

o 58

FIELD MARGINS: INTEGRATING AGRICULTURE AND CONSERVATION

BCPC Monograph No 58

Proceedings of a symposium organised by the British Crop Protection Council in association with the British Ecological Society and the Association of Applied Biologists and held at the University of Warwick, Coventry on 18-20 April 1994

Edited by: Nigel Boatman

BCPC Registered Office:
49 Downing Street
Farnham
Surrey GU9 7PH, UK.

49 Downing Street
Farnham, Surrey GU9 7PH

British Library Cataloguing in Publication Data

British Crop Protection Council
Field Margins: Integrating Agriculture and Conservation
(Monograph Series, ISSN 0306-3941; No 58)
I Boatman, Nigel
II Series 630.2

ISBN 0-948404-75-2

Cover design by Major Design & Production Ltd, Nottingham
Printed in Great Britain by Major Print Ltd, Nottingham

Contents

Page

Posters

SESSION 5:
POLITICAL ASPECTS

Posters

Preface

Field margins mean different things to different people. To the livestock farmer, the field margin is a barrier to prevent stock wandering, which may also provide some shelter. Field margins have often been viewed by arable farmers as a source of weeds, pests and diseases, best removed or at least kept in check by annual applications of herbicide. Nowadays, attitudes are changing and the benefits of marginal habitats as reservoirs of beneficial invertebrates, predators of pest species or crop pollinators, are becoming more widely appreciated. To the general public, the more obvious field boundary structures such as hedges, shelter belts and stone walls are major elements in the landscape, defining features of the countryside we have come to regard as "traditional". To the landscape ecologist, field margins are corridors forming a network through which organisms can move between larger habitat patches. To the conservationist, field margins may represent the last haven for some types of wildlife in an otherwise hostile environment created by intensive modern farming, whilst to the agricultural historian boundary structures can give a clue to the practices of a former age.

The role and management of field margins in agriculture has changed in recent years. Many formerly mixed farms have become entirely arable, and hedges or other field boundaries have lost their previous purpose. The resulting large scale removal of hedges in some areas has caused widespread public concern. At the same time, increased labour costs have led to a decline in the practice of traditional labour-intensive maintenance techniques such as hedge-laying and dry stone-walling, causing the gradual dereliction of many remaining field boundaries. This has been accelerated in stock rearing areas by increased stocking rates, resulting in intensive grazing pressure on hedge bases. Such hedges may eventually lose their effectiveness as stock-proof barriers, and be replaced by post and wire fences. Increased use of inorganic fertilisers and pesticides on crops may have effects on the fauna and flora of field edges via drift, surface run off or leaching into drainage ditches.

The changing status of field boundaries in agriculture has coincided with an increased awareness amongst the wider public of the conservation potential of field margins in the widest sense. Commodity surpluses have shifted agricultural support policy away from productivity orientated incentives to production stabilising mechanisms, with environmental benefits becoming an increasingly prominent factor in policy and spending decisions at national and European level. There is now a wide range of grants and incentives available to farmers to establish, manage or maintain field margins to provide wildlife, landscape and public amenity benefits. Over the last decade, research into all aspects of field margin ecology and management has expanded considerably, providing a sound scientific basis to underpin policy decisions. This symposium provides an international synthesis of this research set in its wider social, political and economic context.

N D Boatman

Symposium Programme Committee

Chairman:	Dr N D Boatman	The Game Conservancy Trust/ Allerton Research and Educational Trust, Loddington House, Loddington, East Norton, Leicestershire LE7 9XE

Session Organisers and Chairmen

Session 1		
Organiser and Chairman:	Dr E J P Marshall	Department of Agricultural Sciences, University of Bristol, Institute of Arable Crops Research, Long Ashton Research Station, Britsol BS18 9AF
Session 2		
Organiser and Chairman:	Dr N W Sotherton	The Game Conservancy Trust, Fordingbridge, Hampshire SP6 1EF
Session 3		
Organiser and Chairman:	Dr R J Froud-Williams	Department of Agricultural Botany, University of Reading, 2 Earley Gate, Whiteknights, PO Box 239, Reading RG6 2AU
Organiser and Chairman:	Dr H M Carnegie	Land Resources Department, Scottish Agricultural College, 581 King Street, Aberdeen AB9 1UD
Session 4		
Organiser and Chairman:	Dr T A Watt	Wye College, University of London, Wye, Ashford, Kent TN25 5AH
Organiser:	Dr H Smith	Wildlife Conservation Research Unit, Department of Zoology, University of Oxford, South Parks Road, Oxford OX1 3PS
Session 5		
Organiser and Chairman:	Dr S Webster	Department of the Environment, Room 917b, Tollgate House, Houlton Street, Bristol BS2 9DJ

Poster Session Organiser:	Mr M A Lainsbury	BASF plc, Lady Lane, Hadleigh, Ipswich, Suffolk IP7 6BQ
Field Visit Organiser:	Dr N D Boatman	The Game Conservancy Trust/ Allerton Research and Educational Trust, Loddington House, Loddington, East Norton, Leicestershire LE7 9XE
Field Visit Organiser:	Dr R Feber	Wildlife Conservation Research Unit, Department of Zoology, University of Oxford, South Parks Road, Oxford OX1 3PS

Acknowledgements

The organisers are grateful for donations received from the following organisations:

AgrEvo UK Crop Protection Ltd
BASF plc Agricultural Division
Monsanto Agricultural Group
Willmot Pertwee Ltd
Zeneca Crop Protection

Field Margin Terminology

(Adapted from Greaves & Marshall, 1987)

Field boundary

Hedge, grass bank, fence, wall, plus hedge bank if present with its herbaceous vegetation, plus ditch or drain if present.

Boundary Strip

Area of ground between boundary and crop. It may include a farm track, a grass strip, an unsown cultivated strip with naturally regenerated flora and/or a "sterile strip" of bare ground, maintained by cultivation or herbicide.

Crop margin (headland)

The outer part of the crop itself, usually considered as the area between the edge of the crop and the first tramline (tractor wheeling). The term "headland" is often used to describe this region, though strictly speaking this refers to the turning area used by agricultural machinery, and therefore only applies to two sides of a field. The crop margin is often managed differently in certain ways from the rest of the field.

Reference

Greaves, M P; Marshall, E J P (1987). Field margins: definitions and statistics. In: *Field Margins*, J M Way and P W Greig-Smith (Eds), *BCPC Monograph No 35*, Thornton Heath: BCPC Publications, pp. 85-94.

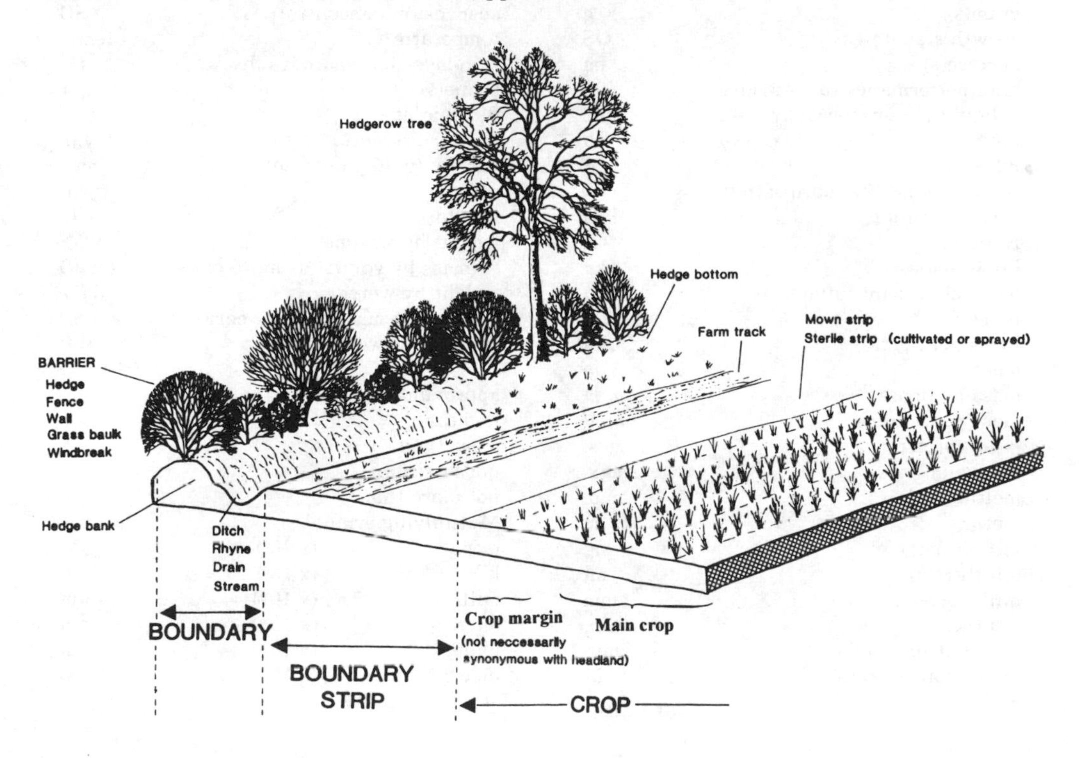

Abbreviations

acid equivalent	a.e.
active ingredient	AI
boiling point	b.p.
British Standards Institution	BSI
centimetre(s)	cm
concentration x time product	ct
concentration required to kill 50% of test organisms	LC50
correlation coefficient	*r*
cultivar	cv.
cultivars	cvs.
day(s)	d
days after treatment	DAT
degrees Celsius (centigrade)	°C
dose required to kill 50% of test organisums	LD50
dry matter	d.m.
Edition	Edn
Editor	Ed
Editors	Eds
emulsifiable concentrate	EC
freezing point	f.p.
gas chromatography-mass spectrometry	gcms
gas-liquid chromatography	glc
gram(s)	g
growth stage	GS
hectare(s)	ha
high performance (or pressure) liquid chromatography	hplc
hour	h
infrared	i.r.
International Standardisation Organisation	ISO
Kelvin	K
kilogram(s)	kg
least significant difference	LSD
litre(s)	Litre
litres per hectare	l/ha
mass	*m*
mass per mass	*m/m*
mass per volume	*m/V*
mass spectrometry	m.s.
maximum	max.
melting point	m.p.
metre(s)	m
milligram(s)	mg
millilitre(s)	ml
millimetre(s)	mm
minimum	min.
minute (time unit)	min
molar concentration	M
nuclear magnetic resonance	nmr
number average diameter	n.a.d.
number median diameter	n.m.d.
organic matter	o.m.
page	p.
pages	pp.
parts per million by volume	mg/l
parts per million by weight	mg/kg
pascal	Pa
percentage	%
post-emergence	post-em.
power take off	p.t.o.
pre-emergence	pre-em.
probability (statistical)	*P*
relative humidity	r.h.
revolutions per minute	rev./min
second (time unit)	s
standard error	SE
standard error of means	SEM
soluble powder	SP
species (singular)	sp.
species (plural)	spp.
square metre	m^2
subspecies	ssp.
surface mean diameter	s.m.d.
suspension concentrate	SC
temperature	temp.
thin-layer chromatography	tlc
tonne(s)	t
ultraviolet	u.v.
vapour pressure	v.p.
variety (wild plant use)	var.
volume	*V*
weight	*W*
weight by volume (mass by volume is more correct)	*W/V* (m/*V*)
weight by weight (mass by mass is more correct)	*W/W* *(m/m)*
wettable powder	WP

approximately	c.
less than	<
more than	>
not less than	<
not more than	>

Multiplying symbols-		Prefixes
mega	(x 10^6)	M
kilo	(x 10^3)	k
milli	(x 10^{-3})	m
micro	(x 10^{-6})	µ
nano	(x 10^{-9})	n
pico	(x 10^{-12})	p

Session 1

The Role of Field Margins in the Landscape

Session Organiser & Chairman DR E J P MARSHALL

FIELD MARGINS - AN HISTORICAL PERSPECTIVE

JOHN CHAPMAN

Department of Geography, University of Portsmouth,
Buckingham Building, Lion Terrace, Portsmouth PO1 3HE

JOHN SHEAIL

Institute of Terrestrial Ecology (Natural Environment Research Council), Monks Wood, Huntingdon, Cambridgeshire PE17 2LS

INTRODUCTION

According to The Times Countryside Correspondent, there is no more potent symbol of rural England than its hedgerows, the green sinews of the countryside that Wordsworth called 'little lines of sportive wood run wild' (The Times, 17 August 1993). And yet, according to a survey recently undertaken by the NERC Institute of Terrestrial Ecology, for the Department of the Environment, barely half the 500,000 miles of hedgerow estimated to have been in existence at the end of the war, now survive. A third of the entire loss was sustained between 1984 and 1991. In England, the survey estimated that, whilst some 2,000 miles might be planted each year, about 4,000 miles had been destroyed (Department of the Environment, 1993).

The Hedgerow Incentive Scheme launched by the Countryside Commission, and other endeavours to secure some form of statutory protection, are likely to stimulate further appraisals as to the character and importance of this diminishing resource. As with previous studies of the hedgerow and field margins, as published, for example, in the Collins New Naturalist series (Pollard et al., 1974) and by the British Crop Protection Council (Greaves & Marshall, 1987), there will continue to be reference to the origins and development of this landscape phenomenon, one taken so much for granted in Lowland England and found so rarely in other parts of the world, beyond Western Europe, New England and Tasmania.

THE UTILITY OF HEDGES

In a paper, published in the Journal of the Royal Agricultural Society in 1985, Dr E.J.T. Collins took conservationists and ecologists to task for regarding the inauguration of the ploughing-up campaign of the second world war as necessarily the benchmark from which to measure change in the agricultural environment. Collins (1985) argued that many of the changes since 1940 had been in effect a resumption of a trend that had begun in the seventeenth and eighteenth centuries. Through the destruction of the commons, reclamation, and the adoption of tighter systems of crop and animal husbandry, the British landscape had become, by the time Victoria ascended the throne, second only to the Dutch in western Europe, in being so tamed and intensively worked. If the years of agricultural depression after 1870 are regarded as the 'middle past', when progress was interrupted and in some instances reversed, the decades since 1940s may be seen as a time when lost ground was regained and surpassed. .

If that longer time perspective is adopted, it comes as no surprise to discover that the utility of hedgerows was keenly debated during the period of 'High Farming' in the mid-nineteenth century. In 1845, the Royal Agricultural Society offered a prize for the best essay on hedges. The winning essay was simply entitled, 'On Fences' (Grigor, 1845). Another focused 'On the Advantages of Reducing the Size and Number of Hedges' (Cambridge, 1845), and a third more dogmatically 'On the Necessity for the Reduction or Abolition of Hedges' (Turner, 1845). All were united in their condemnation of hedges, which took up so much land, made the use of machinery difficult, acted as weed magazines and asylums for pests, impoverished the soil, and prevented the free circulation of air. The prize winner, James Grigor, estimated that, on the basis of a sample survey of four arable districts of Norfolk, there were 25 miles of hedgerow per square mile, covering over 10% of the surface area. Applying the formula to 'the forty divisions of England', the total area occupied was equivalent to 'two of the largest counties'.

To others, it seemed extraordinary that a feature, so widely admired on the Continent, should be threatened with destruction. Richard Jefferies protested at the 'modern agricultural endeavours', in arable districts, to cut down trees and grub up hedges', on the pretext that crops were shaded by the foliage and damaged by their roots. They afforded abundant shelter to sparrows and other pest species (Jefferies, 1879). Claims that the hedges had to be removed so as to allow English farmers to compete more successfully with foreign producers reflected the wisdom of city counting-houses, and scientific lecture-rooms, which looked upon the land as little more than 'a manufactory of agricultural produce' (Johnston, 1851). Fortunately the mercantile spirit and middle-class devotion to profit did not always prevail. In the words of William Johnston, the countryside was held in such affection by the great body of people.

The debate as to the utility of hedgerows was likely to become even fiercer, in the last quarter of the nineteenth century, as the increasing competition from the growing volume of imported American grain called for the greatest economies in the use of capital and labour. On the premise that £1 per acre had to be invested in farm boundaries, and a further 3 shillings per acre in their annual upkeep, one authority estimated that the fencing required for some 45 million acres of enclosed farmland in the United Kingdom represented an investment of nearly £50 millions of capital. A further £6,750,000 were required for annual upkeep. Account had to be taken of both the thickness and layout of the field boundaries. A hedge of only 2 feet in width, with a margin of 1 foot left on each side, might occupy a 55th part of the area of a holding of 250 acres, where laid out in 25 fields of 10 acres in size (Scott, 1883). The optimal size of fields was generally reckoned to be between 20 and 25 acres - fields should not be smaller than 5 acres nor larger than 40 acres. A square field saved frequent turnings on short ridges. Where there were long ridges, horses became fatigued and the soil badly washed by the strong currents which heavy rains formed in the long furrows. In determining the exact size and shape, account had to be taken of the nature and use of the land. Whilst on light, sandy soils, the value of hedges in retaining moisture might increase with their number, they could be injurious on naturally damp and wet soils (Stephens 1890).

Whatever the conclusions drawn from such computations, the rate of hedgerow loss remained small, compared with that recorded in surveys some hundred years later. Whilst some might be destroyed when fields were enlarged to accommodate machinery, particularly following the introduction of the steam plough, A.D. Hall, in his Pilgrimage of British farming, written in 1910-12, continued to be astonished at the extent to which hedgerows survived as obstacles to farming. A green sheltered country of little fields might make for a charming property but, to the farming eye, such a spectacle denoted 'the same retail way of business as the endless tiny shops in the suburbs of a manufacturing town' (Hall, 1913).

PRESCRIPTIONS FOR THE MAKING OF BOUNDARIES

The author of a paper on 'Hedges and hedge-making', published in the Journal of the Royal Agricultural Society in 1899, looked for seventeen qualities in a quick hedge (Table 1). No species possessed so many of these qualities as the whitethorn or hawthorn. Since the thorns deterred all forms of livestock, the hedge could be cut into a very compact form, thereby ensuring its branches offered little refuge for birds and insects. Whilst it grew less vigorously on thin soils, it was only at high altitidues that the whitethorn encountered difficulty in establishing itself. The fact that it was the only shrub used as fencing along the entire length of the railways, through a great variety of soils, topography and climates, provided ample evidence of its adaptability (Malden, 1899).

Among the many changes which the construction of the railways had brought to the customs of the countryside, Henry Stephens, in his Book of the farm, identified the most important to be the way that hedgerows were planted and established. The practice had been to plant the hedges on banks composed of material excavated from an adjacent ditch. Whilst this had the effect of immediately providing some kind of barrier, the sides of the bank inevitably fell away, exposing the rooting system. The railway companies had provided an 'excellent object lesson' in how the first consideration should be the welfare of the hedge. The quicks were planted on the level, with a ditch cut only where needed for drainage. However planted, the ground should be at least fallowed, limed and manured. Every effort had to be made over the first few years to stir the soil, so as to prevent the quicks from being choked by weeds (Stephens, 1890).

Some of the most detailed prescriptions were to be found in Farm roads, fences and gates. A practical treatise, published in 1883 by John Scott, the one-time Professor of Agriculture and Rural Economy at the Royal Agricultural College, Cirencester. Whilst quicks of one, two or three years of age were commonly used, they took a long time to develop into a hedge, and would certainly perish unless well fenced and nursed. Where available, Scott recommended the use of stock of at least 6 years, and ideally over 10 years, in age. Rather than mixing them up (as often happened), it was easier to give more attention to the weaker plants, where they were segregated from the stronger. Opinion was sharply divided as to the advantages of planting in one or two rows. Optimally, the distance between each quick was 6 to 10 inches. Whilst closer planting would help establish the hedge more quickly, and help compensate for any gaps that might occur, say, through the browsing of hares and rabbits, the longer-term effect was to encourage the plants to be drawn upwards, with less lateral growth. The pernicious practice of planting trees within the line of the hedge should be resisted. Most trees grew faster than the whitethorn. With roots and branches spreading in all directions, they would soon overshadow and deprive the young quicks of nourishment (Scott, 1883).

TABLE 1. The seventeen most important attributes of a hedge according to W.J. Malden, in his paper, 'Hedges and hedge-making', published in 1899.

The hedge would:

1 develop in a reasonably short time,
2 be long-lived,
3 be easily repaired, if neglected,
4 be uniform in growth,
5 be easily kept within suitable bounds,
6 present a compact front,
7 prevent animals from escaping, ideally by having thorns,
8 be easily grown from seed,
9 be adaptable to most soils,
10 be able to withstand severe weather,
11 afford shelter to livestock in cold winds,
12 produce shoots close to ground for containing small animals,
13 afford little harbourage for insects,
14 be able to withstand fungal and other diseases,
15 have reasonably compact rooting systems,
16 be able to withstand browsing by livestock or game,
17 be able to regenerate, when cut down, to or near its stump

Even where pruned, hedges might be woefully mismanaged. Whilst the height and severity of cutting might vary, the overriding object was to promote new growth. If too much old wood was left, the heart of the hedge tended to become hollow as the younger growth on the outside smothered that of the inside. Wherever practicable, the wood was best cut with an upward stroke. Water would then run easily off the smooth surface. The vibration caused by a downward cut would cause the wood to splinter, leading to dampness and often considerable decay. Over time, it might become a chief cause of gaps in the hedge. In broad terms, one of two management systems might be adopted. The more common was to train the hedge into an upright, triangular section, that followed closely the natural form of the hawthorn tree. It might reach a height of 4 or 5 feet, without ceasing to be thick and well-clothed at the very bottom. Not only did this produce an effective barrier for livestock, but whitethorn shoots were seriously damaged by the shade cast by growth above them. Since lateral growth had the effect of curbing the natural tendency of sap to flow to the upper shoots, the bottom of the hedge, once weakened, rapidly became weaker. The alternative method of management, and the one preferred by railway companies, was to cut, rather than grow, the hedge into shape. It was first allowed to grow to a height of 6 to 8 feet, and then 'wattled' at an angle of about 40 degrees, stakes being left at 2 feet intervals, the wattling rods being hacked close to the ground, and woven in between the live stakes. The hacking encouraged a strong growth of young shoots from the base.

HEDGEROW-NEGLECT AND REPLACEMENT

A particular feature of the hedgerow survey, carried out by the Institute of Terrestrial Ecology in 1990, was the significance attached to hedgerows that might have survived removal, but had nevertheless been abandoned through neglect. As Malden noted, almost a century earlier, hedgerows had always been subject to neglect even in pastoral areas. They were obvious targets for economies in the straitened circumstances of Agricultural Depression.

One of the more obvious signs of agricultural depression was the neglect or skimping of management work. Whilst, on the one hand, the hedges might be overgrown with 'every weed that gets leave to shed its seed for miles around', there were, on the other hand, so many gaps as to render them useless for containing stock, unless infilled with slabs, paling or loose stones. Observers commented on how, between Oxford and Thame, the hedgerows, once kept 'so painfully low and well-trimmed', had been allowed by the 1890s to grow high. Through neglect, the hedges in parts of Essex and Suffolk had taken on the appearance of 'shaws', or lines of woodland, growing up to 25 feet in height, and encroaching onto fields and roadside wastes (Hissey, 1891; Collins, 1985).

Whilst understandable, commentators stressed the short-sightedness of neglect. Once a weakness or gap developed, the whole purpose of the hedge was lost and remedial action might be costly. Re-planting on the site of a thoroughly-neglected or worn-out hedge was rarely successful. Whilst there was no actual evidence that injurious matter accumulated in the soil, it was usually presumed so. The only course was to remove and replace the top soil with fresh soil from nearby, mixed with well-rotted dung. On thin or barren soils, well-rotted turf or sod was useful. The better course was to ensure the hedgerow never reached the point where it needed replacing. Much of the work could be performed when there was little alternative employment for the farmer's best labourers. The most effective method of dealing with overgrown, yet gappy, hedges was to plash them, the long rods being suitable for wattling. Any decaying stumps should be cut level with the ground, so as to encourage regeneration. If laid well, some judicious thinning and keeping the ground clear at the base should be enough to keep the hedge in shape for 20 to 40 years. All too often, however, it was again allowed to become large and straggly.

The cost of establishing and maintaining hedges was an obvious incentive to finding substitutes. From the 1840s, factory-made iron railings and posts, and wire for strengthening fences, became increasingly available. Whilst too expensive for ordinary farm purposes, the iron-bar fencing was in much demand for use around parks and pleasure grounds, and along roads. It combined great strength with good appearance. Increasing quantities of galvanised wire-netting were used for protecting paddocks and turnip fields, and for fencing rabbit-warrens, poultry yards and pheasantries.

According to Scott (1883), wire fences had, in late years, become the most convenient and profitable of boundary-materials. They were relatively cheap, durable and easy to erect. A drawback was the way in which the wire, fixed to straining and intermediate wooden-posts, soon gave way to pressure from livestock, particularly as the posts began to decay. Scott commended the far superior fencing developed by New Zealand colonists, that only required posts at intervals of as much as 12 to 22 yards. 'Droppers' were fixed at intervals of 6 feet, so as to prevent the wires being pushed apart. Since the 'droppers' did not reach the ground, the fence retained a degree of elasticity.

The future was, however, with steel barb, or barbed wire, fencing. Because livestock soon learnt to keep well clear of the barbs, perhaps only a quarter to a half of the number of fence posts might be needed. Since its introduction in 1873, the fencing had come to consist of at least two barbs, of no more than 5 or 6 inches apart, twisted not around, but between, two strands of wire so as to prevent their being loosened. As Scott remarked, one or two lengths of this barbed wire, within an ordinary, plain-wire, fence, or entwined in the line of a hedge, could have a magical effect on the efficiency of the whole. Costing perhaps under one penny per yard, barbed wire was the only certain way of keeping full-sized horses, oxen and cattle within, and the enemies of sheep, including prowling dogs, outside an enclosure.

Although unequalled in its efficiency and cheapness, many owners of stock soon abandoned the use of barbed wire. Whilst ordinary wire, lodged on the top, or stapled to the sides of the posts, presented little hazard to fox-hunting, barbed wire might seriously injure a horse. Arrangements were made by most Hunts to remove the strands during the hunting season. The Fernie Hunt set up a special Wire sub-committee in 1895, and contemplated the appointment of a wire inspector in 1904. Whilst there was no possibility of a general curb being placed on the use of barbed wire, a Bill was introduced by a group of members of parliament in 1893, making it easier and cheaper to obtain legal redress where injury was caused by 'any wire with jagged projections' beside a public highway. The preamble to the draft Bill recalled how there had been many accidents, as well as 'danger, injury and cruelty to animals'. The Act, as amended by the Local Government Board, enabled local authorities to seek the removal of such wire, where it constituted a nuisance to users of the highway (Public Record Office, HLG 29,41; Parliamentary Debates, 4th series, XII, 302-7; Barbed Wire Act, 1893, 56 & 57 Victoria, c.32).

LANDSCAPE ECOLOGY OF FIELD MARGINS

Beyond providing confirmation that hedgerows have always been the subject of contention as to their utility, however prosperous farming might be, what significance may be attached to these earlier prescriptions as to how hedgerows might be established and maintained, and the merit of replacing them with, say, a barbed-wire fence? The paper concludes by noting two initiatives currently being developed by the agricultural historian and ecologist, and their possible implications for interpreting, say, the changing species composition of the rural mosaic.

Historians have tended to understate the importance of the enclosure movement to landscape change in the eighteenth and nineteenth centuries. They have concentrated on the Parliamentary aspects, namely those schemes carried out by private act or under the auspices of the public general enclosure acts of 1836 and 1845. Such Parliamentary enclosures were certainly of considerable significance. The 5,500 individual Acts or Orders affected some 25% of the total surface area of England, and 12 to 15% of Wales (Chapman, 1987; 1992). There is, however, growing evidence of a contemporaneous and substantial amount of enclosure by private agreement, by the actions of the lord of the manor, and by piecemeal withdrawal of land from the open fields and common wastes. A recent study of south-central England has indicated that 42% of open-field enclosures were achieved by non-Parliamentary methods. Although it would be unwise to project such figures to the country as a whole, it seems likely that a third, rather than a quarter, of the English landscape may have been enclosed during that period (Chapman & Seeliger, 1993).

The environmental impact of the movement depended on the type of land involved. For Parliamentary enclosure, almost 60% of the land was 'common waste', in other words, moorland, heath, downland, or fen. The aim here was to convert the land to more productive use. It did not always succeed. Extensive areas of 'waste' survived in upland areas, such as Central Wales and the Pennines. But for the most part, the grass, heath and scrub were replaced by fields of arable and improved pasture, divided by hedgerows. A further 33% of the land enclosed consisted of open arable, where reallotment was intended to increase productivity of both soil and labour. Although land-use change was less dramatic, open landscapes of some 200 to 300 acres might be converted into hedged fields of no more than a tenth of that size. The remaining 7% consisted of common meadow, or of land already held individually but which was exchanged so as to secure a better layout of estates. The proportions of land involved in non-Parliamentary enclosures are not known with any certainty, but it seems likely the proportion of open field was somewhat higher. The changes were also more piecemeal. Without an overall plan, the new fields were almost invariably smaller, often of no more than two or three acres. The length of hedgerow was consequently far higher.

Beyond the aspirations of individual landowners, enclosures were justified as being the only means of feeding a growing population. Famine could only be avoided through greater efficiency. This single-minded policy, and the loss of the common 'wastes' for public recreation was only challenged in the late nineteenth century, when parliament placed restrictions on further enclosures. By that time, the new hedged-landscapes were reaching some degree of maturity on so large and intimate a scale as to be mistaken, by later generations, for an integral part of 'our natural heritage'.

Far from being the product of countless numbers and generations of farmers, the enclosure landscapes were the responsibility of a comparatively few individuals. Whilst over 4,000 people were employed as enclosure commissioners under the various acts, most of the work was done by some 50 of these, who became effectively full-time professionals. Their power to replan the landscape was enormous. They re-drew property boundaries, realigned roads and streams, and dictated the type of hedge or fence to be used. It is to their overwhelming insistence upon hedges as the means of separating properties, often even in districts where walls or ditches were the norm, that so much of our landscape owes its form. Men such as George Barnes of Andover or John Outram of Burton Agnes are obscure figures compared with, say, Capability Brown, but they planned a far greater acreage than any of the landscape gardeners (Chapman, 1989). The rudiments of their landscapes survive relatively intact, and, indeed, those with Parliamentary sanction are subject to a degree of legal protection, for the awards usually specify that the owners and their successors shall maintain the hedges in perpetuity.

As historians provide a better understanding of the complexity and scale of change in recent agricultural landscapes, so a new discipline has begun to emerge in ecology. Whereas attention once focused on the conceptual beauty of well-balanced, homogeneous ecosystems, the ecologist now attaches increasing importance to the differing properties and behaviour of the patches, or mosaic, that make up the ecological systems that constitute the landscape. The first step in determining how patch dynamics work is to define more fully the patches that comprise the landscape, how they are bonded together, the relative importance of geometry and other landscape characteristics, and how far the boundaries might influence communication and interaction between the component patches. As Turner remarked, 'clever empirical studies' are required to answer such questions. 'Like looking at the world through a keyhole', ecologists were already peering through small spatial and temporal windows to understand what was, in effect, large-scale dynamics (Turner, 1987; Hansen & Castri, 1992).

A landscape of hedgerows of varied density, size and composition provides an obvious testbed for such explorations in landscape ecology. Drawing on both historical and ecological insights, a better understanding might emerge as to how animal and plant life might have reacted to the evolution of the hedged landscapes of the eighteenth and nineteenth centuries. There is some documentary evidence to suggest that the rabbit, an alien species, may have first become established and abundant as a wild animal in many parts of the country, during the period some historians call the Agricultural Revolution of the eighteenth and nineteenth centuries. As the hedges developed, they provided an ideal retreat for the animals, which had to move only a few yards from their burrows to graze the vetches and winter corn. When disturbed in the harvest fields, the animals found refuge close by (Sheail, 1971). Clearly much else was also happening. Game preservation and fox-hunting became important over extensive parts of the country. Whilst the rabbit shared the same natural predators, and benefitted from their large-scale slaughter by game keepers, the status of the animal may also have been enhanced through further changes in the layout and texture of the landscape, as game and fox coverts were established within a hedged landscape.

As the ecologist and agricultural historian come, from their separate perspectives, to take fuller account of the complex spatial and temporal trends in the countryside, and seek more rigorous ways of recording and interpreting the dynamics involved, a clearer understanding is beginning to emerge as to the processes that determine how the landscape is occupied by so large, varied and mobile an array of plant and animal life.

REFERENCES

Cambridge, W. (1845) On the advantages of reducing the size and number of hedges. Journal of the Royal Agricultural Society of England, 6, 333-42

Chapman, J. (1987) The extent and nature of Parliamentary enclosure. Agricultural History Review, 35, 25-35.

Chapman, J. (1989) Saviour of our agriculture. Geographical Magazine, 61 (2), 22-5.

Chapman, J. (1992) A guide to Parliamentary enclosures in Wales, Aberystwyth, University of Wales Press, 23.

Chapman, J., and Seeliger, S. (1993) Non-Parliamentary enclosure in Southern England. University of Portsmouth, Department of Geography Working Papers, 28.

Collins, E.J.T. (1985) Agriculture and conservation in England: an historical overview, 1880-1939. Journal of the Royal Agricultural Society of England, 146, 38-46.

Department of the Environment (1993) Countryside survey 1990 London, HMSO for Department of the Environment, 55-66

Greaves, M.P.; Marshall, E.J.P. (1987) Field margins: definitions and statistics, In Field margins J.M. Way and P.W. Greig-Smith (ed.), Thornton Heath, BCPC Monograph 35, pp.3-10.

Grigor, J. (1845) On fences. Journal of the Royal Agricultural Society of England, 6, 194-228.

Hall, A.D. (1913) A prilgrimage of British farming, 1910-1912 London, Murray.

Hansen, A.J.; di Castri, F. (1992) Landscape boundaries. Consequences for biotic diversity and ecological flows New York, Springer Verlag, pp.v-vii.

Hissey, J.J. (1891) Across England in a dog-cart from London to St Davids and back London, Bentley, p. 383

Jefferies, R. (1879) Wild life in a southern county (reprinted London, Lutterworth, 1949), p.312.

Johnston, W. (1851) England as it is, political, social, and industrial, in the middle of the nineteenth century, London, Murray, pp. 1-9.

Malden, W.J. (1899) Hedges and hedge-making. Journal of the Royal Agricultural Society of England, 6, 87-115.

Pollard, E., Hooper, M.D., & Moore, N.W. (1974) Hedges, London, Collins.

Scott, J. (1883) Farm roads, fences and gates. A practical treatise, London, Crosby Lockwood.

Sheail, J. (1971) Rabbits and their history, Newton Abbot, David & Charles.

Stephens, H. (1890) The book of the farm, Edinburgh, Blackwood, 4th edition revised.

Turner, J.H. (1845) On the necessity of the reduction or abolition of hedges. Journal of the Royal Agricultural Society of England, 6, 479-88.

Turner, M.G. (1987) Landscape heterogeneity and disturbance New York, Springer-Verlag, pp.v-vi.

THE CURRENT STATUS OF FIELD MARGINS IN THE UK

R G H BUNCE, C J BARR, D C HOWARD & C J HALLAM

Institute of Terrestrial Ecology, Merlewood Research Station, Grange-Over-Sands, Cumbria LA11 6JU

ABSTRACT

A review is presented of the available ecological information on field margins in Britain. Hedgerows and streams are well covered, with information on both botanical and faunistic composition. Other features such as walls and grass strips, are not so well covered, although the 1990 Countryside Survey carried out by the Institute of Terrestrial Ecology will provide some basic information. It is concluded that field margins not only contain a major resource of botanical capital in British landscapes, but also represent a potential source of biological diversification.

INTRODUCTION

Throughout Europe, at the present time, there is a trend in agriculture towards intensification and a contrasting environmental pressure towards the maintenance, and recently enhancement, of diversity. The debate about the expenditure under the Common Agricultural Policy emphasises the link between policy and what will happen on individual farms and their associated semi-natural areas and field margins. These must first be defined in order to understand how they fit into the overall ecology of different types of landscape.

Hedges have generally been regarded in Britain as the most ecologically significant field margins. Recently, however, the botanical significance of other features has been assessed (Bunce & Hallam, 1993). Not only did they contain species not present elsewhere in the open landscape, but they also contained a wide range of variation in terms of the vegetation types present. It was concluded that linear features still contained much botanical capital in comparison with the surrounding landscape, and were especially important in the lowlands where the vegetation is often impoverished.

The objective of the present paper is to summarise the available information on field margins in Britain and then to indicate their overall ecological significance.

TYPES OF FIELD MARGIN

Hedgerows

Hedgerows were identified as important linear features at an early date, not only for their ecological content, but also because of their visual contribution to the landscape. This importance was recognised in the book by Pollard *et al.* (1974), which summarised the available information at that time. Losses of hedgerows were reported, but it was not until

1986 in the Monitoring of Landscape Change project (Huntings Surveys and Consultants (1986)) that they were quantified for England and Wales and later for Great Britain by Barr *et al.* (1986). Further losses were reported by Barr *et al.* (1991) and changes in the species composition by Cummins *et al.* (1992). There was also a major conference at Wye College in 1992 summarising the available information, the proceedings of which are currently being published. In ecological terms the botanical contribution of hedges can be divided into the woody and the herbaceous hedge bottom flora. The majority of hedges are of hawthorn (*Crataegus monogyna*) but there are also mixed hedges typical of both western and eastern Britain which contain other woody species usually only found elsewhere in ancient woodland. In addition there are hedges typified by single species, such as *Ulex* spp. and *Fagus sylvatica.* The hedge bottoms contain almost 300 species, some of which are from woodlands, e.g. *Digitalis purpurea,* but also from surrounding habitats such as damp grassland, e.g. *Filipendula vulgaris*. Overall therefore, as Bunce & Hallam (1993) point out, hedges are of major significance to botanical diversity in Britain, except in the uplands.

In terms of management, hedges were maintained by traditional methods, such as cutting, laying and then trimming the regrowth. This practice has now declined because of labour costs and most hedges are now trimmed by flail cutters. In many respects therefore, the hedge is maintained in a coppice type cycle, with regular openings of light which help to maintain diversity. Cummins *et al.* (1992) point out that dereliction leads to the domination of the hedge base by relatively few shade-tolerant species and the current trend for a decline in management therefore is likely to reduce diversity. Furthermore, once a hedge has lost its role as a barrier, it seems more likely to be removed from the landscape.

Streams

The vegetation of streams and rivers has long been celebrated in literature but has only recently begun to be recorded quantitatively. Holmes (1983a) describes an appropriate methodology for recording riverside vegetation and its application to different systems. Haslam (1978) describes the vegetation growing actually in the water. The Institute of Freshwater Ecology have also developed a procedure for recording the vegetation in the water, because of its importance to other river life (Bolton & Dawson, 1992). Bunce & Hallam (1993) describe the first comparisons of riverside vegetation throughout Great Britain, concluding that it contributes much to diversity in all major landscape types. As with hedgerows, there are specialist species, especially those growing in the water, for example *Sparganium* species, that cannot survive anywhere else in the landscape. In addition, there are other species which are able to grow elsewhere, but may have now become restricted. Further details of such species and their contribution to diversity will become available with further analysis of the Countryside Survey 1990 data.

The management of streamsides varies according to the type of landscape. Thus, in the lowlands, the banks are managed separately from the arable fields. In lowland grasslands, the banks are often grazed to the edge of the water, or fenced off to some degree. In both such areas, trees or woodland may grow right up to the waters' edge. In the uplands there is usually no separate management, with the vegetation along the stream being continuous with that elsewhere, unless next to a gorge. The differences in vegetation are thus often due to nutrient enrichment and water level, rather than management.

Roadsides

As with streams there are many references in the literature to the attractiveness of British roadsides, especially country lanes, many of which border farmland directly. In this case the construction of motorways stimulated the first extensive work described by Way (1977). Otherwise there is much information on management techniques, for example Parr & Way (1988). Bunce & Hallam (1993) compare roadside verges with hedges and streamsides, concluding that overall, they contained more species than either of the previous category. They were especially rich in mesotrophic meadow species. In some areas the verges have maintained fragments of formerly widespread grassland types, e.g. chalk grassland on the South Downs. There are many types of management, with flail mowers most widely used on approximately the first two metres; behind this area there is often scrub invasion or coarse competitive species. Many minor roads are no longer cut at all, with consequent changes in species composition. Few modern cutting methods remove the dead material, whereas the previous regimes were similar to hay cutting, leading to a rich flora.

Walls

Walls have received much attention for their visual appeal but little from a botanical point of view. General texts are available, e.g. Rackham (1986), as well as some local surveys, but a literature search revealed no papers. There is, firstly, the flora growing actually on the wall itself, mainly lichen and bryophytes, but species of fern and flowering plants may grow in the wall if there is soil between the rocks. Secondly, there is the area at the base of the wall, which may be unmanaged with shrub development or residual unimproved grassland because of protection from fertiliser application.

The ecological significance of walls therefore remains to be assessed in terms of vegetation. As with other linear features, they may also act as barriers or corridors to species in the landscape.

Grass strips/fences/banks

The former occur mainly in arable landscapes between crops, whereas the latter are either to divide grass paddocks or along hedges and walls to restrict stock movement. Their botanical significance has not been assessed in Britain, although Smith *et al.* (1993) have produced some general conclusions. They mainly consist of different assemblages of grassland species. There has, however, been work on their zoological significance, e.g. Thomas *et al.* (1992). The ITE Countryside Survey will provide an overview of the botanical composition in Britain.

Boundary margins

Narrow strips are present around many arable fields. Bunce & Hallam (1993) showed that they can contain unusual ruderal species lost from the surrounding fields. Marshall (1989a; 1989b) has also studied them from a weed invasion viewpoint and summarised their management implications. Other gaps can be quite wide and converge with headlands at the corners of fields where tractors turn. Further information would be useful on their extent and species composition. The work by the Game Conservancy on headlands is relevant to these margins, there being a close relationship between them.

Green roads

These are sunken old roads that have now fallen out of use. Richard Mabey has lectured about them and Raistrick (1978) has described them in the Pennines. They are a rich source of shady woodland plants absent elsewhere in many lowland landscapes. Other unsurfaced or grassed tracks on farms may support a range of plant and animal species, if disturbance is not too severe. Rackham (1986) describes the significance of trees which are often present along green roads or any of the above margins.

THE ECOLOGICAL VALUE OF FIELD MARGINS

As mentioned above, some linear features have a sufficient level of botanical capital to be important in their own right. Even the poorer features contain species either not present in the surrounding landscape or of restricted distribution. In the present context, these features are of particular significance in their potential for the expansion of their constituent species into the wider countryside, if agricultural management pressure declined.

The evidence of the rate and pattern of such colonisation is incomplete, since most studies until recently have assumed agricultural expansion. Concerning the process of colonisation from linear features the proceedings of the conference on setaside (Clark 1992) gives an up-to-date summary of recent work. Baudry & Bunce (1991) also summarise abandonment. Otherwise, Marshall (1989b) has carried out detailed studies of the movement of species from hedgerows, especially weeds, because of their agricultural significance. Movement into crops was relatively limited, with only a small of a number of species; but this was into crops. Otherwise evidence of movement tends to come from observation, with motorway verges providing some good examples eg the expansion of *Primula veris* and *Chrysanthemum leucanthemum* from field margins onto embankments. An important principle is that individuals move rather than assemblages and whilst the movement of some species eg indicators of ancient woodland from hedgerows can be predicted, many others cannot. Rackham (1980) suggests that such movement is likely to be very slow. The richness of the feature is also of importance since a wider complement of species is available for expansion. The actual seed supply is also a limiting factor and the subsequent composition of vegetation following colonisation is usually determined by the propagules available.

Different linear features have different potential for expansion eg a roadside verge of chalk grassland species are unlikely to expand into an abandoned field because of competition from ruderal species, whereas hedges have greater potential. The processes are also different for example van Dorp & van Groenendael (1991) have shown how species can spread from river banks by flooding.

The value of linear features as corridors for movement of species has only been proven in a few cases. Verkaar (1990) for example describes the movement along streambanks. Holmes (1983) also points out that few aquatic aliens have been successful although some exotics, eg *Mimulus* species have colonised fast flowing rivers. Observational evidence exists eg oil seed rape along motorways and *Senecio squalidus* along railways. In hedges the true hawthorn content of many post enclosure hedges shows virtually no colonisation by woody species for almost 200 years.

RELATIVE FREQUENCY OF FIELD MARGINS

During the Countryside Survey 1990 programme (Barr *et al.* 1993) ITE surveyed 508 1 x 1 km squares throughout Great Britain in a four month period between June and October 1990. In each square, as part of the broader work programme, the land cover and landscape features were recorded and mapped. At a more detailed level, vegetation was recorded in up to 27 plots in each square. Five 200 m^2 quadrats in five random locations were recorded within each square and 1 x 10 m linear plots were placed beside field margins where these were within 100 m of the plot. These data therefore give an estimate of the relative frequency of the different field margins in Britain, although more accurate estimates could be made in due course by further analysis of data now available. The results are presented in Table 1.

Table 1 presents the boundary plot types by Land Class groups (Barr *et al.* 1993) showing the dominance of the hedge series in the two lowland groups, whereas fences extend into the marginal uplands and uplands; walls are present throughout, whereas most of the other types are dominated in the two lowland groups.

This Table characterises the four groups which each show different patterns reflecting their ecological character. Thus the uplands have few boundaries reflecting the continuous nature of the semi-natural vegetation, the marginal uplands are dominated by fences and walls, showing a degree of dissection. The lowland grass series is dominated by fences but with many hedgerows and contributions from other types. The lowland arable series also has fences most commonly but with a high proportion of hedges and water - perhaps surprisingly, as it is often considered that the arable areas have a lower number of hedgerows.

Table 1 Relative occurrence of different types of field margins in linear plots placed within 100 m in 5 random 200 m plots within 508 1 km^2 in Britain. The four landscape types are described by Barr *et al.* (1993)

Land class group	Hedge	Fence	Wall	Water	Grass strip	Bank	Verge	Other	Total
Lowland arable	200 26%	265 35%	39 5%	127 17%	24 3%	15 2%	79 10%	8 1%	757
Lowland grass	168 23%	333 46%	82 11%	47 7%	12 2%	43 6%	23 3%	10 1%	718
Marginal upland	18 9%	126 60%	51 24%	8 4%	0 0%	2 1%	5 2%	1 0%	211
Upland	0 0%	85 70%	28 23%	6 5%	0 0%	0 0%	2 2%	0 0%	121
Total GB	386 21%	809 45%	200 11%	188 10%	36 2%	60 3%	109 6%	19 1%	1807

CONCLUSIONS

Although there is a wide range of data now available on field margins the above review shows that there is much work yet to be carried out.

Linear features are not only important for flora but also for fauna. Some species such as the phytophagous insects are directly dependent upon plants; others depend upon them for shelter and movement. Examples of these inter-relationships have been presented by Schreiber (1988).

The role of many linear features, as a refuge for species which have not been able to survive in the fields, is important in terms of the response of vegetation to changes in agricultural practices, such as set-aside. For example, a decline in management of grasslands could lead to the expansion of species from field boundaries, as seen in the derelict areas in the Pyrenees. The mobility of different species and their ability to colonise existing vegetation will determine which are successful.

Field margins still retain much botanical capital that can be utilised to replace species lost from open landscapes. They need, therefore, to be incorporated into the development of landscape design, for maintaining and enhancing ecological diversity. While it is recognised that designated conservation areas for specific habitats are essential, even intensively farmed landscapes still contain many species and offer many opportunities for conservation.

REFERENCES

Barr, C.J., Benefield, C.B., Bunce, R.G.H., Ridsdale, H.A. & Whittaker, M. (1986) Landscape changes in Britain. Abbots Ripton: Institute of Terrestrial Ecology.

Barr, C.J., Howard, D.C., Bunce, R.G.H., Gillespie, M.K. & Hallam, C.J. (1991) *Changes in hedgerows in Britain between 1984 and 1990.* NERC contract report to the Department of the Environment. Institute of Terrestrial Ecology: Grange-Over-Sands.

Barr, C.J., Bunce, R.G.H., Clarke, R.T., Fuller, R.M., Furse, M.T., Gillespie, M.K., Groom, G.B., Hallam, C.J., Hornung, M., Howard, D.C. & Ness, M.J. (1993) *Countryside Survey 1990.* Main Report. London: Department of the Environment.

Baudry, J. & Bunce, R.G.H. (Eds.) (1991) Land abandonment and its role in conservation. (Options Mediterraneennes. Serie A. Seminaires Mediteraneennes no. 15). Zaragoza: Centre International de Hautes Etudes Agronomiques Mediterraneennes.

Bolton, P. & Dawson, F.H. (1992) The use of a check-list in assessing possible environmental impacts in planning watercourse improvements. In: *Proceedings of the International Symposium on Effects of Watercourse Improvements: assessment, methodology, management assistance*, 29-42. (Joint Keynote presentation).

Bunce, R.G.H. & Hallam, C.J. (1993) The ecological significance of linear features. In: *Landscape ecology and agroecosystems*, edited by R.G.H. Bunce, L. Ryzkowski & M.G. Paoletti, 11-19. Boca Raton: Lewis.

Clarke, J. (Ed.) (1992) *Set-aside.* BCPC Monograph No. 50. Farnham: BCPC Publications.

Cummins, R.C., French, D., Bunce, R.G.H., Howard, D.C. & Barr, C.J. (1992) *Diversity in British hedgerows.* (NERC contract report to the Department of the Environment). Institute of Terrestrial Ecology: Grange-Over-Sands.

Haslam, S.M. (1978) *River plants; the macrophytic vegetation of watercourses.* Cambridge: Cambridge University Press.

Holmes, N. (1983a) *Typing British rivers according to their flora.* Focus on Nature Conservation, No. 4. Peterborough: Nature Conservancy Council.

Holmes, N. (1983b) Garden plants along our rivers. *Living Countryside*, **124**, 2468-2471.

Huntings Surveys & Consultants Ltd. (1986) *Monitoring landscape change*. Borehamwood: Hunting.

Marshall, E.J.P. (1989a) The ecology and management of field margin floras in England. *Outlook on Agriculture*, **17**, 178-182.

Marshall, E.J.P. (1989b) Distribution patterns of plants associated with arable field edges. *Journal of Applied Ecology*, **26**, 247-257.

Parr, T.W. & Way, J.M. (1988) Management of roadside vegetation: the long-term effects of cutting. *Journal of Applied Ecology*, **25**, 1073-1087.

Pollard, E., Hooper, M.D. & Moore, N.W. (1974) *Hedges*. London: Collins.

Rackham, O. (1980) *Ancient woodland: its history, vegetation and uses in England.* London: Edward Amola.

Rackham, O. (1986) *The history of the countryside*. London: Dent.

Raistrick, A. (1978) Green tracks in the mid-Pennines. Ashbourne: Moorland.

Schreiber, K.F. (1988) Connectivity in landscape ecology. *Proceedings of the 2nd International Seminar of the International Association of Landscape Ecology*. Paderborn: Schoningh.

Smith, H., Feber, R.E., Johnson, P.J., McCallum,. K., Jensen, S.P., Younes, M. & Macdonald, D.W. (1993) *The conservation management of arable field margins.* Peterborough: English Nature.

Thomas, M.B., Sotherton, N.W., Coombes, D.S. & Wratten, S.D. (1992) Habitat factors influencing the distribution of polyphagous predatory insects between field boundaries. *Annals of Applied Biology*, **120**, 197-202.

Van Dorp, D. & Van Groenendael. (1991) Dispersal of seeds, a key factor in the regeneration of species rich grasslands. In: *Proceedings of the European IALE Seminar on Practical Landscape Ecology*, 99-105. Roskilde: Roskilde University Centre.

Verkaar, H.J. (1990) Corridors as a tool for plant species conservation? In: *Species dispersal in agricultural habitats*, edited by R.G.H. Bunce & D.C. Howard, 82-97. London: Belhaven Press.

Way, J.M. (1977) Roadside verges and conservation in Britain: a review. *Biological Conservation*, **12,** 65-74.

FIELD MARGINS - AN AGRICULTURAL PERSPECTIVE

M B HELPS

ADAS Drayton, Alcester Road, Stratford-upon-Avon, Warwickshire CV37 9RQ

SUMMARY

Field boundaries require periodic maintenance which has been and is now generally carried out by the occupiers of the farmland.

Hedges provide shelter for some distance into the field, but on balance these benefits are limited in all but relatively specialised situations.

The greatest area of controversy on field margins is in arable farming landscapes. The problems here are the relative neglect of the hedges and the farmers concern that weeds that develop under the hedge or on the headland can spread into the field.

In mixed or livestock farming districts, fields tend to be smaller, the length of hedges/unit area greater and the hedge itself is seen as more useful as a livestock barrier. However, there are some reservations about the value of hedges in livestock farming systems.

INTRODUCTION

This is very much an agricultural perspective representing, I hope, the reasonably well-informed farmer's view of field margin management. It is recognised that hedges are a valuable ecological resource and the extent of that resource will be emphasised in the later research papers. Field margins, with their associated farm hedges, ditches, headlands and field corners are a characteristic feature of much of lowland England. In our uplands, the place of the hedges can be taken by stone walls in all their varied forms.

In many of our landscapes these field boundaries and field margins are the only semi-natural feature. Even these have to be subjected to some degree of management if the landscape is to survive in its present form. Even so, hedges and associated field margins are subject to constant changes, either improving or deteriorating. The changes are inevitably going to please some interest groups and displease others. Carter (1983) estimated that the 500,000 miles of hedges in the country stood on an area of land that exceeded that of all this country's nature reserves put together. The annual cost of hedge maintenance of these hedges was estimated at £2m. a year (Hall, 1978).

Hedges at ADAS Drayton

ADAS Drayton research centre near Stratford-upon-Avon extends to 200 ha, with soils derived from the Lower Lias, and is typical of a large proportion of the Midland

Clays. The farm is divided into 33 enclosures, ranging from 0.8 to 13 ha and includes approximately 16.6 km of hedges. Average field size is 5.6 ha and we have 89 m hedge/ha.

Recommendations as to an ideal length of hedge/unit area are limited. Records on this are available from surveys of specific areas (e.g. Pollard et al 1974). More recently a range of 60-80 m/ha has been suggested to retain a high density of birds and a broad range of species (Lack, 1992). This means an average field size of between 4 and 7 ha. He comments that with more hedge than this there will be a higher overall density of birds but perhaps fewer species, with those that prefer open areas dropping out. With fewer hedges the overall bird density and the number of species both fall.

Table 1. Length of hedge per hectare for different agricultural situations

		m/ha
Arable area	1969	57.25
Grassland area	1969	89.84
ADAS Drayton	1993	89.10

Land use on Drayton is mixed, with combinable crops (mainly winter wheat), long leys and some permanent grass. The livestock includes both cattle and sheep.

Environmental issues, including some work on farm hedges and field margins, have been a part, but only a part, of the centres experimental programme in recent years, but all our work has been carried out in the context of productive farming systems appropriate to our geographic situation.

Hedges are mechanically trimmed to approximately 1½ m high x 1½ m wide, most of them annually - some in late August/early September and the remainder in December/January. A proportion are left to grow taller with periodic side trimming. Some of the hedges have obviously been laid, at least once, in the distant past, but there is no record of hedgelaying in recent years. Hedgerow trees in the district were decimated by Dutch Elm disease but some Oak (*Quercus* spp) and Ash (*Fraxinus excelsior*) remain and new plantings in field corners and some shelter-belts are developing. Our current species lists include 209 plant species and 63 bird species, 28 of which are breeding on site.

It is my brief to raise topics and issues from an agricultural perspective so I will deal with the subject under the following headings and leave other speakers to deal with issues from the conservationist's point of view.

1. Frequency of hedge trimming and cost
2. Shelter
3. Arable cropping
4. Livestock husbandry

FREQUENCY OF HEDGE TRIMMING AND COST

Hedges were originally planted by farmers/landowners for two purposes.

1. To mark boundaries
2. To prevent livestock from straying.

If they are to exist at all any field boundary has to be managed. For any hedge this will mean some form and frequency of cutting, for a ditch some form and frequency of cleaning and for a drystone wall some necessary repairs. Without this management, margin structures will sooner or later become dilapidated. The fact that our landscapes look as they do is because those who have been managing the land in the past have devoted time and money to the maintenance of field boundaries. In general terms I suggest that the farming community must have managed their hedges and field boundaries reasonably over a long period of time, otherwise there would be fewer hedges in the landscape than there now are.

Field size governs the length of hedge per unit area and also the proportion of land covered by hedges and their associated field margins. It can also have an important influence on the work rate of field machinery, through its effect on the amount of turning per unit area and therefore the amount of unproductive time. The proportion of compacted headland will be greater in smaller fields. The relationships between field size and hedge and headland area is illustrated in Fig. 1. From the logistical point of view there is a lot to be said for 20 ha fields!

Figure 1. Area of headland as a percentage of field area

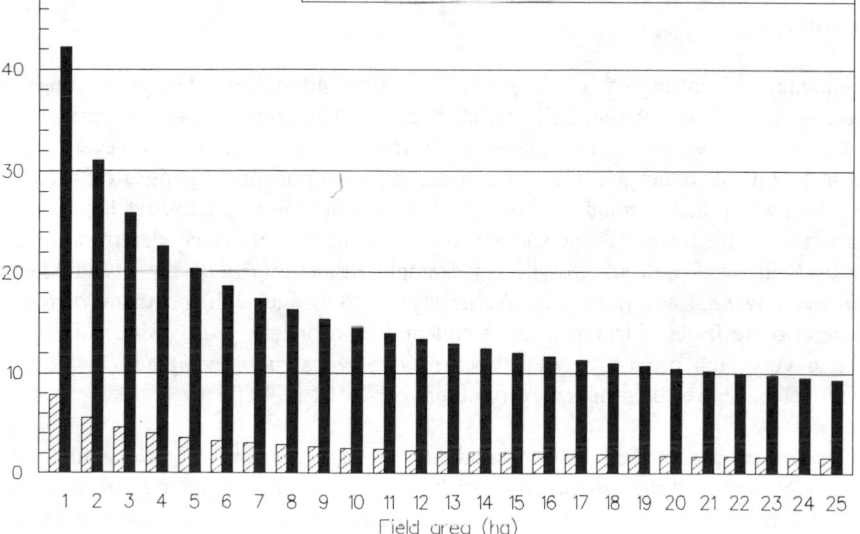

Perhaps, once upon a time, it was possible to trim hedges by hand between haytime and harvest and lay a proportion of hedges each winter. This is now almost impossible. Hedge trimming between late July and early August would nowadays be discouraged because of likely damage to nesting birds.

Over the last 50 years the workforce on farms has roughly halved and farming systems have changed considerably. Not only are fewer staff employed, but farm systems have changed and there is little "spare" time. The usual method of hedge management today is trimming with a flail cutter - often done by a local contractor. Even when this machinery is being used the job competes for a tractor and driver with other essential work. Hedge laying is now very expensive, relative wages of farm staff have increased and this job has not benefited very much from mechanisation. In addition to the problem of cost and availability of labour, hedges to be laid would have to be allowed to grow for some years before the job was done. This would allow climbing plants to develop, encourage gaps in hedges and reduce their value as stock-proof barriers.

The range of costs today is about
£2/m - £3/m for laying and
£0.07/m - £0.17/m for machine trimming

The desirable form or shape of hedge gives rise to much disagreement. Even people who share a strong interest in ornithology can be divided on this point, according to their relative interest in song birds, gamebirds or raptors. From the agricultural point of view, the shape and size of hedge that needs the fewest passes with the hedgecutter has much to recommend it. Subtleties of hedge trimming can be difficult to put into practice if using a contractor. The timing of hedgecutting can also give organisational problems to arable farmers. It is desirable to leave seed and berries on the hedges as late as possible, as a food supply for birds and small mammals. However, if the hedge is only accessible from a ploughed field, the hedge may have to be trimmed before the autumn crop is drilled in mid October, to avoid difficult travelling conditions on the cultivated and drilled headland during the winter.

Frequency of cutting is a subject of potential difficulty. Hedge cutting is undoubtedly easier if it is done annually on all hedges. The softer growth is easier and neater to cut with less visible damage and possibly less risk of die back. However, there are points in favour of longer periods between cutting a proportion of hedges on a farm. Before deciding which hedge could or should be left, it is desirable to consider the species mix and structure of the hedge. Some species will, if left unchecked, cause deterioration of the hedge by their over-vigorous growth, for example Bramble (*Rubus* spp.), Old Man's Beard (*Clematis vitalba*) and Elder (*Sambucus nigra*). A hedge with too many of these may well deteriorate faster if trimmed less frequently than once a year. Alternatively, a hedge with a very high proportion of hawthorn (*Crateagus monogyna*) and few other species may well be able to be trimmed every other year without ill effect.

Most farmers accept the need for variability in shape and form of field boundaries and their management, and the benefits this variability can have for the ecology of the area. On most farms much variability occurs automatically as a result of differences in slope, aspect, accessibility etc., which give variation in hedge height and shape, as well as botanical variation in hedges and field margins.

SHELTER

A great deal of work has been done in the past on the influence of an individual hedge on the immediately adjoining field. The primary effect of a hedge is to alter the wind speeds in the area immediately adjoining the hedge. We should also note the quite separate shading effect which is normally less important than wind shelter although perhaps more obvious to the casual observer. It is this shading effect which, combined with the growth of roots from the hedge itself into the field, can be partly responsible for the poor growth of crops often seen in the few rows closest to the hedge.

The shadow cast by a hedge may be very long when the sun is low but significant shading effects are limited to a distance of one or two times the height of the hedge into the field. Many shading/shelter effects are proportional to the height of the hedge.

Influence of shelter on wind speed will have marked effects on the micro climate immediately adjoining the hedge (Fig. 2).

Figure 2. Shelter effect of hedges

SOLID

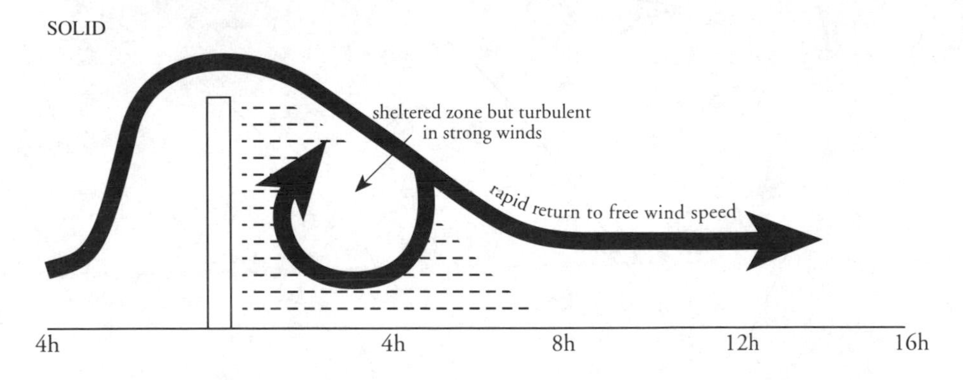

PERMEABLE

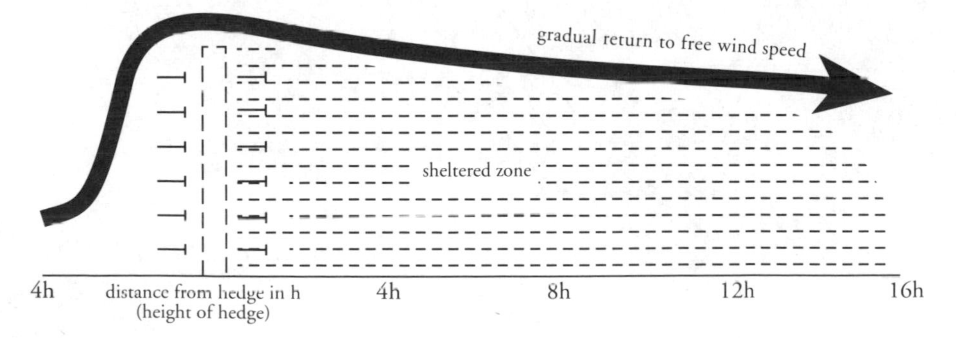

Farm hedges do in fact provide practically all gradations from an almost solid barrier to an almost completely permeable one. It is perhaps rather surprising that a single permeable hedge provides more shelter in proportion to height than a substantial block of woodland, because the established woodland acts as a solid barrier.

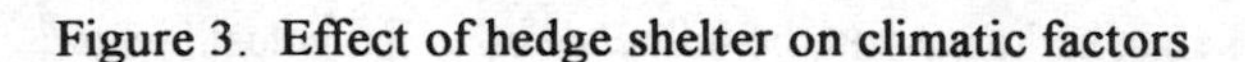

Figure 3. Effect of hedge shelter on climatic factors

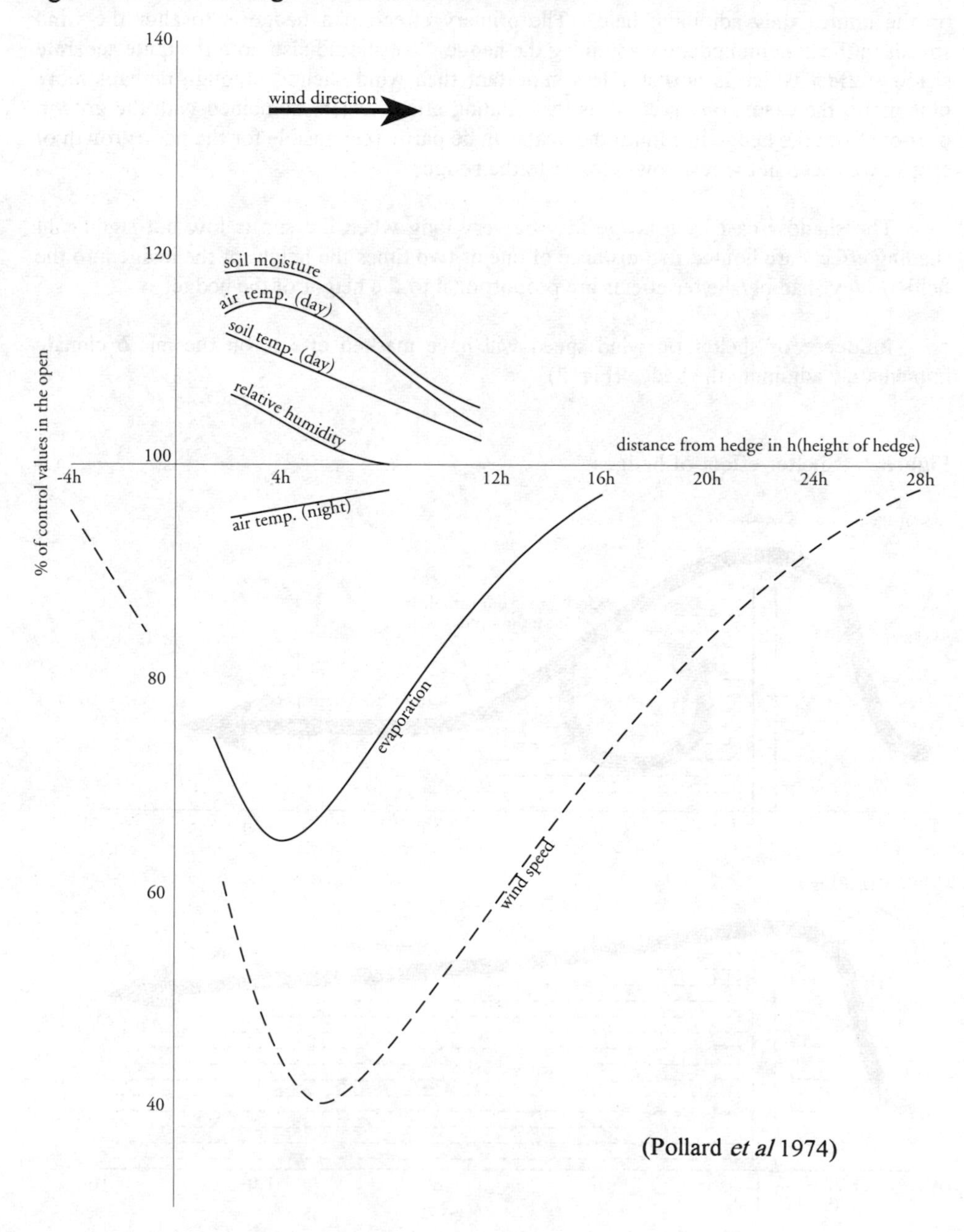

(Pollard *et al* 1974)

This summary diagram (Fig. 3.) is a generalisation based on a large number of different experiments with many types of shelter. There are many variables involved but this does provide a basis for discussion of the way in which the physical presence of the hedge can affect soil, plants and animals in the adjoining fields.

It can be said that most of the physical parameters recorded show very little variation beyond the normal headland adjoining a well trimmed, typical lowland hedge, 1.5 metres high. Shelter from wind would appear more important in certain crops (e.g. fruit) and on certain soils than on others, e.g. root crops on light soil subject to wind erosion. Similarly, shelter would be more important for certain classes of livestock, e.g. newborn lambs or young calves at turnout. Shelter can in fact be harmful if the field boundary creates snowdrifts or frost pockets. It follows that shelter is much more important at some times of the year than at others. Modern farm systems have to some extent reduced the general importance of hedges as a source of shelter, but increased the importance of specialised forms of shelter in horticultural or arable cropping. For example, specialised husbandry techniques to reduce the risk of soil blowing, or high hedges to protect orchards.

ARABLE CROPPING

Much controversy on farm hedges is based on perceptions of hedges in the context of today's specialised and well-mechanised arable cropping systems. In these, hedges are generally tolerated, but the intensity of hedge management has decreased and many hedges have deteriorated - some to the point of being difficult to classify as hedges any longer (Barr *et al.,* 1991).

In the context of intensive cereal production hedges can be perceived as of little practical benefit. To a farmer restricted growth and later ripening of crops close to a tall dense hedge can be all too obvious. This may be due to the rabbits living in the hedge, or the Couch grass, Cleavers and Sterile Brome invading the adjoining field from the hedge bottom, or to the hedge itself. Either way, the neglected hedge becomes a candidate for removal.

Management decisions about the hedge-bottom, the crop headland and any gap between these are less clear. The farmers concerns about this area are generally concentrated on weeds and the risk of spreading them from the field margin into the field. In practice these risks need not be very great. The problem weeds have to be there in the first place. The best defence is likely to be a strong close natural vegetation in the hedge bank and the farmer should minimise the risks of damaging the natural vegetation close to the hedge for this reason.

The bromes (particularly *Bromus sterilis* and *Bromus commutatus*) are a particular hedgerow problem, as potentially is Couch (*Elymus repens*). Cleavers (*Galium aparine*) is probably the only dicotyledonous weed species which appears as a widespread problem in field margins. It has, however, been suggested that hedgerow populations do not pose a major threat to arable crops (Froud-Williams 1984). Despite this, *G. aparine* is widespread in many field margins and fields

Neither fertiliser or herbicides should be applied into the hedge. They can both directly or indirectly favour weed species at the expense of the natural flora of the hedgebottom so the practise is both wasteful and environmentally damaging. Successful hedge and field margin maintenance and weed control depends very much on the details of workmanship involved. Weed species are encouraged by any bare soil left in the field margin. When ploughing the field margins the furrow is best turned towards the centre of the field to give a sharp cut off to the cultivated area. This will also reduce the low spots a few metres in from the field margin where waterlogging frequently occurs. Normal cultivation will tend to fill the furrow but should work to this clearly marked edge. If the line of cultivation of a field margin is allowed to vary from year to year closer to or further from the hedge, this may hasten the spread of weeds or weed seeds into the field. Similarly, undrilled or uncultivated bare soil between the crop and the hedge-bottom will encourage growth of weeds.

Ditches are part of the farm drainage system and, like the rest of the field margin, need maintenance. Grass banks need regular trimming and silt and sludge must be removed and deposited on the adjoining headland rather than on the ditch-bank itself where it will smother the existing vegetation, encourage weed problems and more rapidly slip back into the ditch.

There are several advantages in having a narrow managed gap between the hedgebottom and the crop headland.

1. Physical spread of weeds from hedgebottom can be controlled by rotary cultivation, spraying or mowing repeated as necessary during the growing season.

2. A clear gap reduces the risk of applying fertiliser or agro-chemicals into the hedge-bottom.

3. This gap provides a clear track for the divider of the combine.

The crop headland can be an important proportion of the field - in a square 10 ha field a 12 m headland is about 14% of the area.

Turning on the headlands in difficult soil conditions can result in surface compaction and a poor seedbed. Conditions on the headlands may require different cultivations from the rest of the field. Differential sowing dates between headland and the remainder of the crop are not a practical proposition because of wheeling problems, but some farmers do say they use heavier seedrates. Herbicide performance can be reduced if there is a poor quality seedbed and headlands with poorer crops offer less competition to weeds. Wild Oats and Blackgrass are particularly difficult to eradicate from headlands. If fields are large enough to accommodate them, set aside headlands may be a suitable proposition.

Soil borne pest problems are not known to be more common in field margins than elsewhere in the crop. In fact a compacted headland could inhibit the free movement of slugs and reduce their feeding. Other pests known to be more serious on headlands tend to be those of non-cereal crops.

Defenders of hedges and field margins can make impressive lists of predators and parasites of pests, and of pollinating insects which can also move between hedges and field margins and the crops. It is not easy to strike a balance between the beneficial and negative aspects of field margin flora and fauna. It is unlikely that there is any benefit to agricultural crop production beyond a certain distance from the hedge. Similarly, hedges do not usually cause pest problems in cereal crops.

GRASS AND LIVESTOCK

Hedges are frequently claimed to provide a stock-proof barrier to grazing livestock. Indeed many were originally established for this purpose. This is reflected in the greater importance of hedges in mixed farming and livestock areas. Unfortunately stock management has changed since the hedge lines were planted. As a result of intensification, the size of individual livestock units and animal group size has increased, as has grass production/unit area. Grazing systems have been developed which result in high grazing pressures. These have risen to as much as 17-21 ewes/ha, possibly with twin lambs, set stocked until weaning, or 5 dairy cows/ha on a rotational grazing system in early summer. These levels are not unusual on productive grassland.

At these high grazing pressures, the livestock will browse hedges to which they have access. When this happens alongside long-term leys or permanent swards the damaged hedge does not remain a reliable stock-proof barrier. Any accessible ground flora beneath the hedges is eaten or damaged by the stock and after several years the hedge takes on a typical tufted appearance. The hedge thus deteriorates. This problem suggests a need to double fence hedges in these situations. It is of course also necessary to fence stock out of ditches at the field margin of fields used for grazing, to avoid stock treading in the bank.

Late trimming of hedges is considered the environmentally friendly option. This carries with it the risk that small pieces of debris will still be hard and sharp in the spring and there is a risk that they can cause lameness in young lambs. This risk is increased if the lambs can get in underneath a hedge, as they like to do in play and to get shelter. Shade and shelter are attractive to all types of grazing livestock.

The hedges and/or the livestock sheltering there are attractive to flies, some of which can be carriers of infection, e.g. summer mastitis in dairy heifers and dry cows, or New Forest Eye disease in cattle. In sheep, particularly in lambs on lowland farms, the risk of blow fly strike has been a reason for the use of a persistent summer dip. Again, blow flies prefer living close to hedges rather than in the middle of a field.

CONCLUSIONS

The overall perception of farmers would be

1. There are landscape and amenity benefits to be derived from the diversity and variability of the British landscape. Hedge and field boundaries give character to this diversity.

2. There are possible, but at this stage unquantified, macro-climate benefits from a hedged landscape, as opposed to a prairie.

3. There are ecological benefits to be derived from having a proportion of semi-natural vegetation within any landscape.

4. All field boundaries need periodic maintenance and this costs both time and money, both of which are becoming harder to find.

REFERENCES

Barr, C.; Howard, D.; Bunce, R.; Gillespie, M.; Hallam, C. (1991). Changes in hedgerows in Britain between 1984 and 1990. NERC Contract Report to Department of Environment. Institute of Terrestrial Ecolocy. 13 pp.

Carter, E. S. (1983) Management of hedgerows and scrub. *BCPC Monograph No. 26,* In: *Management of natural and semi-natural vegetation,* J. M. Way (Ed.), Croydon, pp. 177-187.

Froud-Williams, R. J. (1984) In: *Understanding Cleavers* (*Galium aparine*) *and their control in cereals and oilseed rape.* Association of Applied Biologists, Oxford, pp.1-20.

Hall, J. (1978) Management of hedges and hedgerows. *Big Farm Management* 2, 29-32.

Lack, P. (1992) Birds on Lowland Farms. London. HMSO.

Pollard, E.; Hooper M. D.; Moore, N. W. (1974) Hedges. London : Collins

THE ROLE OF FIELD MARGINS IN THE LANDSCAPE

G. L. A. FRY

Norwegian Institute of Nature Research, PO Box 5064, The Agricultural University of Norway, N-1432 Norway.

ABSTRACT

I examine the various roles of field margins at the landscape level. Emphasis is on how the pattern of field boundaries in the landscape affects their functions as wildlife habitat, movement corridors, a recreation resource, and their interactions with agricultural systems. The success of field margins in fulfilling these roles is dependent on a combination of their local site quality, length per unit area of farmland and spatial arrangement in the landscape. Of particular importance is their degree of connectedness, alignment to slope and aspect. I conclude that an approach based on the principles of landscape ecology is essential for developing a theoretical foundation for understanding field margin function.

INTRODUCTION

The appreciation of the role of landscape level processes in determining field margin function is a major change since the last BCPC conference in 1986 (Way & Greig-Smith, 1987). In this collection of papers only one explicitly examines the evidence for larger scale effects (Morris & Webb, 1987). Other papers provide evidence for the increase in species and numbers of individuals with length of hedgerow in a landscape. These papers do not, however, make reference to the network pattern formed by hedges and other margins. Since then, research has focused on whether the spatial arrangement of margins in the landscape determines their value. This focus from site to landscape scale is partly due to the rapid growth of landscape ecology as a branch of ecology (Forman & Godron, 1986; Turner & Gardner, 1991). Landscape ecology has forced us to consider new dimensions and larger scales. For field margins, the studies of Baudry in France (Baudry, 1988; Baudry, 1989; Burel & Baudry, 1990) and Merriam in Canada (Fahrig & Merriam, 1984; Merriam, 1988; Middleton & Merriam, 1983; Wegner & Merriam, 1990) have been very influential in demonstrating the need to consider their spatial pattern in the landscape. My aim in this paper is to examine the importance of a landscape ecological approach to field boundary management. I do not examine the importance of the structure, vegetation and local management of field margins for wildlife, as other chapters address these issues.

SOME LANDSCAPE CONCEPTS

As a basis for my argument, I use some definitions of primary landscape characteristics (Forman & Godron, 1986) and apply them to field margins.

Landscape structure

At the landscape level, structure refers to the relationship between the various components of a landscape; their size, shape, diversity, density, and spatial pattern. Although we tend to concentrate on the biological components of field margins, structure also includes other components such as energy, nutrients, and their physical architecture. Field margins vary in type, width, length, habitat diversity, and their connectedness in the landscape. Their structure may be very simple (fence line) or very complex (old hedgerow). Some field margins are very uniform along their length while others are very variable.

Landscape function

Landscape functions relate to the interactions between components of a landscape. For field margins, these will be determined by the flows of materials, energy and species along them, and between them and agricultural fields or other habitats such as woods, streams, roads, etc.

Landscape change

Landscapes change through time, and so do their structure and function. Field margins originated mainly because they were of direct advantage to farming systems; keeping stock from crops, marking ownership boundaries, as shelter belts, etc. Over time, they undergo successional changes in biota, and management will respond to technology and the needs of farming. If margins fall out of use, they are removed or become neglected and degraded. This evolution affects the structure of the field margin network and all its functions. Restoration schemes need to address specific functions of field margins. This will influence the selection of appropriate management techniques. No single margin can fulfil all potential functions.

FIELD MARGINS AS WILDLIFE HABITAT

Most of the wildlife living on farmland is dependent on fragments of once extensive semi-natural habitat. These fragments maintain populations of plants and animals on farmland that would not otherwise survive there. If we define fragmentation as the disruption of habitat continuity (Lord & Norton, 1990), then clearly most semi-natural habitats on farmland are very fragmented. Even larger habitat units such as woodlands are isolated patches in a matrix of agricultural fields in many European countries. The ecology of fragmented habitats and the species that inhabit them is a major component of modern conservation biology (Hansson, 1992; Saunders *et al.*, 1991). This also provides the most important theoretical argument for understanding the role of field boundaries in conserving biodiversity on farmland. Here I concentrate on the role of the landscape network of field margins in maintaining wildlife populations.

Metapopulation dynamics

The theory of metapopulation dynamics (Gilpin & Hanski, 1991) provides a general model for describing systems of local populations. Hanski (1989) discusses the development of metapopulation theory and its relevance to nature conservation. Many species living on farmland exist as small local populations surviving on fragmented habitat patches. They only manage to persist because their rate of extinction within patches is less than their rate of recolonisation. The important factors in the survival of metapopulations are the size and dynamics of each sub-population, and the rate of dispersal between patches (Wu *et al.*, 1993). Specific metapopulation models have been developed for plants, birds, mammals and frogs; see the reviews by Gilpin & Hanski (1991), Hansson (1992), and Opdam (1991). Field margins are patches of habitat (e.g., grassland or woodland) supporting sub-populations of species which operate as a metapopulation. These species would otherwise be rare or absent on farmland. We need to identify which species have this type of population structure and how we can improve the links between sub-populations.

Source-sink dynamics

In mosaic landscapes where species reproduce in more than one habitat, considerable variation in reproduction rates may occur. Populations in good habitats where reproduction exceeds mortality rates (sources) may contribute to other sub-populations where the reproduction rate is lower than mortality rates (sinks) (Howe *et al.*, 1991; Pulliam, 1988). Field margins are surely the source habitats for many species surviving on farmland. These species persist only if there are sufficient lengths of good quality field margin habitat in a matrix of crop fields. The pattern of field margins in the landscape affects inter-patch movement and is an important component of "source-sink" dynamics. Through source-sink processes, field margins may be supporting populations of declining species on farmland, even though surveys would find most individuals in other, non-sustaining habitats.

Movement corridors

Perhaps the most controversial claim for field margins is their function as movement corridors. There are several reviews of the value of wildlife corridors e.g., (Bennett, 1990; Hobbs, 1992; Saunders & Hobbs, 1991; Simberloff *et al.*, 1992) but there remains little evidence that field margins act as corridors linking habitat patches, enhancing the dispersal of individuals and stabilizing populations. Several studies (Baudry, 1988; Baudry, 1989; Burel, 1989; Burel & Baudry, 1990; Middleton & Merriam, 1983; Samways, 1993; Sustek, 1992; Wegner & Merriam, 1990) clearly demonstrate the extension of woodland species of plants and animals into farmland through field margins. Saunders & Hobbs (1991) also include examples of species dispersal such as birds moving along bush corridors on migration. However there is little empirical evidence of movement corridors being essential for the persistence of populations. In his review, Hobbs (1992) concludes that although we lack experimental data on the beneficial role of movement corridors, observational evidence accumulates. He recommends collecting hard data on corridor function but points to the dilemma of conservation planners who have to make decisions now.

When examining the wildlife corridor concept, we must distinguish between connectedness - an expression of the physical linkages in a landscape, and connectivity - the landscape function describing the degree of inter-connection between the sub-populations of a demographic unit. Even when field margins are linked, they may fail to connect habitat patches because some lengths are unsuitable as movement corridors (Burel, 1989). Obtaining evidence for the corridor function of field margins will take careful planning and design, as described in (Inglis & Underwood, 1992).

Field margins are important habitats for a wide range of animal and plant species found on farmland. Although there are few species found exclusively in field margin habitats, many would have restricted ranges or be absent from intensively farmed land without margins. For each species there are a set of habitat conditions most favourable for growth, and reproduction. In addition, the position of field margins in the landscape will be a major determinant of their role in maintaining species diversity. This will apply to species diversity both within hedgerow networks and on farmland generally.

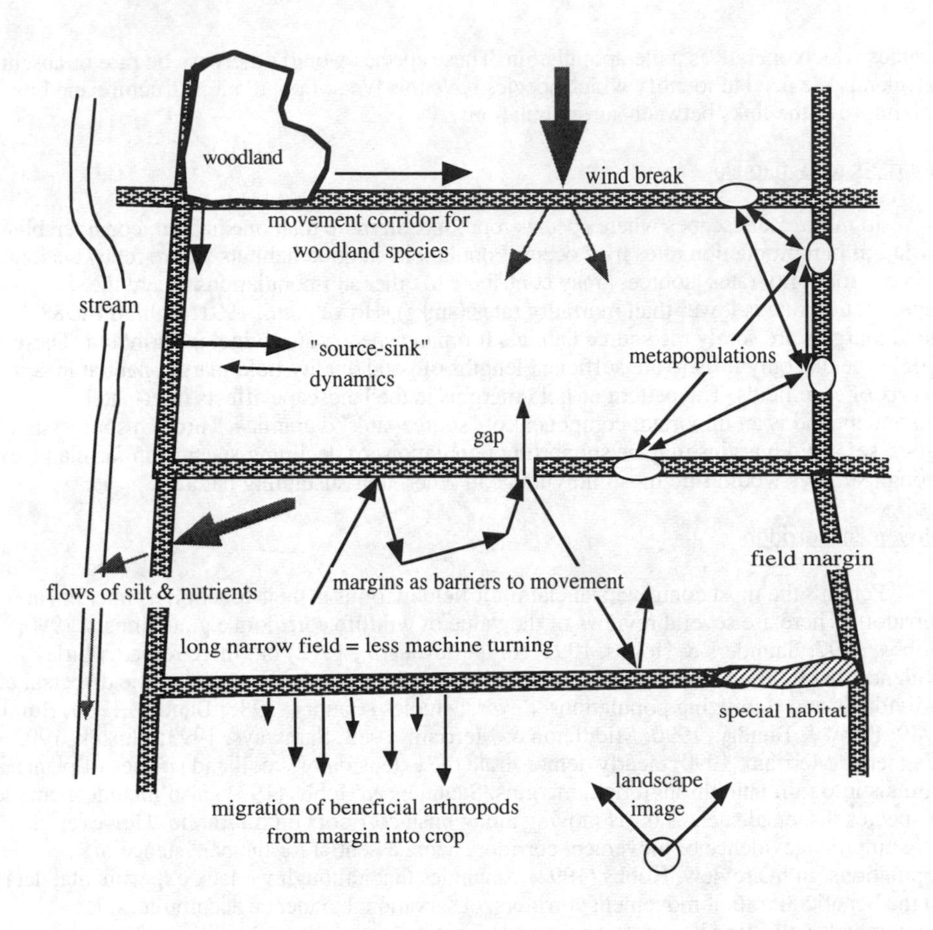

Fig. 1 Schematic view of some functions of field margins which require study at the site, field, and landscape levels.

THE ROLE OF FIELD MARGINS IN INTEGRATED FARMING

It is generally accepted that we need to move agriculture away from over-reliance on external sources of energy and chemicals and towards greater dependence on ecological services. Field margins are often recommended as landscape features worthy of protection for their value to nature and visual quality. We also have a duty to examine their role in crop production and the longer term sustainability of farming systems.

Enhancement of integrated pest management

There is wide acceptance of the importance of field margins as reservoirs of the natural enemies of crop pests (Wratten, 1988). Data is also available on the movement of beneficial arthropods from margins into crops (Coombes & Sotherton, 1986; Dennis, 1991). Many polyphagous predators have significantly higher densities close to margins (15-30m) in the spring. Thomas *et al.* (1991) found similar effects for specially designed margins (beetle banks) planted to enhance predator densities. The potential of these findings for pest control has been of great interest to integrated farming schemes. Until recently, however, few cases of pest control have been planned at the landscape level. The ever increasing power of desktop computers, the widespread availability of geostatistical software, and data on species dispersal have increased the ease of analysis and modelling of pest control at the field, farm and landscape levels (Liebhold *et al.*, 1993). The distances that natural enemies penetrate the crop (at the pest growth stage where they can reduce economic losses) could become the basis for optimizing the spacing of field margins in the landscape. To improve pest control models we need more information on the factors affecting natural enemy migration out into crops, its rate and timing. The spatial pattern of margins in the landscape clearly affects the within-field dispersal of beneficials, and, therefore, the costs and benefits of margins to agriculture.

Reducing the apparancy of crops

Crops grown in large monocultures suffer greater pest damage than when grown as polycultures or in rotation (Paoletti *et al.*, 1992). Part of the reason is that pests have to find their host plants (in space and time) and monocultures make this easy for them. Field margins can have an impact on the host searching behaviour of many pests, especially those with poor dispersal ability. The same argument supports the use of inter-cropping, but margins are more permanent strips of non-crop vegetation which may be managed to enhance natural enemies and to fragment the crop into patches, reducing its apparancy.

Shelter and wind breaks

The benefits of field margins as wind breaks and shelter are one of the few reasons why they remain as an integral part of some farming systems (Russel & Grace, 1979). The correct positioning of shelter belts in a landscape is based on simple physical attributes of the landscape such as wind direction, slope, and aspect. In hilly and mountainous areas, wooded field margins across slopes also hinder the downward flow of heavy cold air into the productive valleys.

Reducing environmental problems and landscape restoration

The ecological basis for landscape restoration, especially of degraded agricultural land, is a very active field of research (Saunders *et al.*, 1993). Both the structure of individual margins and their landscape pattern are integral components of restoration schemes to combat soil erosion, salination (Saunders *et al.*, 1993), and nutrient run-off (Haycock *et al.*, 1993). Since field margins influence the flows of materials in landscapes, they play an important role in any restoration strategy.

Reducing the impact of pesticides on non-target organisms also requires us to examine the properties of individual field margins and their spatial relationships, as well as pesticide application regimes (Sherratt & Jepson, 1993). If species move freely from field to field they may be vulnerable to correlated extinctions. In contrast, if margins have low permeability, they may prevent species escaping from pesticide application or cultivation. In ameliorating environmental problems, the arrangement of field margins in the landscape, as well as their quantity and quality, will necessitate careful planning to achieve success.

FIELD MARGINS, LANDSCAPE IMAGES, AND RECREATION

Recreation is an increasingly important product of agricultural land. Urban populations have increasing access to the countryside for a wider range of recreational activities. Field margins are important visual elements determining aesthetic appreciation of the countryside. They are associated with most footpaths, bridal paths and other access routes providing shelter and natural history experience. The work of the Game Conservancy has demonstrated that margins can enhance the game value of farms (e.g., for partridge) while improving their visual quality through wild flower and butterfly diversity. The chocolate box or calendar images, portraying traditional patch-work field structures divided by woody hedges, are highly dependent on the spatial pattern of landscapes. Appreciation of the countryside is linked to previous experience and other social factors. This causes problems when people are asked to accept change. We can expect a storm of protest if new types of margin do not resemble this visual ideal, even when rich in wildlife and aiding agricultural sustainability. We therefore need a better understanding of how the pattern of field margins interacts with people's enjoyment of the countryside.

The type and pattern of field boundaries are also important factors in selling landscape restoration schemes to farmers. In the USA, farmers resisted take-up of soil conservation plans even when offered financial incentives. Landscape architects discovered, through interview and landscape perception experiments, that the visual impression of conservation measures conveyed an image of poor stewardship. Areas of semi-natural vegetation including buffer strips and margins were given low ratings by farmers. However, if pictures of the same farms were 'landscaped' by image processing to look more tidy and cared for, e.g., addition of white board fences, objections turned to praise (Nassauer, 1992). In countryside planning we must be continuously aware of the powerful emotional feelings evoked by landscape and that field margins are very important visual components of farms.

SOME DIFFICULT MANAGEMENT QUESTIONS

Field margins as barriers

For good or bad, field margins are also semi-permeable barriers in the landscape. This is a benefit in reducing pesticide drift, hindering the flow of nutrients and silt in run-off, reducing the spread of pests and disease, and for providing shelter from wind. However, the consequences of barrier effects on species conservation is poorly understood (see also corridors above). Field margins may significantly reduce the dispersal of carabid beetles between fields (Mauremootoo & Wratten, 1993) and increase isolation between populations of meadow butterflies (Fry & Main, 1993). It may therefore be necessary to create functional gaps in hedgerow networks to allow for inter-field movement of weak dispersers. Since such gaps may be insurmountable barriers for the movement of other species along woody hedgerows, a clear conflict of interests arises. It may be possible to resolve this through better understanding of the needs of different ecological groups of species (Bink, 1989; Duelli *et al.*, 1990; Hodgson, 1993). It is, however, likely to remain a difficult management choice.

What about costs ?

The economy of scale has been a major argument for the destruction of thousands of kilometres of field margins. Increased mechanisation and the scale of farm machinery are often quoted as the rationale for the loss of hedges, walls and other margins. Recent research in Norway and Finland (Sky, 1992) has questioned the validity of this argument. Agricultural engineers found the major costs involved in cereal production are related to the time it takes for machinery to work a crop. All operations from preparation through to harvest take their toll in wear on expensive equipment, fuel consumption and manpower costs. Modelling and field trials showed that field shape plays a key role in determining the magnitude of these costs. These studies found the economy of scale argument applied only to very small field systems (at least for mechanisation), with only marginal benefits gained from fields over 5 ha (see Fig 2).

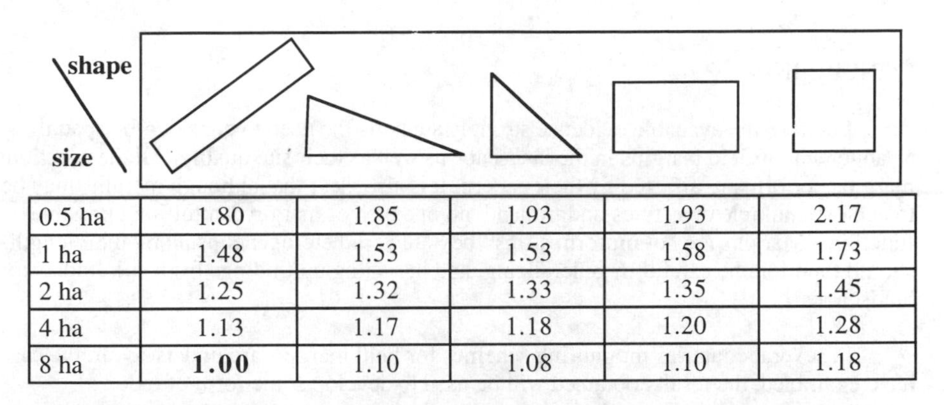

size \ shape					
0.5 ha	1.80	1.85	1.93	1.93	2.17
1 ha	1.48	1.53	1.53	1.58	1.73
2 ha	1.25	1.32	1.33	1.35	1.45
4 ha	1.13	1.17	1.18	1.20	1.28
8 ha	**1.00**	1.10	1.08	1.10	1.18

FIG. 2 An example of time costs associated with field shape and size for cereal production (from Sky, 1992). Comparisons are shown against the time cost for the most efficient shape, shown in **bold** type (1.0 = c.10 h/ha).

Large fields are associated with the most serious environmental problems in Norway and probably elsewhere. If the costs caused by these problems were taken into account, it would reduce even further the benefits of scale. It would seem that it is more important to have an optimal shape of field (long and narrow) to reduce turning large tractors and harvesters than to increase the size of field with the consequent loss of margin habitat.

Land ownership and landscape management

Although there is a strong emotional bond between a farmer and the boundaries of his holding, landscape management will call for greater co-operation between farms. This may require a new look at the responsibilities of land ownership. Otherwise, the introduction of environmental measures on farmland will be unfair. Costs will be unevenly distributed, leaving some farmers economically disadvantaged while neighbouring farms benefit from those actions. This is not new thinking; farmers have shared resources, such as access to water or isolated pastures for many thousands of years.

Various schemes of official land consolidation have evolved to redistribute land to increase farming efficiency, especially where holdings are fragmented and difficult to manage as a unit. Around 20% of farmland in Europe have been reorganized in this way in recent times - often with no consideration of the environmental consequences and resulting in major losses of hedgerows (Baudry & Burel, 1984). The scope of official land consolidation is being broadened to explore the potential for adjusting farm boundaries to meet environmental objectives. Similar processes also occur in most countries without formal organization. Farmers work together to tackle the threats posed by flooding, avalanches, soil erosion or pollution. For example, in the wheatbelt of Western Australia, Land Care groups have introduced catchment management plans to combat salination and soil erosion. These co-operative projects across ownership boundaries aim to even out the costs and benefits to participating farmers (Saunders *et al.*, 1993). Voluntary approaches, where whole catchments or valleys work as a single management unit, benefit the environment and farming efficiency. To manage field margin networks, co-operation between farmers and planning at a larger scale than the individual farm will be essential.

CONCLUSION

I believe the available evidence strongly supports the need to consider the spatial arrangement of field margins in the landscape as well as their site quality. All the functions that margins perform are affected by their pattern. It is also clear that although margins may be able to achieve multiple objectives on farmland, no one type of margin can fulfil all possible functions. Margins are not uniform strips, they are very heterogeneous along their length. We should therefore be careful in generalizing, and in scaling up findings from site studies to landscapes.

In several countries monitoring schemes for field margins are underway. In the cases I have examined, the results obtained will be used to develop some form of index of environmental quality. Few schemes include a range of field margin types (emphasis being on hedges); some, but not all, include an assessment of quality; and very few make any attempt to assess the landscape pattern of margins. Classification into type, assessment of quality (against clearly defined criteria) and the patterns of margin networks are all needed for a comprehensive assessment and to make comparisons between areas.

We are only beginning to understand the effects of different spatial patterns of field margins on their function. The available data suggests that long narrow fields, following the contours of the land would give the best return for investment. Rectangular fields are the most efficient for machinery, for the dispersal of beneficial insects and for reducing silt and nutrient run-off. If we combine this knowledge with data on the merits of different types of margin, we should be able to optimize the value of margins to both agriculture and nature conservation.

ACKNOWLEDGEMENTS

I thank the BCPC for their support. Jon Marshall and Wendy Robson provided helpful comments on the manuscript. This review was funded by NINA and the Directorate for Nature Conservation.

REFERENCES

Baudry, J. (1988) Hedgerows and hedgerow networks as wildlife habitat in Europe. In: *Environmental Management in Agriculture.* Park, J. R. (Ed.), London: Belhaven Press. pp. 111-124.

Baudry, J. (1989) Interactions between agricultural and ecological systems at the landscape level. *Agriculture, Ecosystems and Environment*, **27**, 119-130.

Baudry, J. ; Burel, F. (1984) Landscape project "remembrement": landscape consolidation in France. *Landscape Planning*, **11**, 235-241.

Bennett, A. F. (1990) *Habitat Corridors - Their role in wildlife management and conservation.* Melbourne: Department of Conservation and Environment.

Bink, F. A. (1989) The butterflies of the future, their strategy. In: *Future of butterflies in Europe, strategies for survival.* Pavlicek-van Beek, T. ; Ovaa, A. H. ; Van der Made, J. G. Agricultural University Wageningen. pp. 134-138.

Burel, F. (1989) Landscape structure effects on carabid beetles spatial patterns in western France. *Landscape Ecology*, **2**, 215-226.

Burel, F. ; Baudry, J. (1990) Structural dynamic of a hedgerow network landscape in Brittany France. *Landscape Ecology*, **4**, 197-210.

Coombes, D. S. ; Sotherton, N. W. (1986) The dispersal and distribution of polyphagus predatory Coleoptera in cereals. *Annals of Applied Biology*, **108**, 461-474.

Dennis, P. (1991) The temporal and spatial distribution of arthropod predators of the aphids *Rhopalosiphum padi* W. and *Sitobion avenae* (F.) in cereals next to field-margin habitats. *Norwegian Journal of Agricultural Sciences*, **5**, 79-88.

Duelli, P. ; Studer, M. ; Marchland, I. ; Jakob, S. (1990) Population movements of arthropods between natural and cultivated areas. *Biological Conservation*, **54**, 193-207.

Fahrig, L. ; Merriam, G. (1985) Habitat patch connectivity and population survival. *Ecology*, **66**, 1762-1768.

Forman, R. T. T. ; Godron, M. (1986) *Landscape Ecology*. New York: John Wiley & Sons.

Fry, G. L. ; Main, A. R. (1993) Restoring seemingly natural communities on agricultural land. In: *Reconstruction of fragmented ecosystems.* Saunders, D. A., Hobbs, R. J.; Ehrlich, P. R. (Ed.). Chipping Norton: Surrey Beatty & Sons. pp. 225-241.

Gilpin, M. ; Hanski, I. (Ed.), (1991) *Metapopulation Dynamics: Empirical and Theoretical Investigations.* London: Academic Press.

Hanski, I. (1989) Metapopulation dynamics: does it help to have more of the same? *Trends in Ecology and Evolution*, **4**, 113-115.

Hansson, L. (Ed.), (1992) *The Ecological Principles of Nature Conservation.* New York: Elsevier Science Publishers Ltd.

Haycock, N. E. ; Pinay, G. ; Walker, C. (1993) Nitrogen retention in river corridors: European perspective. *Ambio*, **22**, 340-346.

Hobbs, R. J. (1992) The role of corridors in conservation - solution or bandwagon. *Trends In Ecology And Evolution* , **7**, 389-391.

Hodgson, J. G. (1993) Commonness and rarity in British butterflies. *Journal of Applied Ecology*, **30**, 407-437.

Howe, R. W. ; Davis, G. J. ; Mosca, V. (1991) The demographic significance of 'sink' populations. *Biological Conservation*, **57**, 239-255.

Inglis, G. ; Underwood, A. J. (1992) Comments on some designs proposed for experiments on the biological importance of corridors. *Conservation Biology*, **6**, 581-586.

Liebhold, A. M. ; Rossi, R. E. ; Kemp, W. P. (1993) Geostatistics and geographical information systems in applied insect ecology. *Annual Review of Entomology*, **38**, 303-327.

Lord, J. M. ; Norton, D. A. (1990) Scale and the spatial concept of fragmentation. *Conservation Biology*, **4**, 197-202.

Mauremootoo, J.R. ; Wratten, S.D. (1993) Permeability of field boundaries to predatory carabid beetles. *IOBC/WPRS Bulletin* (in press).

Merriam, G. (1988a) Landscape dynamics in farmland. *Trends in Ecology and Evolution*, **3**, 16-20.

Middleton, J. ; Merriam, G. (1983) Distribution of woodland species in farmland woods. *Journal of Applied Ecology*, **20**, 625-644.

Morris, M. G. ; Webb, N. R. (1987) The importance of field margins for the conservation of insects. In: *Field Margins, BCPC Monograph 35* . Way, J. M. & Greig-Smith, P. W. (Ed.), Thorton Heath: BCPC Publications, pp. 53-65

Nassauer, J. I. (1992) The appearance of landscape as matter of policy. *Landscape Ecology*, **6**, 239-250.

Opdam, P. (1991) Metapopulation theory and habitat fragmentation: a review of holarctic breeding bird studies. *Lanscape Ecology*, **4**, 93-106.

Paoletti, M. G. ; Pimentel, D. ; Stinner, B. R. ; Stinner, D. (1992) Agroecosystem biodiversity: matching production and conservation biology. *Agriculture, Ecosystems and Environment*, **40**, 3-23.

Pulliam, H. R. (1988) Sources, sinks, and population regulation. *The American Naturalist*, **132**, 652-661.

Russell, G. ; Grace, J. (1979) The effect of shelter on the yield of grasses in southern Scotland. *Journal of Applied Ecology*, **16**, 319-330.

Samways, M. (1993) *Insect Conservation Biology*. London: Chapman & Hall.

Saunders, D. A. ; Hobbs, R. J. ; Ehrlich, P. R. (Ed.), (1993) *Nature Conservation 3: Reconstruction of fragmented ecosystems*. Chipping Norton: Surrey Beatty & Sons.

Saunders, D. A. ; Hobbs, R. J. ; Margules, C. R. (1991) Biological consequences of ecosystem fragmentation: a review. *Conservation Biology* , **5**, 18-32.

Saunders, D. A. ; Hobbs, R.J. (Ed.), (1991) *Nature Conservation 2. The Role of Corridors*. Chipping Norton, Australia: Surrey Beatty & Sons.

Sherratt, T. N. ; Jepson, P. C. (1993) A metapopulation approach to predicting the long-term effects of pesticides on invertebrates. *Journal of Applied Ecology*, **30**, (in press).

Simberloff, D. ; Farr, J. A. ; Cox, J. ; Mehlman, D. W. (1992) Movement corridors: conservation bargains or poor investments? *Conservation Biology* , **6**, 493-504.

Sky, P. K. (1992). The use of geographic information systems in the Land Consolidation Service of Norway (in Nowegian). *Kart og Plan*, 51, 307-313.

Sustek, Z. (1992) Windbreaks and line communities as migration corridors for carabids (*Col.Carabidae*) in the agricultural landscape of South Moravia. *Ekologia*, **11**, 259-271.

Thomas, M. B. ; Wratten, S. D. ; Sotherton, N. W. (1991) Creation of 'island' habitats in farmland to manipulate populations of beneficial arthropods: predator densities and emigration. *Jounal of Applied Ecology* , **28**, 906-917.

Turner, M. G. ; Gardner, R. H. (Ed.), (1991) *Quantitative Landscape Ecology*. New York: Springer-Verlag.

Way, J. M. ; Greig-Smith, P. W. (Ed.), (1987) *Field Margins*. British Crop Protection Council Monograph No. 35. Thornton Heath, BCPC Publications.

Wegner, J. ; Merriam, G. (1990) Use of spatial elements in a farmland mosaic by a woodland rodent. *Biological Conservation*, **54**, 263-276.

Wratten, S. D. (1988) The role of field margins as reservoirs of beneficial insects. In: *Environmental Management in Agriculture* . Park, J. R. (Ed.), London: Belhaven Press. pp. 144-150.

Wu, J. ; Vankat, J. L. ; Barlas, Y. (1993) Effects of patch connectivity and arrangement on animal metapopulation dynamics: a simultation study. *Ecological Modelling*, **65**, 221-254.

Session 2

Field Margins as Wildlife Habitats

Session Organiser
& Chairman DR N W SOTHERTON

BOTANICAL DIVERSITY IN BRITISH HEDGEROWS

R G H BUNCE, D C HOWARD & C J BARR

Institute of Terrestrial Ecology, Merlewood Research Station
Grange-Over-Sands, Cumbria LA11 6JU

R C CUMMINS & D FRENCH
Institute of Terrestrial Ecology, Banchory Research Station, Hill of Brathens, Glassel
Banchory, Kincardineshire AB31 4BY

ABSTRACT

The Institute of Terrestrial Ecology conducted surveys on the countryside of Great Britain in 1978 and 1990, collecting detailed botanical data on the composition of hedgerows. These data were analysed to define the botanical diversity currently present in British hedgerows and the changes that have taken place over that period. Classifications of hedgerows are described based on woody and herbaceous species separately. No particular type was lost preferentially but diversity declined, mainly in pastural landscapes.

INTRODUCTION

Although the botanical interest in hedgerows has concentrated on the woody species, there has also been much interest in the vegetation growing at the base of the hedge. Pollard *et al.* (1974) and Rackham (1986) provide the ecological background on the hedges in Britain. Of particular interest has been the work of Hooper (1970) who showed that generally the older the hedge the more woody species it contained. Thus, he reported that some sites where hedges had been established for over six hundred years, there may be 6-10 woody species in a 30 m length, whereas hedges less than one hundred years old usually have fewer than four species and often tended to be mono-typic with hawthorn predominant. However, although this relationship is broadly correct the actual ratio, ie number of woody species to age of hedge will vary regionally. Thus, old hedges in the north of England would be likely to have fewer species than similar aged hedges in the south, due to geographical limitations on the distribution of some species, eg *Acer campestre* and *Viburnum lantana*. Few regional studies have been carried out to support the general relationship expressed by Hooper and this is even more so in relation to the ground vegetation. Marshall (1989) has described details of hedge floras around fields in the south and east of Britain, but otherwise only fragmentary information is available. Bunce & Hallam (1993) described the difference between the flora of the hedgerows and the surrounding fields, emphasising that many species in lowland Britain in particular are only present in hedgerows, rather than the open fields surrounding. Thus, in the south, species like *Digitalis purpurea* and *Teucrium scorodonia* would be found only in hedgerows and not in the surrounding fields. The National Vegetation Classification describes hedgerows in the woodland context.

The results described in the present paper were derived from the surveys of the countryside of Great Britain conducted by the Institute of Terrestrial Ecology (ITE) in 1978 and 1990 (Barr *et al.*, 1993). Data were collected for a wide range of features of the countryside but the present paper describes those relating to the vegetation and management of hedgerows and the adjacent land use. The Department of the Environment and the former Nature Conservancy Council part-funded the 1990 survey but the previous survey was carried out by the Natural Environment Research Council.

METHODS

The sampling strategy was the same in both the 1978 and 1990 surveys, ie a random sample of 1 km squares throughout GB, stratified according to ITE Land Classes (Barr *et al.*, 1993). In 1978, 256 squares were sampled, which was increased to 508 squares in the 1990 survey. The field survey methods were similar in both years and included:

a. mapping the hedges within a square, and
b. describing the hedge floristics using quadrats.

In 1978, two plots were placed along hedges in each square using a set procedure to ensure that the plots were well separated in the 1 km square. Plots were relocated if there was not a clear metre between the centre of the hedge and another linear feature. At each plot a 10 x 1 m quadrat was set out running parallel to the hedge with its 10 m base along the centreline of the hedge. The species and percentage cover, where greater than 5%, were recorded for all plants rooted in the quadrat. In 1990, 259 of the plots recorded in 1978 were reliably relocated and, if the hedge was still present, the recordings were repeated. In 1990 only, further hedge plots were recorded beside a random selection of boundaries. Altogether, 278 plots from 1978 and 918 from 1990 were used in the analysis. Initial TWINSPAN analysis of all the species together, ie woody and herbaceous, gave a classification which was confused by the large information present. It was therefore decided to separate the woody and herbaceous species and create separate classifications, which was also convenient for the interpretation of the data. Percentage cover was therefore used to classify the woody species and presence/absence data to classify the herbaceous species within the hedge bottom. The relationships between species, regardless of the hedgerow classes, were examined using the statistical procedure DECORANA. The scores from that analysis were then used within nearest neighbour analysis (woody species) or Wards Minimum Variance Analysis (herbaceous species) to determine species associations. Full details of these analyses and the groupings that they produced are provided in Cummins *et al.* (1992). The following terminology is used in the paper:

- the woody species component will be referred to as the hedge
- the herbaceous component as the hedge bottom
- the two components together as the hedgerow
- Hedges are referred to by their woody species class and hedge bottoms by their herbaceous groups.

RESULTS

Hedgerow classifcation

The classification of woody species initially divided a series of less comon hedges, all but one having fewer than 20 plots in the sample (Table 1). The amount of hawthorn present was a strong factor in sub-dividing the remainder core hedge types. Eleven classes of hedges are summarised in Table 1. The four species profiles of the hedgerows are described in Cummins *et al.* (1992) but the most widespread type was the hawthorn dominant type 4a, probably associated with the extensive planting of hawthorn during the enclosures. The two most diverse types are those of 4b, mixed hawthorn type, more typical of eastern Britain and the mixed hazel type, 5b, more typical of the west. Certain types, eg type 8, was dominated by a single species, in this case, gorse. The TWINSPAN analysis of herbaceous species produced 70 interpretable classes. However, these were summarised at Level 3 of the hierarchy to give four composite groups which were used subsequently to indicate the types of hedge bottom floras. Many species occur throughout the groups but are present in different quantities. The four herbaceous groups may be described as typical of the following types of land use:

HG1 - arable and crop land
HG2 - other intensively managed ground (mainly lowland)
HG3 - rough grazing and less intensively managed grasslands
HG4 - woodland vegetation

TABLE 1. TWINSPAN hedge classes as described by Cummins *et al.* (1992), types and the number of plots in each class.

Class	Type	No of plots
1a-d	mostly planted non-native species	9
2	wild privet present	5
3	beech dominant	19
4a	hawthorn	552
4b	mixed hawthorn	61
4c	elder/hawthorn	40
5a	willow or rose dominant	6
5b	mixed hazel premoninant	157
6	blackthorn predominant	270
7	elm predominant	49
8	gorse dominant	8

The structure of the TWINSPAN classifications can be used to categorise the characteristics of the different types and these are recognisable in the field. Further work using the discriminant function could produce more robust keys.

The strength of the associations between the woody species and the herbaceous species were assessed by the two classifying techniques described in the analysis procedures. The clustering of the woody species showed two groupings. The first, typical of long-lived woodland species which are all native except for sycamore. The second group was also native with the exception of oak and blackthorn, but do not live as long as those in the first cluster. Apart from that, a group of exotics, the reasons for clustering other species were not obvious, often tending to be incidental, associated with strongly dominant species. Thus, the conclusion from this is that there was a core of native species separated into the longer-lived as opposed to the shorter-lived, more temporary species, with individual other exotics, such as beech, interpolated between them.

Twelve herbaceous species associations were identified, summarised in Table 2 according to the main habitat types which they represent. The full list of the species comprising these types is given in Cummins *et al.* (1992). Despite the physical dominance of the woody components of hedges, it was clear that the vegetation of hedge bottoms is not confined into associations typical of woodlands. The occurrence of species from the woodland clusters, 4b and 6, together accounted for only 15% of the total. The unmanaged nature of many hedgerows was indicated by the 25% occurrence of species typifying abandoned or derelict areas. As emphasised in the report on the Countryside Survey 1990 (Barr *et al.*, 1993) these results emphasised the significance of hedgerows in maintaining species of open habitats within intensively farmed landscapes. The association between the clusters of the herbaceous species and the classification of the overall composition of the hedge bottom species are shown in Table 3. Thus, the arable group was associated with clusters typical of arable land, ie those of nitrogen and phosphorus rich areas and arable weeds. The second group included species from a wide range of habitats, having no strong positive affinities. The third group could be generally summed as grasslands, due to the positive association with all four grassland clusters and negative associations with the clusters more characteristic of arable areas and unmanaged ground. The fourth group was the only group having positive affinities with woodland species. Detailed analysis of the results showed that the vegetation was associated primarily with a gradient in management levels, from high to low intensity. This association with management, rather than with other factors, suggested that the adjacent land use was affecting the species composition of the hedge bottom in many areas. The herbaceous species clusters were therefore likely to be affected more by the adjacent land use than by the hedge type. Therefore, few classes of hedge will have a specific type of ground flora with the possible exception of the mixed hazel hedges and woodland herbaceous plants of the west of Britain. It is therefore important that both the woody and the herbaceous components of the hedgerows are assessed separately, which has important implications in future management proposals for hedgerows.

TABLE 2. The occurrence of herbaceous species in different clusters as a percentage of all herbaceous species records.

Cluster	Habitat type	% of records
1a	meadow	19
1b	short-term grassland	8
2	old meadow	5
3	nitrogen/phosphorus-rich areas	1
4a	acid grassland	3
4b	acid woodland	5
5	abandoned/derelict areas	22
6	lowland wood/scrub	10
7	arable weeks	2
8a	humus-rich basophiles	14
8b	southern derelict areas	3
9	common weeds	7

TABLE 3. Types of hedge-bottom and herbaceous species clusters with at least 1.5 times more (+) or fewer (-) plots than expected statistically.

		Hedge-bottom (HG)			
Species cluster		1	2	3	4
1a	(meadow)	-		+	
1b	(short-term grassland)	-		+	-
2	(old meadow)	---		++	
3	(N/P rich)	+		--	---
4a	(acid grassland)	---		++	++
4b	(acid woodland)	--	-		+++
5	(abandoned/derelict)	+		--	---
6	(lowland wood)				+
7	(arable weeds)	+		-	---
8b	(southern derelict)			-	

+ indicates observed value is at least 1.5 times expected. ++ at least twice expected and +++ at least 3 times expected. Similarly, - -- and --- indicate expected is 1.5, 2 or 3 times observed

Diversity

Bunce *et al.* (1992) described the various approaches which have been used to define diversity and concluded that direct records of species number was one of the most appropriate measures of diversity. The results of the species diversity with hedge class are shown in Table 4. The diversity of woody species was notably low in Class 4a, hawthorn dominant, and high in Classes 4b and 5b. Of the rarer classes, gorse hedges were possibly species-rich, whereas those of the elm were probably the poorest. The diversity of herbaceous species increases by about 80% in passing through the series. This is demonstrated in Table 5. The difference between the woody species values was not as pronounced as Table 4, although still significant, but the woody species contribute comparatively little to the overall diversity because of the far more abundant herbaceous species. There is therefore, little association between the diversity of woody species and herbaceous species within the two classifications separately, ie plots which were rich in woody species will not necessarily be rich in herbaceous species and vice versa. The results suggested that the overall species richness of hedgerows was not determined solely by either the woody component or the herbaceous component, although it is weighted by the latter, simply because there are more herbaceous species than woody species. Furthermore, the numerical diversity of species in hedgerows should not be considered in isolation from the types of vegetation they contain. Thus, two hedge bottoms may contain similar numbers of species, but one that is representative of, say, an old meadow established over many years will be more difficult to recreate than a hedge bottom with a large number of invasive weed species. Analysis of the relationship between management and the diversity of the hedgerows showed that very intensive land management and no management at all were both deleterious to the diversity of herbaceous species. In some situations, the indications of the analysis suggested that limited grazing could be a valuable management tool. Hedgerows adjacent to roads, tracks and wooded ground were particularly rich in woody species. Management of the hedge itself had no significant affect on species diversity either within or between the hedge. However, the hedge should be managed to maintain it as a coherent feature and to prevent its degeneration, eg into a row of trees, because otherwise different species would not be able to maintain themselves within the hedge boundary. At least 60% of the samples of both species-rich type of hedges had more than 10% gaps in their length. Rejuvenation of these types is probably therefore necessary in order to maintain the current mixture of species and the maintenance of their diversity.

TABLE 4. Mean number of species per plot in hedge classes with more than 10 plots. Figures in parentheses are standard errors.

	Hedge class						
	3	4a	4b	4c	5b	6	7
Woody spp	2.8 (0.3)	2.0 (0.04)	3.6 (0.2)	2.7 (0.2)	4.4 (0.1)	3.2 (0.1)	2.9 (0.2)
Herbaceous spp.	15.1 (1.2)	12.8 (0.2)	12.5 (0.7)	12.0 (0.8)	15.3 (0.6)	14.1 (0.4)	11.4 (0.8)
Total	17.9 (1.3)	14.8 (0.2)	16.1 (0.7)	14.8 (0.9)	19.7 (0.6)	17.3 (0.4)	14.3 (0.9)

TABLE 5. Mean number of species per plot in hedge-bottoms (HGs). Figures in parentheses are standard errors.

	HG			
	1	2	3	4
Herbaceous spp	10.2 (0.2)	14.4 (0.3)	18.2 (0.6)	18.1 (0.6)
Woody spp	2.5 (0.1)	3.0 (0.1)	2.2 (0.1)	3.3 (0.1)
Total	12.7 (0.2)	17.4 (0.3)	20.4 (0.6)	21.4 (0.6)

Changes in hedgerows between 1978 and 1990

Two hundred and fify-nine of the plots sampled in 1978 were reliably relocated in 1990 and were used to assess the changes in the hedgerows. Losses of the more common types of hedge between 1978 and 1990 from all causes, including neglect, range from 21% (hawthorn dominated and a blackthorn) to 31% (mixed hawthorn). However, there was no evidence that any one particular type of hedge was being lost to a greater degree than any other type. The same analysis was carried out on the herbaceous hedge classification, with similar conclusions. Loss of hedgerows was therefore proportionately taking place throughout all the major types of hedges. The overall loss was 24%, which compares with a 25% loss of hedge length for the same period calculated from the results given by Barr *et al.* (1991). Both reports showed similar losses due to removal over the 12 years. Full details of these changes are described in Cummins *et al.* (1992).

In relation to the average cover of the different species there were increases and decreases in the average cover of both woody and herbaceous species. The only statistically significant change amongst woody clusters was an increase in the maple/sycamore cluster and a decrease in the gorse cluster. The 10% decrease in the wych elm association (cluster 3) although not significant, may be the result of Dutch elm disease. In contrast, the English elm cluster increased by about 5% which could be due to recovery by its abundant suckers. Three of the changes in herbaceous clusters were statistically significant:

1 There was an increase in the cover of species in cluster 5 (abandoned derelict such as would occur with the removal of grazing or an increase in the size of headlands in arable areas).
2 The cover of cluster 6 (lowland woodland) also increased, reflecting the changes described above.
3 The decrease in cover of species from short-term grasslands is consistent with an increase in arable land.

The change in individual species is best summarised by looking at the different landscape types described by Barr *et al.* (1990). These results are shown in Table 6. There was a decline in diversity in both the landscape types with large samples, but only the pasture was significant. These results were consistent with other results reported in the Countryside Survey 1990 and showed that the biggest changes over the period of time were within pastural landscapes.

TABLE 6. Change (1978-90) in species numbers recorded in paired plots placed along hedgerows, by landscape type.

Landscape type	No of plots	Mean species no 1978	Mean species no 1990	Change in mean species no.	SE of change	P
Arable	116	11.0	10.2	-0.8	0.6	
Pastural	111	14.6	13.1	-1.5	0.6	*
Marginal upland	24	17.0	17.5	0.4	1.3	
GB	251	13.1	12.2	-1.0	0.4	*

(Significance is based on pair t-test. Probability (P) * <0.1).

In relation to changes in the individual species, out of more than 450 species recorded, the percentage cover of 33 species changed significantly between 1978 and 1990 (Table 7). Exactly two-thirds of these changes were decreases. In relation to woody species, sycamore, which is very invasive, was the only species to increase in cover, whereas dogwood and gorse, both relatively uncommon, decreased. Wych elm was the only large tree species to decrease, by almost 5%, probably due to Dutch elm disease. A major increase in the cover of *Hedera helix* and bramble suggested that hedge management has declined in some places, resulting in increased shading of the hedge bottom. Observations suggest that *Clematis vitalba* has spread extensively on the chalk. Most of the other species which increased are aggressive/invasive species. In contrast, the species which have decreased in cover came from a wide range of habitats.

CONCLUSIONS

In conclusion therefore, not only have hedges been lost, but there has also been a general decrease in the number of both woody and herbaceous species per plot. The results do not pinpoint any common type of hedge that was particularly susceptible to loss, except that gorse hedges appeared to be vulnerable. National proportions of the different hedge types were unaltered and therefore there is no reason to target any particular types of hedges solely on the grounds of susceptibility to change. The vegetation of the hedge bottom is strongly influenced by the adjacent land use. Hence, changes associated with land use and the intensity of management are more likely to reflect the hedge flora than other factors.

TABLE 7. Species whose overall cover changed significantly between 1978 and 1990 in paired plots

n + No. of plots. m.d. = Mean Difference between % cover in 1978 and 1990. s.e. = Standard Error of mean difference. sig. = significance: * p < 0.1 ** p < 0.01 *** p < 0.001

	n	m.d.	S.E.	P
Species increasing				
Acer pseudoplatanus	29	10.3	4.6	*
Agrostis stolonifera	91	2.7	0.98	**
Bromus sterilis	60	4.1	2.0	*
Convolvulus arvensis	41	1.4	0.66	*
Festuca rubra	70	2.6	1.2	*
Galium aparine	172	5.0	1.1	***
Hedera helix	108	10.5	2.4	***
Mercurialis perennis	19	3.1	1.8	*
Rubus fruticosus agg.	179	4.1	1.6	*
Sochus oleraceus	12	0.5	0.26	*
Trifolium repens	24	1.6	0.72	*
Species decreasing				
Achillea millefolium	31	-1.1	0.45	*
Agrostis gigantea	10	-0.8	0.2	**
Arrhenatherum elatius	166	-3.3	1.68	*
Arum maculatum	27	-2.0	0.75	*
Campanula rotundifolia	7	-0.6	0.20	*
Cardamine hirs/flex	5	-0.6	0.24	*
Cerastium fontanum	38	-0.4	0.13	**
Cirsium vulgare	50	-0.7	0.14	***
Cornus sanguineus	10	-10.1	3.8	*
Corylus avellana	46	-5.5	2.9	*
Epilobium montanum	21	-1.0	0.52	*
Filipendula ulmaria	15	-1.4	0.46	**
Holcus lanatus	117	-3.7	1.4	**
Plantago lanceolata	19	-0.4	0.19	*
Poa trivialis/nemoralis	107	-5.3	1.1	***
Potentilla erecta	10	-0.5	0.22	*
Silene dioica	38	-1.8	0.93	*
Stellaria media	69	-0.5	0.16	**
Taraxacum agg.	60	-0.2	0.14	*
Ulex europaeus	10	-21.2	8.5	*
Ulmus glabra	18	-4.9	2.5	*
Vicia sepium	19	-0.4	0.21	*

REFERENCES

Barr, C.J.; Howard, D.C.; Bunce, R.G.H.; Gillespie, M.K.; Hallam, C.J. (1991) *Changes in hedgerows in Britain between 1984 and 1990*. NERC contract report to the Department of the Environment. Institute of Terrestrial Ecology: Grange-Over-Sands.

Barr, C.J.; Bunce, R.G.H.; Clarke, R.T.; Fuller, R.M.; (Furse, M.T.); Gillespie, M.K.; Groom, G.B.; Hallam, C.J.; Hornung, M.; Howard, D.C.; Ness, M.J. (1993) *Countryside Survey 1990*. Main Report. London: Department of the Environment.

Bunce, R.G.H.; Hallam, C.J. (1993) The ecological significance of linear features. In: *Landscape ecology and agroecosystems*, R.G.H. Bunce, L. Ryzkowski and M.G. Paoletti (Eds), 11-19. Boca Raton: Lewis.

Bunce, R.G.H.; Howard, D.C.; Hallam, C.J.; Barr, C.J.; Benefield, C.B. (1993) *Ecologial consequences of land use change*. London: Department of the Environment.

Cummins, R.C.; French, D.; Bunce, R.G.H.; Howard, D.C.; Barr, C.J. (1992) *Diversity in British hedgerows*. (NERC contract report to the Department of the Environment). Institute of Terrestrial Ecology: Grange-Over-Sands.

Marshall, E.J.P. (1989) The ecology and management of field margin floras in England. *Outlook on Agriculture*, **17**, 178-182.

Pollard, E.; Hooper, M.D.; Moore, N.W. (1974) *Hedges*. London: Collins, pp256

Rackham, O. 1986. *The history of the countryside*. London: Dent, pp 445

BOTANICAL DIVERSITY IN ARABLE FIELD MARGINS.

P.J. WILSON.

The Farmland Ecology Unit, The Game Conservancy Trust, Fordingbridge, Hampshire. SP6 1EF

ABSTRACT

The plant communities of Britain's arable field margins have evolved over the past 5000 years. There is considerable variation within the arable flora, chiefly related to soil type and climate. Farming practices have however undergone radical changes since 1945, and many species associated with arable cultivation have become extremely rare. Most of the sites where species-rich communities survive are on calcareous soils in the south and east of England. In addition to overall reductions in range and abundance of these species, they are in most cases now confined to the extreme edges and corners of fields, and the removal of field margins in recent years has contributed to the decline of the arable flora.

INTRODUCTION

Arable crops have been grown in Britain since about 3000 BC. The weed communities of these early arable fields are unknown, but it may justifiably be said that these plant communities have an origin of similar antiquity to coppiced woodlands, hay meadows and other anthropogenic habitats.

Farming has undergone great changes during the past 200 years, particularly since 1945. Since then, herbicides have become the major method of weed control, amounts of nitrogen applied to cereal crops have increased by over 600%, and highly nitrogen-responsive crop varieties have been developed. Mechanisation has been almost complete, and fields have been enlarged to facilitate the use of large machinery (Wilson, 1990).

These and other developments have had profound consequences for the flora of Britain's arable field margins, and have imposed an impoverished uniformity on the arable ecosystem. Conservation of the arable flora is therefore an urgent priority. In order to carry out effective conservation however, an understanding of the variation within arable vegetation is necessary, as is an understanding of the processes affecting it.

THE ORIGINS OF BRITAIN'S ARABLE FLORA

H. Godwin (1956) recorded 31 species now characteristic of arable habitats from the fossil record of the last inter-glacial period. While there is no evidence for continuity of habitat since then, Godwin considered that suitable open habitat could have remained between the last glaciation and the start of arable farming. Several arable species including *Petroselinum segetum, Galeopsis angustifolium* and *Ranunculus parviflora* still occur in semi-natural habitats. Arable agriculture would have opened up large areas of land for such species to expand into.

Other species probably arrived with successive waves of human immigration and opening of new trade routes. The majority of these originated around the Mediterranean, are at the climatic limits of their distributions in north-western Europe, and are therefore particularly susceptible to environmental changes (Holzner, 1979). Other species have arrived more recently. *Veronica persica* was first recorded in 1825 and is of Asian origin, while the New World was responsible for *Matricaria matricarioides* (Salisbury, 1961). Other species including *Bromus sterilis* and *Galium aparine* originating in habitats adjacent to arable fields have become prominent in recent years.

VARIATION WITHIN BRITAIN'S ARABLE FLORA

Britain's arable flora includes over 200 species. As arable agriculture extends from the Scilly Isles to the Shetlands, and ranges from the intensive cereal growing of south-eastern England to the machair farming of the Hebrides, there is much potential for variation within the associated flora.

Silverside's phytosociological study of Britain's arable flora detected 15 plant associations (Silverside, 1976)(Table 1). A survey aimed at areas where uncommon arable species were still known to occur (Wilson, 1990) recorded four of the five alliances recorded by Silverside, the *Arnoseridion minimi* having become extremely rare in Britain.

Although many species such as *Veronica persica, Polygonum aviculare* and *Stellaria media*, are ubiquitous throughout Britain, most are restricted either geographically or to particular soil types or farming practices. The majority of sites that still support rich arable floras are to be found on calcareous soils in the south-east of England. At their richest, such communities include *Adonis annua, Papaver hybridum, Papaver argemone, Petroselinum segetum, Galeopsis angustifolium, Buglossoides arvensis, Fumaria parviflora, Fumaria densiflora* and *Valerianella dentata*. *Papaver hybridum, P. argemone, V. dentata* and *F. densiflora* can all be quite common in localised areas of south-east England. Species such as *Scandix pecten-veneris, Euphorbia platyphyllos* and *Torilis arvensis* also occur on chalky soils, but are less typical. Such vegetation corresponds to Silverside's *Caucalidion lappulae* and the *Caucalidion platycarpi* of central Europe (Hüppe and Hofmeister, 1990). Several species extinct in Britain or no longer known from true arable habitats, including *Caucalis platycarpos, Melampyrum arvense, Bunium bulbocastaneum, Bupleurum rotundifolium* and *Ajuga chamaepitys*, are characteristic of the most thermophilous of these communities in Germany (Hüppe & Hofmeister, 1990).

TABLE 1. Classification of arable plant communities (class Stellarietalia) in Britain under the Zürich-Montpellier system (Silverside, 1976).

	Community	Habitat
Order	Polygono-Chenopodietalia.	Roots and spring cereals
1.	Fumario-Euphorbion	
	a. Veronico-Lamietum hybridi	Loams, E. Anglia
	b. Alopecuro-Matricarietum chamomillae	Loams, E. Anglia
	c. Fumarietum bastardii	Loams, Atlantic
2.	Spergulo-Oxalidion	
	a. Spergulo-Chrysanthemetum segeti	Sands
	b. Stachys arvensis community	Loams, south-west
	c. Chenopodio-Violetum curtisii	Hebridean machair
	d. Lycopsietum arvensis	Sands, east England
	e. Descuranio-Lycopsietum arvensis	Sands, Breckland
Order	Centauretalia cyani.	Cereal crops.
1.	Arnoseridion minimi	
	a. Teesdalio-Arnoseridetum minimae	Poor sands, eastern
2.	Aphanion arvensis	
	a. Papaveretum argemonis	Basic sands, southern
	b. Alchmillo-Matricarietum chamomillae	Clays, southern
	c. Euphorbia exigua-Avena fatua community	Basic clays
3.	Caucalidion lappulae	South-eastern
	a. Linarietum spuriae	Chalk
	b. Papaveri-Melandrium noctiflori	Chalk
	c. Adonido autumnalis-Iberidetum amarae	Calcareous loams

Sandy soils can also support rich communities. The most outstanding are in the East Anglian Breckland, where the former low intensity farming systems on marginal land, the continental climate and the mixture of calcareous and acidic sandy soils, combined to favour a flora unique in Britain. The Breckland has many rare arable annuals including *Veronica praecox, Veronica verna* and *Apera interrupta*. Other species which are rare in the rest of Britain can be very common in the Breckland. These include *Descurania sophia* and *Apera spica-venti*, and to a lesser extent *Silene noctiflora, Anthemis arvensis* and *Lamium hybridum*. Elsewhere calcareous sands provide suitable conditions for rare species outside their main ranges. In Cornwall and Norfolk, coastal fields have populations of *P. hybridum, P. argemone* and *Silene noctiflora*. Oolitic sands on the southern side of the North Yorkshire Moors are a northerly limit for *S. noctiflora, Legousia hybrida* and *Stachys arvensis*, and in east Scotland, coastal sands support outlying populations of *F. densiflora* and *P. argemone*.

Most very freely draining sandy soils tend to be acidic, and support several characteristic and uncommon species. Spring barley and root crops are typical. Less common species include *Fumaria bastardii, Fumaria purpurea, Misopates orontium* and *Silene gallica* especially in the south-west, while *Chrysanthemum segetum* is scattered mainly over the west of the country as far north as the Shetlands. *Fumaria occidentale* is an endemic species found only in arable fields and hedgerows in Cornwall.

The few localities where rich arable floras persist on heavy clay soils are highly vulnerable to changes in farming practices. The most well known site is at the Rothamsted Experimental Station, where part of the Broadbalk winter wheat experiment has never received herbicides, and where *Ranunculus arvensis, S. pecten-veneris, T. arvensis* and *Galium tricornutum* still occur. The first three of these occur in other of the richest clay soil sites, one of which, owned by the Somerset Wildlife Trust, is among the three best sites for the arable flora in Britain, with populations of *Euphorbia platyphyllos, Valerianella rimosa* and *Vicia tenuissima*. *Scandix pecten-veneris* is still considered a farming problem in one area of Suffolk, and in its few sites elsewhere can occur in large quantities. Areas where water collects in the winter support a distinctive flora dominated by *Polygonum spp.*, and uncommon species include *Myosurus minimus* and in two locations, *Lythrum hyssopifolia*.

CHANGES IN BRITAIN'S ARABLE FLORA

Many species including some which were once very common have become extremely rare in recent years (Table 2). Improved methods of seed-cleaning were probably responsible for the early declines of some species including *Agrostemma githago* and *Bupleurum rotundifolium* whose seeds have similar dimensions to cereal grains (Salisbury, 1961). The majority have however declined since the 1940s, coincidental with the massive changes that have occurred in farming practices. At the same time, other species which were pre-adapted to modern high-input farming methods, such as *Bromus sterilis* and *Galium aparine* have increased to become the new problems of modern cereal farming.

TABLE 2. The status of some uncommon arable weeds in Britain in terms of the number of 10km. squares in which they have been recorded. Taken from Perring & Walters, 1990; Smith, 1986; Wilson, 1990; A. Smith, pers comm.. Casual records excluded.

	1930-60	1960-75	1976-85	1986-89
Adonis annua	36	34	13	12
Agrostemma githago	>150	14	17	0
Bupleurum rotundifolium	17	8	1	0
Centaurea cyanus	264	<100	<50	3
Galeopsis angustifolia	238	-	-	18
Galium tricornutum	77	16	7	2
Ranunculus arvensis	432	169	71	22
Scandix pecten-veneris	426	86	<20	20
Torilis arvensis	136	35	16	10
Valerianella rimosa	60	17	11	5

Twenty five species in Britain's arable flora are classified as "Nationally Scarce", a further 24 were included in the "Red Data Book" (Perring & Farrell, 1983), and at least five further species are now of "Red Data Book" status. Of these "Red Data Book" species, seven are now extinct, and ten others no longer occur in strictly arable habitats. Nine are now fully protected under the Wildlife and Countryside Act (1981), although only two of these still occur at all in arable habitats (Table 3).

Table 3. "Nationally scarce" and "Red Data Book" (Perring & Farrell, 1983) species in Britain.

Red Data Book status species. e extinct, n no longer in arable habitats, p legally protected, * not included in Perring & Farrell.		Nationally scarce species (Strictly arable species only.)
Adonis annua	*	*Anagallis foemina*
Agrostemma githago	e	*Apera interrupta*
Ajuga chamaepitys	n*p	*Briza minor*
Alyssum alyssoides	np	*Bromus arvensis*
Anthoxanthum puelii	n	*Bromus secalinus*
Arnoseris minima	e	*Buglossoides arvensis*
Bromus interruptus	e	*Euphorbia platyphyllos*
Bunium bulbocastaneum	n	*Fumaria densiflora*
Bupleurum rotundifolium	e	*Fumaria parviflora*
Caucalis platycarpos	e	*Fumaria bastardii*
Centaurea cyanus	*	*Fumaria vaillantii*
Echium plantagineum		*Galeopsis angustifolium*
Filago apiculata	np	*Geranium columbinum*
Filago pyramidata	p	*Misopates orontium*
Fumaria martinii	np	*Myosurus minimus*
Fumaria occidentale		*Papaver argemone*
Galeopsis segetum	e	*Papaver hybridum*
Galium spurium		*Petroselinum segetum*
Galium tricornutum	*	*Ranunculus arvensis*
Gastridium ventricosum		*Ranunculus parviflorus*
Lolium temulentum	*e	*Scandix pecten-veneris*
Lythrum hyssopifolia	p	*Silene gallica*
Melampyrum arvense	np	*Silene noctiflora*
Rhinanthus serotinus	np	*Valerianella dentata*
Teucrium botrys	np	*Vicia tenuissima*
Valerianella rimosa		*Lathyrus aphaca*
Veronica praecox		
Veronica triphyllos	np	
Veronica verna		

DISCUSSION

Developments in farming practices are thought to have been responsible for the changes in the composition of arable weed communities. The richest communities have survived on calcareous soils where the climate is more favourable to species whose centres of distribution are around the Mediterranean. This has also been observed from the rest of Europe (Holzner, 1979), and it is possible that with the loss of species such as *Arnoseris minima* or *Caucalis platycarpos* from Britain we have lost whole communities of our most thermophilous annual plants. The species most sensitive to agricultural change seem to have retreated to refugia where the effects of modern agriculture are less extreme. These refugia are not simply areas of optimum climate or soil, but may also be places where agricultural activities are less intensive and crop performance is poorer, such as around the edges or in corners of fields. A survey of weed distribution in relation to distance from field margins (Wilson, in press) demonstrated a concentration of uncommon species and diversity within four metres of the field edge. It therefore appears that the more sensitive arable species are not only becoming rarer on a national scale, but also are becoming restricted to the most favourable areas of the fields in which they still occur. The extreme edge of the field between the regularly cultivated area and the hedge bottom is of vital importance for a number of threatened species including *Ajuga chamaepitys* and *Teucrium botrys* that now rarely grow within arable crops. Large-scale field boundary removal in recent years may have contributed greatly to the decline of many species.

Although the origins of the arable flora are in many cases obscure, and some species may not be truly "native" to Britain, arable weed communities nevertheless represent an unique document of man's agricultural activities. They form well-defined and stable communities similar to those in Europe, and can show considerable persistence at sites. They also contain some of Britain's most endangered plants, many of which are experiencing more rapid declines than any others in the British flora. Field margins are therefore extremely important areas for the conservation of some of Britain's most threatened species, and plant communities of great aesthetic value.

REFERENCES

Godwin H. (1956). The History of the British Flora. Cambridge University Press, Cambridge.

Holzner W. (1978). Weed species and weed communities. Vegetatio, **38**. 13-20.

Hüppe J. & Hofmeister H. (1990). Syntaxonomische Fassung und Ubersicht uber die Ackerunkrautgesellschaften der Bundesrepublik Deutschland. Ber. d. Reinh. Tuxen-Ges. 2 61-81.

Perring F.H. & Farrell L. (1983). British Red Data Books: 1. Vascular Plants. RSNC, Lincoln.

Perring F.H. & Walters S.M. (1990). Atlas of the British Flora. BSBI

Salisbury E. (1961). Weeds and Aliens. Collins, London.

Silverside A. (1976). A phytosociological survey of arable weed and related communities in Britain. Unpublished PhD thesis, University of Durham.

Smith A. (1986). Endangered species of disturbed habitats. NCC, Peterborough.

Wilson P.J. (1990). The ecology and conservation of rare arable weed species and communities. Unpublished PhD thesis, University of Southampton.

Wilson P.J. (In press). The distribution of dicotyledonous arable weeds in relation to distance from the field edge. Journal of Applied Ecology.

ARABLE FIELD MARGINS: FACTORS AFFECTING BUTTERFLY DISTRIBUTION AND ABUNDANCE

J.W. DOVER

The Game Conservancy Trust/Myerscough College, Bilsborrow, Preston PR3 0RY

ABSTRACT

The current status of butterfly ecology on arable farmland with specific reference to field margins is reviewed, including major studies with a substantial potential for butterfly conservation. The potential role of pesticides, adjacent habitats, and biotic and abiotic influences on butterfly distribution and abundance are considered, and possible future avenues of investigation discussed.

INTRODUCTION

Butterflies have an intrinsic appeal, even pest species. Historically, farmland was a rich habitat for butterflies but, following the intensification of agriculture since the late 1940's, it is now regarded as impoverished (Thomas, 1984). Farmers themselves have recognised the need to reduce the impact of farming practices on wildlife, a particularly successful example being the farmer-funded development of Conservation Headlands (Sotherton *et al.*, 1989). The change in emphasis in the management of farmland, particularly the permanent vegetation of the field margin, provides a substantial opportunity for the conservation of butterflies. This paper reviews the progress to date.

SPECIES FOUND IN FIELD MARGINS

British butterflies currently comprise sixty-one species, excluding rare migrants. Thirty-one species (51% of the British list) have been recorded from arable field margins, together with the rare migrant *Colias hyale* (Table 1). The majority of these species probably breed in field margins, with species such as *Argynnis aglaja*, *Argynnis paphia* and *Ladoga camilla* flying in from adjacent habitats to nectar or bask. Simple species lists do not, however, give any indication of the relative abundance of species, and many will be, at best, only occasional sightings (Table 1).

LARVAL HOSTS

Full details of the larval host plants of butterflies may be found in Dennis (1992), Emmett & Heath (1989) and recent work on some satyrids in Feber (1993). Marshall (1989) gives details of the flora of field margins. Not all larval hosts present in field margins will be exploited because of biotic and abiotic influences on adults such as shelter (Dover, 1990a), shade (Courtney, 1982), weather (Courtney & Duggan, 1983), micro-scale influences on oviposition cues (Dennis, 1983), size of hosts (Dennis, 1985), the propensity for many species to lay on the outside of clumps (the 'edge-effect') (Courtney & Courtney, 1982; Dennis, 1984), and the physiological and nutritional status of the plants (Myers, 1985; Pullin, 1987). The cutting of verges can have catastrophic effects on larval survival (Courtney & Duggan, 1983), although this is most likely to be a problem in the margins of fields adjacent to roads.

TABLE 1. Butterfly species recorded in arable field margins in specific English counties, larval hostplants known to grow in field margins, probable breeding status, and relative abundance. Taken from Dover (1991), Feber (1993), Emmet & Heath (1989), Pollard *et al.* (1986)

Hesperiidae		**Pieridae**	
Ochlodes venata	G,Ha,Hu,Ox,*,L	*Anthocharis cardamines*	G,Ha,Hu,Ox,*,L
Thymelicus lineola	Ha,*,L	*Colias croceus*	G,Ha,*,#,I
Thymelicus sylvestris	G,Ha,Hu,Ox,*,L	*Colias hyale*	G,*,#,I
Lycaenidae		*Gonepteryx rhamni*	G,Ha,Hu,Ox,*,L
Aricia agestis	G,Ox,#,I	*Pieris brassicae*	G,Ha,Hu,Ox,*,A
Callophrys rubi	Ha,*#,I	*Pieris napi*	G,Ha,Hu,Ox,*,R
Celastrina argiolus	G,Ha,Hu,Ox,*,L	*Pieris rapae*	G,Ha,Hu,Ox,*,A
Lycaena phlaeas	G,Ha,Hu,Ox,*,L	**Satyridae**	
Polyommatus icarus	G,Ha,Hu,Ox,*,L	*Aphantopus hyperantus*	G,Ha,Hu,Ox,*,R
Quercusia quercus	G,Ha,Ox,*,#,I	*Coenonympha pamphilus*	G,Ha,Hu,Ox,*,L
Strymondia w-album	G,Ox,*,I	*Lasiommata megera*	G,Ha,Hu,*,L
Nymphalidae		*Maniola jurtina*	G,Ha,Hu,Ox,*A
Aglais urticae	G,Ha,Hu,Ox,*,A	*Melanargia galathea*	G,Ha,Ox,*,L
Argynnis aglaja	Ha,*,#,I	*Pararge aegeria*	G,Ha,Ox,*,L
Argynnis paphia	G,Ha,*,#,I	*Pyronia tithonus*	G,Ha,Hu,Ox,*,A
Inachis io	G,Ha,Hu,Ox,*,R		
Ladoga camilla	Ha,*,#,I		
Polygonia c-album	G,Ha,Hu,Ox,*,L		
Vanessa atalanta	G,Ha,Hu,Ox,*,L		
Vanessa cardui	G,Ha,Hu,Ox,*,L		

G - Gloucestershire, Ha - Hampshire, Hu - Huntingdonshire, Ox - Oxfordshire; * - hosts found in field margins; # - unlikely to be breeding in field margins; adult abundance, subjective score: I - infrequently found, L - low numbers, R - reasonable numbers, A - abundant

ADJACENT HABITATS AND GAME COVER

Field margin habitats such as hedgerows and verges are typically quite narrow and discrete entities, although they may form part of a wider hedgerow network. Other margin habitats such as woodland and railway embankments are frequently of considerably greater area. The transition between two habitat types, known as the 'ecotone' is where species characteristic of both habitats may be found. Hence, woodland specialist butterflies such as *Q. quercus*, *A. paphia* and *L. camilla* and grassland species such as *A. aglaja* and *M. galathea* are found associated with arable field margins exploiting some resources of the ecotone, but ultimately returning to their primary habitat. The field margin ecotone may be important to habitat specialists, especially during dispersal or after failure of their principal resources.

On farms with a significant game interest some fields may have a strip of game cover between the field margin and crop. Cover crops on a north-Hampshire farm composed of *Helianthus tuberosus* (Jerusalem artichoke) and/or *Phalaris tuberosa* (canary grass) were found to be of high value for butterflies (Dover *et al.*, 1992). The cover crops studied were subject to relatively low levels of management, were in place for several years, and had been colonised by perennial and annual 'weeds' (Rew, 1988) providing both larval and adult resources.

CONSERVATION HEADLANDS AND EXTENDED FIELD BOUNDARIES

The crop margin management technique known as 'Conservation Headlands' (Sotherton *et al., 1989)*, where pesticide inputs to the outer 6m of cereal fields are prescriptively reduced and selective, has been shown to have a significant benefit for butterflies on farmland. Butterfly transects carried out over a five year period showed significantly more butterflies in field margins which had Conservation Headlands compared with crop margins sprayed according to normal farm practice (Dover, 1991). On average 68% of all butterflies seen on transects were found in Conservation Headlands. Of 57 statistical comparisons between the two experimental regimes, 45 significant differences were identified, of which only one showed significantly more butterflies in field margins sprayed according to normal farm practice (Dover, 1991).

Observations of butterflies revealed significant changes in the temporal and spatial distribution of flight, feeding, resting and interacting (mating) behaviours between field margins with Conservation or fully sprayed headlands; in the Conservation regime flight activity reduced, feeding increased as did resting and interactive behaviour. The shift in behaviour was principally due to the additional nectar resources present in Conservation Headlands and, for some pierid species, additional larval host plants (Dover, 1989a,b,1991, 1992). Nectar is a significant factor in the potential fecundity and longevity of adult butterflies (Watt *et al.*, 1974; Murphy *et al.*, 1983) and may be a limiting factor for butterflies in modern arable farmland.

Population trends of three satyrid and three pierid species on a study farm with half the cereal fields managed with Conservation Headlands were compared with regional data from the National Butterfly Monitoring Scheme (Pollard *et al.*, 1986) over a five year period. The 'open' (continual dispersal) population structure of the pierid species prevented the identification of differing population trends unlike the 'closed' (colony forming) structure of the satyrid species (*A. hyperantus*, *M. jurtina* and *P. tithonus*) which all showed increases in population trends at the study farm compared with regional data (Dover, 1991, 1992).

The impact of increasing the uncropped area of arable field margins by the inclusion of an additional 2m wide boundary strip, created by sowing with a grass and wildflower mixture or by allowing the existing seed-bank to develop (unsown), was examined by Feber (1993; this volume). The two sward types were manipulated by the use of different cutting and hay removal regimes; additionally, some unsown plots were sprayed annually with glyphosate. This latter regime provided the poorest habitat for butterflies with fewer species and lower abundance. The sown sward performed better than the unsown sward as butterfly habitat; cutting in the summer decreased butterfly numbers compared with a spring and autumn cut, or no cut. Comparison of the expanded field margins with a nearby commercial farm with 'normal' margins showed increased abundance of butterflies at the experimental farm. Both the sown and unsown swards provided host plants for some butterfly species, but the sown sward in particular provided enhanced adult nectar resources.

Feber (1993) also compared two grass leys on wider strips (7.2 to 9.6 m) of boundary; a conventional grass/clover mix and a more diverse mix including wildflowers. More butterflies were found on the margins with the diverse sward compared with the conventional ley, and fertiliser use diminished butterfly abundance in both. Neither sward type was as attractive to butterflies as the 2m margins described above.

BUTTERFLY DISTRIBUTION BETWEEN AND WITHIN FIELD MARGINS

Dover (1990a) carried out a mark-release-recapture experiment with *A. hyperantus, P. tithonus* and *M. jurtina,* in the field margins of a 66.5ha block of arable farmland. The field margins principally consisted of hedgerows and grass verges, some with associated farm tracks, and wood-edges. Records of biotic and abiotic parameters were made in a subset of the fields under study during the years 1988 and 1989. Butterfly captures and habitat data for 30m lengths and whole field margins were analysed by stepwise multiple regression. Factors affecting the distribution of butterflies between field margins included: shelter, a small uncropped area between a narrow copse and hedgerow field boundary traversed by a farm track, and nectar sources. Within field margins, distribution was affected by the *degree* of shelter, insolation, nectar sources, farm tracks and variables reflecting different aspects of habitat quality.

BUTTERFLY MOVEMENT AND COLONY COMPACTNESS

Dover, (1990a) and Dover *et al.* (1992) showed that *A. hyperantus, P. tithonus* and *M. jurtina* were capable of moving considerable distances, in excess of 1km, between marking stations in farmland habitats, although the frequency of such movements was low. Dover (1990b) demonstrated that pierid butterflies made use of field margins as flight corridors, apparently in preference to overflying crops. This may be due to the shelter associated with hedgerows and wood-edges, and the presence of adult and juvenile resources. Males of the territorial species *A. urticae* and *I. io* use field margins as territorial habitats in order to maximise their encounters with females searching for oviposition sites along them (Baker, 1972). Dennis (1986), Dover (1991), and Munguira & Thomas (1992) noted that wind buffeting may reduce the passage of species across an unsheltered area such as a gap in a hedgerow or a road. The flight tracks of butterflies are also affected by shade, and individuals can be observed flying along the contours of the shadows cast by hedgerows and hedgerow trees (Dover, personal observation), potentially reducing the time available for oviposition and nectaring.

Movements of *A. hyperantus, P. tithonus and M. jurtina* between field margins were studied by Dover (1990a) and Dover *et al.* (1992). Plots of flights between capture points in MRR studies demonstrated substantial interchange between nearby field margins within colonies. Movement between adjacent colonies of species was evident, suggesting that satyrids exhibit a metapopulation structure on arable farmland. The area of land required to support colonies of butterflies on farmland can be quite small; Dover (1991) showed that in one colony the butterflies were utilising just 0.67ha of field margin, approximately one-third of the total field margin habitat in the study area, or just 1% of the total study area. Thomas (1984) gives details of the minimum habitat area for British species.

Many farms are traversed by road systems, which may act as bottlenecks for movement between field margins. Dennis (1986) in a study of *A. cardamines* in the Bollin valley in Cheshire showed the M56 motorway to reduce overflights by 92%. However, Munguira & Thomas (1992) recorded nineteen species crossing 'A' class roads; these were not considered to impose significant levels of mortality or to prevent butterfly dispersal, although they restricted it in some species.

SPRAY DRIFT

Dover *et al.* (1990) speculated on the impact that reducing pesticide drift into field margins would have on butterflies, either through lower juvenile mortality from compounds with insecticidal activity, or reduced herbicide damage to larval host plants and adult nectar resources. Work on pesticide drift into field margins (Cuthbertson, 1988; Cuthbertson & Jepson, 1988) showed that, during autumn spraying of cereal crops, Conservation Headlands reduced pesticide levels on hedgebank vegetation by over 50%. The presence of a mature crop during summer spraying sheltered the basal vegetation of hedgerows from spray drift, reducing the loading by 75% compared with vegetation above the crop canopy.

Cilgi & Jepson (in press) using fourth instar *P.brassicae* and *P.rapae* larvae demonstrated that exposure for two hours to 1/16th of the field dose rate of the synthetic pyrethroid deltamethrin sprayed onto leaf surfaces resulted in mortality. Further experiments exposing larvae to low dose rates over long time periods demonstrated mortality and sub-lethal effects (eg. anti-feedant response and smaller size of adult at eclosion) at concentrations down to 1/640th of the full field dose rate.

Sinha *et al.* (1990) examined the impact of eight insecticides on two-day old *P. brassicae* larvae and demonstrated a x700 difference in toxicity levels. Davis *et al.* (1991) showed that *P.brassicae* was particularly suitable as an indicator species in bioassays, being more sensitive than *P. napi, P. tithonus* and *P. icarus*. Subsequent trials with the most toxic compound diflubenzuron, at the maximum field dose rate, and *P. brassicae* demonstrated high mortality (>95%) 16 m downwind of the spray boom with a windspeed of 5.3 m/second, but even at low speeds (2.5 m/second) 24% of larvae were killed at 24 m down wind (Davies *et al.*, 1991).

The host-plants of many British butterflies (Dennis, 1992; Emmet & Heath, 1989; Feber, 1993) may be subject to herbicide drift. No information is currently available on the impact of such drift on butterfly populations. Marshall & Birne (1985) in a study of seven herbicides showed a significant impact on the graminaceous host-plants of Hesperiidae and Satyridae, and the broadleaved host-plants of the Pieridae and Nymphalidae, as well as to some adult nectar plants. Death of a host-plant requiring the larva to find a new host, or premature scenescence of a plant, or part of a plant, may be sufficient to cause larval starvation resulting in mortality or the eclosion of less-fit adults (Courtney, 1981). Marrs *et al.* (1989) recommend that buffer zones of 5-10m be left between areas of sensitive vegetation when using herbicides. Most lethal effects occur within 2.5m of the spray boom, although transient effects and flowering suppression may occur on sensitive species up to 20m away from the sprayer (Marrs *et al.* (1989)). Although plants were found to recover from such sub-lethal effects, the impact on larvae and nectivorous adults may be substantial.

DISCUSSION

Information on the ecology of butterflies on arable farmland has increased considerably since the first paper on the subject by Rands & Sotherton (1986) and approaches such as Conservation Headlands and extended field margins have clear demonstrable, benefits for butterflies. However, much information is of a preliminary nature and begs further questions. For example, is the relatively poor abundance of some lycaenid species in field margins due to pesticide and cultural effects (including fertiliser drift) on the host plants and juvenile stages, disruptive impacts on mutualistic associations with ants, or a combination of the two? Several studies have shown the importance of nectar in butterfly longevity, fecundity and microdistribution, but precisely how important is it? Is nectar a limiting factor on arable farmland and if so what is the threshold value and does it differ between butterfly species? Can annual nectar sources in Conservation Headlands fully compensate for degraded perennial nectar sources in field margins? The information available on insecticide and herbicide drift is incomplete. There appear to be substantial differences in the toxicity of the main classes of insecticide to butterflies, herbicides have sub-lethal effects on non-target plants; what are the impacts on larval survival, how do the various factors interact? Information on the use of field margins as flight corridors by butterflies is sparse, but there appears to be a huge potential for the use of this group to explore the concept in a landscape ecological approach.

Using closed population species such as the satyrids it should be possible to develop a simulation model of the impact of habitat and crop management on butterfly distribution and abundance allowing the impact of current and future management options to be explored.

ACKNOWLEDGEMENTS

I would like to thank Dr N.W. Sotherton and Dr N.D. Boatman for their help and support. I am grateful to Dr R. Feber for generous access to the results of her doctoral research.

REFERENCES

Baker, R.R. (1972) Territorial behaviour of the nymphalid butterflies, *Aglais urticae* (l.) and *Inachis io* (L.), *Journal of Animal Ecology,* **41**: 453-69

Cilgi, T & Jepson, P. (in press) Pesticide spray drift into field boundaries and hedgerows: toxicity to non-target Lepidoptera, *Environmental Pollution,*

Courtney, S.P. (1981) Coevolution of pierid butterflies and their cruciferous foodplants III *Anthocharis cardamines* (L.) survival, development and oviposition on different hostplants, *Oecologia (Berl),* **51**: 91-96.

Courtney, S.P. (1982) Coevolution of pierid butterflies and their cruciferous foodplants IV Crucifer apparency and *Anthocharis cardamines* (L.) oviposition, *Oecologia (Berl),* **52**: 258-265.

Courtney, S.P. & Courtney, S. (1982) The 'edge-effect' in butterfly oviposition: causality in *Anthocharis cardamines* and related species, *Ecological Entomology,* **7**: 131-137

Courtney, S.P. & Duggan, A.E. (1983) The population biology of the orange tip butterfly *Anthocharis cardamines* in Britain, *Ecological Entomology,* **8**: 271-281

Cuthbertson, P. (1988) The pattern and level of pesticide spray drift into conservation and fully sprayed arable crop headlands, *Aspects of Applied Biology,* **17**: 273-276

Cuthbertson, P.; Jepson, P. (1988) Reducing pesticide drift into the hedgerow by the inclusion of an unsprayed field margin. *1988 Brighton Crop Protection Conference - Pests and Diseases,* **2**: 747-751.

Davis, B.N.K.; Lakhani, K.H.; Yates, T.J. (1991) The hazards of insecticides to butterflies of field margins, *Agriculture, Ecosystems and Environment,* **36**: 151-161.

Davis, B.N.K.; Lakhani, K.H.; Yates, T.J. & Frost, A.J. (1991) Bioassays of Insecticide Spray Drift: the effects of wind speed on the mortality of *Pieris brassicae* larvae (Lepidoptera) caused by diflubenzuron, *Agriculture, Ecosystems and Environment,* **36**: 141-149.

Dennis, R.L.H. (1983) Egg laying cues in the wall brown butterfly *Lasiommata megera*(L.) (Lepidoptera: Satyridae), *Entomologist's Gazette,* **34**: 89-95.

Dennis, R.L.H. (1984) The edge effect in butterfly oviposition: batch siting in *Aglais urticae* (L.) (Lepidoptera: Nymphalidae), *Entomologist's Gazette,* **35**: 157-173

Dennis, R.L.H. (1985) Small plants attract attention,: Choice of egg-laying sites in the green-veined white butterfly *(Artogeia napi) (L.)* Lep. Pieridae, *Bulletin of the Amateur Entomologist's Society,* **44**: 77-82

Dennis, R.L.H. (1986) Motorways and cross-movements: an insect's 'mental map' of the M56 in Cheshire, *Bulletin of the Amateur Entomologist's Society,* **45**: 228-243

Dennis, R.L.H. (1992) *The Ecology of Butterflies in Britain,* Oxford, Oxford University Press

Dover, J.W. (1989a) A method for recording and transcribing observations of butterfly behaviour, *Entomologist's Gazette,* **40**: 95-100.

Dover, J.W. (1989b) The use of flowers by butterflies foraging in cereal field margins,*Entomologist's Gazette,* **40**: 283-291

Dover, J.W. (1990a) *Butterfly Ecology on Arable Farmland with special reference to agricultural practices* NERC contract report G2/044; Fordingbridge, Game Conservancy.

Dover, J.W. (1990b) Butterflies and wildlife corridors, *The Game Conservancy Review of 1989,* **21**: 62-64

Dover, J.W. (1991) The conservation of insects on arable farmland. In: *The Conservation of Insects and their Habitats,* N.M. Collins and J.A. Thomas (Eds), London:

Academic Press, pp. 293-318.

Dover, J.W. (1992) Butterflies and conservation headlands, In: *The Future of Butterflies in Europe: Strategies for Survival,* T. Pavlicek-van Beek; A.H. Ovaa; J.G. van der Maade (Eds), Wageningen, Agricultural University Wageningen, pp. 327-336

Dover, J.W.; Clarke, S.A.; Rew, L. (1992) Habitats and movement patterns of satyrid butterflies (Lepidoptera:Satyridae) on arable farmland, *Entomologist's Gazette*, **43**: 29-44.

Dover, J.W.; Sotherton, N.W.; Gobbett, K. (1990) Reduced pesticide inputs on cereal field margins: the effects on butterfly abundance, *Ecological Entomology*, **15**: 17-24.

Emmet, A.M.; Heath, J. *The Moths and Butterflies of Great Britain and Ireland, 7:Hesperiidae - Nymphalidae The Butterflies,* Colchester, Harley Books

Feber, R.. (1993) *The Ecology and Conservation of Butterflies on Lowland Arable Farmland.* Unpublished D.Phil. Thesis, University of Oxford.

Marrs, R.H,; Frost, A.J.; Plant, R.A. (1989) A preliminary assessment of the impact of herbicide drift on plant species of conservation interest, *1989 Brighton Crop Protection Conference - Weeds*, **2**: 795-802

Marshall, E.J.P. (1989) Distribution pattern of plants associated with arable field margins, *Journal of Applied Ecology*, **26**: 247-257

Marshall, E.J.P; Birnie, J.E. (1985) Herbicide effects on field margin flora, *1985 British Crop Protection Conference - Weeds* , **x**:1021-1028

Munguira, M.L. & Thomas, J.A. (1992) Use of road verges by butterfly and burnet populations, and the effect of roads on adult dispersal and mortality, *Journal of Applied Ecology*, **29**: 316-329

Murphy, D.D.; Launer, A.E.; Ehrlich, P.R. (1983) The role of adult feeding in egg production and population dynamics of the checkerspot butterfly, *Euphydryas editha*, *Oecologia*, **56**: 257-263.

Myers, J.H. (1985) Effect of physiological condition of the hostplant on the ovipositional choice of the cabbage white butterfly, *Pieris rapae*, *Journal of Animal Ecology,* **54**: 193-204.

Pollard, E; Hall. M.L.; Bibby, T.J. (1986) *Research and Survey in Nature Conservation No. 2: Monitoring the Abundance of Butterflies 1976-1985,* Peterborough, Nature Conservancy Council.

Pullin, A.S. (1987) Changes in leaf quality following clipping and regrowth of *Urtica dioica*, and consequences for a specialist insect herbivore, *Aglais urticae*, *OIKOS*, **49**: 39-45

Rands, M.R.W. & Sotherton, N.W. (1986) Pesticide use on cereal crops and changes in the abundance of butterflies on arable farmland, *Biological Conservation,* **36**:71-82

Rew, L.(1988) *The Importance of Game Cover for the Conservation of Butterflies on Farmland.* Unpublished BSc. Thesis, University of Southampton.

Sinha, S.N.; Lakhani, K.H. ; Davis, B.N.K. (1990) Toxicity of insecticidal drift to the large white butterfly, *Annals of Applied Biology*, **116**: 27-41

Sotherton, N.W.; Boatman, N.D.; Rands, M.R.W. (1989), The 'Conservation Headland' experiment in cereal ecosystems, *Entomologist,* **108**:135-143.

Thomas, J.A. (1984) The conservation of butterflies in temperate countries: past efforts and lessons for the future, In: *The Biology of Butterflies*, R.I. Vane-Wright & P.R. Ackery (Eds.), London, Academic Press, pp.333-353

Watt, H.B.; Hoch, P.C.; Mills, S. (1974) Nectar resource use by *Colias* butterflies: chemical and visual aspects, *Oecologia*, **14**: 353-374.

FIELD MARGINS AS HABITATS, REFUGES AND BARRIERS OF VARIABLE PERMEABILITY TO CARABIDAE

P.C. JEPSON

Department of Biology, School of Biological Sciences, Southampton University, Biosciences Building, Bassett Crescent East, Southampton, SO9 3TU. UK.

ABSTRACT

Carabidae may be sensitive indicators of the ability of fragmented landscapes to support invertebrate wildlife. There is evidence that natural populations undergo frequent extinctions and that the rate of extinction increases in cultivated areas. Little is known however, about the effects of landscape features such as field boundaries on population persistence. Interactions between carabids and field boundaries are complex and variable across the family but there is some evidence that they delay the inter-field movement of species that overwinter in the field. A simple model, incorporating diffusion and population growth rates for Carabidae demonstrates that the environmental resistance of the habitat, which is contributed to by field boundaries, could have a significant effect on recolonisation rates of depleted habitats. The potential of Carabidae to be exploited as indicators of habitat quality for invertebrates may only be realised when some key research questions have been answered.

INTRODUCTION

Linear features in the farming landscape may provide connections between non-crop habitats and assist colonisation and movement by wildlife (although the evidence that this occurs is patchy; Hobbs, 1992). To those organisms that complete their life-cycles within agricultural fields however, associations with the linear features such as field boundaries may be far more complex: the boundary may act as a habitat, a refuge or even a barrier, effectively restructuring the population into sub-units. Habitat fragmentation by field boundaries may have a variety of consequences. Small, sub-populations may suffer an increased risk of local extinction through amplification of stochastic effects. These populations may alternatively, benefit from fragmentation because they are partially protected from the consequences of random catastrophic events by being out of synchrony with other sub-populations (Kareiva, 1991). The degree to which the structure of the farming landscape affects the density and persistence of organisms inhabiting farmland remains one of the more interesting unanswered questions of agro-ecology.

Despite their widespread distribution and abundance as members of the soil invertebrate community, populations of Carabidae (Coleoptera) are susceptible to local extinction in cultivated landscapes (den Boer, 1977; 1990a). Evidence to support this has arisen from an examination of the frequency distributions of population sizes: the form in which certain carabid species deviate from the expected distributions in cultivated areas has been taken to indicate that small populations are being lost (den Boer, 1977). Carabidae within farmland may be more resilient than species that occupy uncultivated areas (den Boer, 1977; 1990a): recently however, cases of local extinction of species from the farmland community have been reported (e.g. Basedow, 1991; Burn, 1992). These events have been associated with intensive pesticide treatments however, the additional role that fragmentation of the agricultural landscape has played in the process of extinction is unknown.

Carabids are potential indicators of the side-effects associated with excessive pesticide use (Jepson, 1988; 1993). Their importance as ecological indicators may extend beyond this however: they may also be indicators of those levels of habitat fragmentation and types of field boundary that characterise habitats which are intrinsically less suited to support invertebrate populations in general. The first theoretical investigation to explicitly examine the importance of field boundary permeability in the local persistence of Carabidae (Sherratt and Jepson, 1993) revealed that there may be optimum field sizes, boundary permeabilities or rates of movement that maximise the chance that some species will persist in farmland with a given level of disruption by pesticides. The low reproductive rates and slow, cursorial dispersal of many Carabidae may make them good general indicators of the influence that landscape structure has upon population processes. The presence or absence of certain carabid species might be used to classify farming landscapes and patterns of land use in terms of their ability to support diverse invertebrate populations and withstand disruption by agricultural practices.

Given the increasing speculation that the fragmentation of cultivated habitats by roads, hedges and ditches has a major impact on the persistence of carabids and other invertebrates (Mader, 1990; den Boer, 1990; Sherratt and Jepson, 1993; Thomas, 1992), what evidence in support of this hypothesis can be derived from the literature?

TYPES OF CARABID INTERACTIONS WITH FIELD BOUNDARIES

It is known that some substrates and habitat types may directly impede carabid movement (den Boer, 1971; Speight and Lawton, 1976) and the rate of progress will be reduced if they cross rugged or densely vegetated terrain. Hedgerows, wooded strips and grass banks that border fields are likely to present more severe obstacles. These sub-divide the habitats of Carabidae that colonise arable cropping systems and are therefore the most commonly encountered impediments to movement. Do carabids coincide with field boundaries or complete parts of their life-cycles within them though?

There is a long history of research, into the associations between Carabidae and field boundaries (Thiele, 1977). Initially, the fauna of dense, wooded strips bordering fields were investigated. These were found to contain an impoverished forest community that rarely entered agricultural crops (Tischler, 1958). Carabidae from adjoining agricultural habitats were at much lower levels of activity/density in the wooded strips. It may be inferred from data presented in Theile (1977) for example, that the activity/densities of *Pterostichus melanarius* Ill. in wooded strips was 20%

of that in the adjoining fields.

Later investigations have found a complex array of possible interactions between Carabidae and the more conventional hedgerow environment. In the spring to late summer, some of the most commonly occurring arable crop Carabidae are associated with the hedgerow, but not restricted to it (eg. *Nebria brevicollis* (F.), *Agonum dorsale* Pontoppidan): others however, only have limited associations with it (*P. melanarius, P. madidus, Harpalus rufipes* (DeGeer), *Bembidion lampros* (Herbst), *B. obtusum* Serville, *Trechus quadristriatus* (Schrank), *Loricera pilicornis* (F.) and *Notiophilus biguttatus* (F.)) (Pollard, 1968b). Pollard (1968a) concluded that the hedge or border zone impinged more upon the life-cycles of nocturnal Carabidae, active when crops are absent (eg. autumn-active *N. brevicollis*) than to diurnal species, active at the same time (eg. *B. obtusum*) or nocturnal species with a late summer peak, when crop cover was good (eg. *P. melanarius* or *H. rufipes*). This picture is complicated however by the finding that a sub-group of commoner species overwinter in the field boundary zone, penetrating the field in the spring (Greenslade, 1965; Fuchs, 1969; Sotherton, 1984). In addition, structural and vegetational properties of the field boundary are important in determining beetle composition and densities because of the narrow habitat preferences of many species (Sotherton, 1985).

With this array of possible interactions with field boundaries, it therefore seems likely that the presence of at least some boundary types might alter the rate of diffusion of Carabidae through farmland.

FIELD BOUNDARIES AS BARRIERS TO MOVEMENT

What are the consequences of the sub-division of fields by hedgerows therefore likely to be for dispersal rate? In agricultural habitats, several investigations have attempted to quantify the extent to which movement between habitats is impeded in the short-term, by features such as dirt, gravel and tarred roads, grass strips and railway tracks (Duelli, Studer, Marchand and Jakob, 1990; Mader, Schell and Kornacker, 1990). The species most commonly investigated represented the larger Carabidae such as *P. melanarius* and *H. rufipes,* rather than the complete spectrum of body sizes however, a range of permeability levels (defined as the proportion of individuals that cross the hedgerow when incident with it) can be derived from these studies (Table 1). In addition, it can be assumed that certain barriers such as irrigation ditches, rivers and canals and brick walls will have zero permeability to epigeal species.

The reduction in recolonisation and recovery rates generally caused by field boundaries will tend to extend the duration of population reductions after harmful interventions such as pesticide spraying (Jepson and Thacker, 1990). Whether or not this has positive or negative long-term implications for population persistence in an area depends upon how adverse the conditions are in the field that might be entered and upon the dispersal rate of the species in question (Sherratt and Jepson, 1993).

MEASURING INTER-FIELD DIFFUSION AND RECOLONISATION RATES

For those species that overwinter in grass banks or hedgerows, there may be a high level of exchange between neighbouring crops between seasons: the importance of the physical features of the boundary in determining the likelihood of exchanging crops is unknown for these species. Up to 50% of the population in a given field might transfer to another field as a result of the reassortment that takes place after the use of hedgerows as an overwintering refuge. There will be also however, be a reduction in the overall displacement rate of these species, because these species preferentially colonise the open field in the spring and summer. This reduction in diffusion rate may however, be ameliorated by the apparently rapid diffusion along the hedgerow/crop junction by members of the boundary-overwintering guild (Jensen, Dyring, Kristensen, Nielsen and Rasmussen, 1989).

TABLE 1. Estimated habitat boundary permeabilities to certain Carabidae, calculated from data in the literature or guessed for extreme cases.

Ref. number	Boundary Type	Percentage permeability	Carabid species
1.	10m woodland strip	20	*P.melanarius*
2.	3m dirt road	51	smaller spp.
	6m tarred road	40	"
	grass strip	141	"
3.	1.2m grass track	55	*P.melanarius* and others
	1m gravel track	15	"
	0.5m paved road	23	"
	5.7m railway embankment	9.8	"
	"	17.4	*P.melanarius*
	"	0	*N.brevicollis*
guesses	canal or irrigation ditch	0	all species?
	crop:same crop interface (strip cropping)	100	"
	overwintering boundary	50	boundary overwintering spp.

References:1 (Thiele, 1964); 2 (Duelli *et al.*, (1990);3 (Mader *et al.*, 1990). For 1-3, % permeabilities are estimated rates of entry or crossing the given boundary types for Carabidae from arable crops: see these papers for details.

For field-overwintering species, the effects of field boundaries upon movement depend upon the level of interaction with the field boundary over the insect's life-cycle and upon the structure of the boundary itself. The level of interaction with the boundary is dependent upon prevailing environmental conditions (Fuchs, 1969), food availability (Williams, 1959) and the habitat requirements over the complete life cycles of different species. Some Carabidae seem to exploit a secondary habitat as part of their life cycle, thus *P. melanarius,* a field-active species, may enter hedgerows during hot, dry, periods (Fuchs, 1969) and may spend part of it's life-cycle in other habitats adjoining wheat crops (Wallin, 1985). This species is however preferentially associated with the centre of the crop and larvae are mostly found in this zone (Wallin and Ekbom, 1988): a distribution that may reflect avoidance of competition and facilitate early exploitation of the crop habitat each season. Use of secondary habitats, requiring field boundaries to be crossed, may only therefore be for those adults that survive the reproductive phase in the field (Lyngby and Nielsen, 1988).

To estimate the degree of inter-habitat diffusion empirically, it is not therefore sufficient to measure boundary permeability over short time-intervals as was done in the recent studies reported above. The degree of interchange must me measured over a whole generation and is likely to be a product of intrinsic phenological characteristics and behavioural responses to conditions in the field, as well as the physical permeability of the boundary itself. A simple 'permeability' term may be adequate for general models, designed to explore the possible significance of field boundaries for local population persistence in Carabidae: it will not be sufficient however for predictions that relate to individual species. This problem is considered further in Jepson (1994).

A SIMPLE MODEL OF 'ENVIRONMENTAL RESISTANCE'

A general population model for exploring the role of differing boundary permeabilities on the local population persistence of Carabidae has been developed by Sherratt and Jepson (1993). This model does not however, incorporate detailed population dynamics, phenological patterns or habitat requirements that might determine the likelihood of entering and crossing a field boundary. Some of this detail would be required if the model was to be used to generate specific predictions. At present, research into movement patterns of epigeal species over farmland is still lacking however, some general predictions of the way in which different types of field boundary might affect the displacement of Carabidae can be made by building simple mathematical models based upon the small amount of information that is available (Jepson, 1994). The example below concerns the special case of the rate at which Carabidae might enter an area from which populations have become extinct. It avoids the complexity of considering competition for food resources by colonists entering habitats that are already occupied.

On theoretical grounds, it may be argued that the rate of radial expansion of a population moving by random diffusion approaches an asymptote over long time intervals (Skellam, 1951). Andow, Kareiva, Levin and Okubo (1991) have developed a tractable test of this hypothesis for a range of invertebrates and vertebrates. The asymptotic velocity of the advancing wave-front of organisms is estimated from measurements of diffusion coefficients and intrinsic rates of population increase when resources are not limiting (hence the need to consider Carabidae entering a carabid-depleted habitat). Within certain limitations of the phenology of the population in question, the asymptotic velocity of the reinvading population front may be estimated as:

$V_F = \sqrt{4\alpha D}$ (from Andow *et al.*, 1991)........... function 1

Where V_F is velocity (distance/time), α is the intrinsic rate of population growth and D is the coefficient of diffusion.

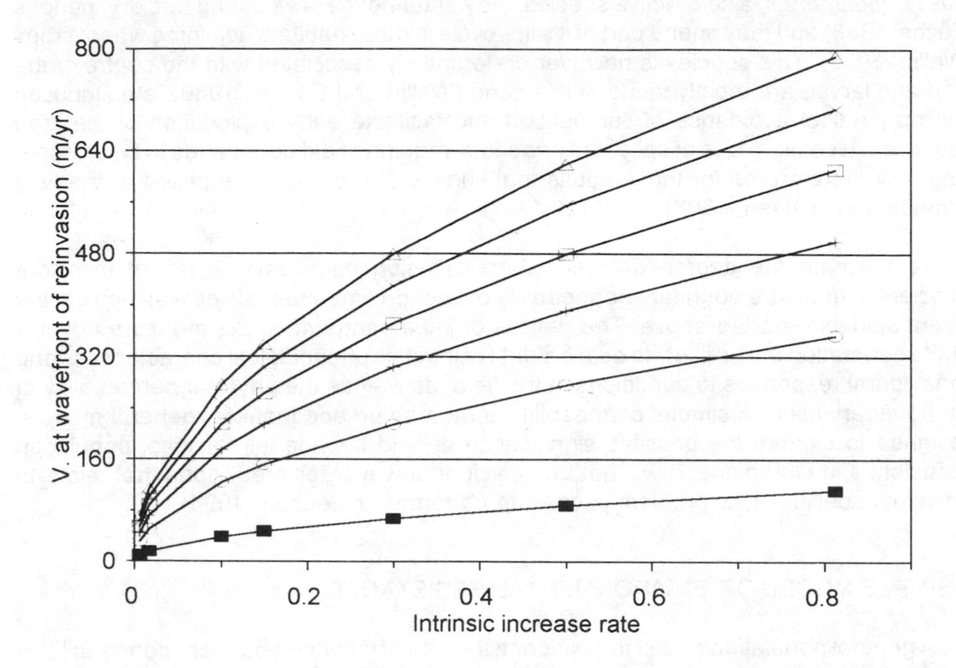

Fig. 1: Velocity of advance of the wave front of carabid colonists re-entering a system where pesticides have caused extinctions. Increase rates represent the range measured in cultivated land for 24 common species. Velocity is calculated for a range of net values of boundary permeability (see text). The upper line is for a net permeability of 100%. The other lines are 80%, 60%, 40%, 20% and 2% respectively.

Data is available to satisfy the requirements of this expression assuming that the rate of progress of the carabid population is effectively a random walk that can be predicted from the maximum daily displacement (see Jepson, 1994). Accepting these assumptions, the diffusion coefficient can then be derived from:

$$D = \frac{Ms\ (t)}{4t}$$ (from Andow *et al.*, 1991) ....function 2

Where D is the coefficient of diffusion, Ms is the mean of the squared displacements and t is the number of time intervals. From the records of maximum displacement in the literature (N=8) (given in Jepson, 1994), the estimate of mean of squared displacement is 4806 and D is therefore 1202 m^2/day (function 2). Den Boer (1990b) gives mean population increase rates for established populations of 24 species of Carabidae. The range of ln (R) (where R was the total trap catch in year N+1 divided by the total catch in year N) calculated from the mean R values for the 24 species, was 0.0065 to 0.8122. These were taken to represent the range in intrinsic increase rates attainable by carabid populations in farmland.

The effects of introducing impediments to movement such as hedgerows, may then be estimated by varying D (function 2) in proportion to the known permeability of field boundaries. This assumes, for a radially dispersing wave front, that the encounter rate with hedgerows will be the same in all directions. Figure 1 expresses the velocity of the reinvading wave front as a function of increase rates (across the full range measured by den Boer (1990b)), varying boundary permeability. Here, the measure of boundary permeability may be read as a net value, independent of the number of encounters with boundaries per annum. A value of 0.02 could be arrived at by a single encounter with a substantial obstacle or by several encounters with less severe features (eg. two encounters with boundaries of permeability 0.14 or three with a permeability of approximately 0.27). The figure may be interpreted as indicating that the rate of recolonisation of a particular habitat by Carabidae could be strongly affected by the net rate of displacement over the range of population increase rates reported in the literature. Thus, in a habitat with relatively impermeable field boundaries from which Carabidae have been rendered extinct or substantially reduced, the likelihood that a population might be reestablished will be reduced. This is discussed further in Jepson (1994) and Sherratt and Jepson (1993) however, the values given in Figure 1 represent the first estimates of the effects that field boundary permeability has on the overall scale and rate of movement by ground beetles.

AGENDA FOR RESEARCH

For an invertebrate family that is so sensitive to landscape characteristics, the potential for exploitation of Carabidae as indicators of adverse features, that might increase local extinction rates in the invertebrate fauna in general, should be explored. The ecological justification for this is that although the Carabidae are frequent colonists of disturbed land, their low reproductive rates and limited dispersal powers may make them amongst the most sensitive organisms to anthropogenic effects such as habitat fragmentation or excessive pesticide use (Jepson, 1988; Jepson, 1993). This hypothesis however needs to be tested by devising testable predictions concerning the patterns of presence and absence of Carabidae in different landscapes. These predictions would be based upon comparisons of the expected species composition in different habitats (determined largely by the physical characteristics of each habitat) with the composition predicted once anthropogenic activities have been imposed.

Several key research questions need to be answered in order to make these predictions possible:

1. Investigations are required of the components of field boundary permeability, over the life-cycles of selected species. These investigations would have to include representatives from the carabid guild that overwinters in the field boundary, as well as the guild that overwinters within the crop.

2. Mathematical models are then required to explore the consequences of differing levels of field boundary permeability for the local population persistence of carabid species, within agricultural habitats subjected to different patterns of land use (ie. differing degrees

of fragmentation, cultivation or pesticide use). These might be based upon the basic model presented by Sherratt and Jepson (1993).

3. Research is also needed to furnish predictions of the composition of the carabid community in different agricultural habitat types from the physical characteristics of that habitat. Physical factors determine the distribution and composition of carabid assemblages over long time scales (Hengeveld, 1985), over large spatial scales (Luff, Eyre and Rushton, 1989) and even within a given habitat type (Eyre, Luff and Rushton, 1990; Gardiner, 1991) and are likely to be the most important factor that underlie the distribution and abundance of different species.

4. Predictions of the likely carabid assemblages in habitats with differing degrees of fragmentation and disruption by agricultural practices could then in theory, be made. Sites could then be selected to test these predictions and investigate the degree to which species were lost in different systems. For the first time in might then be possible to properly determine the role of field boundaries on the distribution and abundance of Carabidae and also the degree to which loss of carabid species is a good general indicator for sites with depleted invertebrate faunas.

CONCLUSIONS

Most ecological research takes place on a small scale and investigates small numbers of organisms, commonly individual species. If we are to be able to understand the role that field boundaries play in the ecology of farming landscapes however, research must focus on much larger spatial scales and consider larger assemblages of organisms, even whole communities. This does not however imply vagueness or a departure from scientific rigour. Detailed mechanistic studies are needed, that permit testable predictions, supported by relevant ecological theory, to be made and to justify the assertion that some ecological processes can only be properly understood by considering this larger perspective. It has perhaps been too tempting to assume that field boundaries act as essential corridors for wildlife and that the most diverse and resilient farmland flora and fauna can only be maintained in habitats with the maximum quantity of field boundary per unit area. This assumption has ignored the possibilities that field boundaries could have negative consequences for some organisms or that qualitative characters, that determine permeability for example, might be important.

REFERENCES

Andow, D.A.; Kareiva, P.M.; Levin, S.A.; Okubo, A. (1991) Spread of invading organisms. *Landscape Ecology*, **4**, 177-188.

Basedow, Th. (1991) Effects of insecticides on Carabidae and the significance of these for agriculture and species number. In: *The role of ground beetles in ecological and environmental studies*, Stork, N.E. (Ed.), Andover: Intercept, pp. 115-125.

Burn, A.J. (1992) Interactions between cereal pests and their predators and parasites. In: *Pesticides and the environment: the Boxworth study*, Greig-Smith, P.; Frampton, G.H.; Hardy, A. (Eds), London: HMSO, pp. 110-131.

den Boer, P.J. (1971) On the dispersal power of carabid beetles and its possible significance. In: Dispersal and dispersal power of carabid beetles. *Miscellaneous papers* **(8)** *Lanbouwhogeschool, Wageningen*, pp.199-238.

den Boer, P.J. (1977) Dispersal power and survival: carabids in a cultivated countryside. *Miscellaneous papers* **(14)** *Lanbouwhogeschool, Wageningen,* pp.1-190.

den Boer, P.J. (1990a) The survival value of dispersal in terrestrial arthropods. *Biological Conservation*, **54,** 175-192.

den Boer, P.J. (1990b) Density limits and survival of local populations in 64 carabid species with different powers of dispersal. *Journal of Evolutionary Biology*, **3**, 19-48.

Duelli, P.J.; Studer, M.; Marchand, I.; Jakob, S. (1990) Population movements of arthropods between natural and cultivated areas. *Biological Conservation*, **54,** 193-207.

Eyre, M.D.; Luff, M.L.; Rushton, S.P. (1990) The ground beetle fauna of intensively managed agricultural grasslands in Northern England and Southern Scotland. *Pedobiologia*, **34**, 11-18.

Fuchs, G. (1969) Die okoligische Bedeitung der Wallhecken in der Agrarlandschaft Nordwestdeutschlands, am Beispiel der Kafer. *Pedobiologia,* **9,** 432-458.

Gardiner, S.M. (1991) Ground beetles (Coleoptera: Carabidae) communities on upland heath and their association with heathland flora. *Journal of Biogeography,* **18**, 281-289.

Greenslade, P.J.M. (1965) On the ecology of some British carabid beetles with special reference to life histories. *Transactions of the Society for British Entomology*, **16**, 149-179.

Hengeveld, R. (1985) Dynamics of Dutch beetle species during the 20th century (Coleoptera: Carabidae). *Journal of Biogeography,* **12**, 389-411.

Hobbs, R.J. (1992) The role of corridors in conservation: solution or bandwagon? *Trends in Ecology and Evolution*, **7**, 389-392.

Jensen, T.S.; Dyring, L.; Kristensen, B.; Nielsen, B.O. (1989) Spring dispersal and summer habitat distribution of *Agonum dorsale* (Coleoptera: Carabidae). *Pedobiologia*, **33,** 155-165.

Jepson, P.C. (1988) Ecological characteristics and the susceptibility of non-target invertebrates to long-term pesticide side-effects. In: Field methods for the study of the environmental effects of pesticides. *BCPC Monograph,* **40,** Thornton Heath, Surrey: BCPC Publications, pp. 191-200.

Jepson, P.C. (1993) Ecological insights into risk analysis: the side-effects of pesticides as a case study. *The Science of the Total Environment,* **134** (Suppl.), 1547-1566.

Jepson, P.C. (1994) Rates and patterns of movement by Carabidae and their consequences for populations and communities. *Acta Jutlandica*, submitted.

Jepson, P.C.; Thacker, J.R.M. (1990) Analysis of the spatial component of pesticide side-effects on non-target invertebrate populations and its relevance to hazard analysis. *Functional Ecology*, **4**, 349-358.

Kareiva, P.M. (1991) Population dynamics in spatially complex environments: theory and data. *Philosophical Transactions of the Royal Society London, Series B,* **330**, 175-190.

Lyngby, J.E.; Nielsen, H.B. (1988) The spatial distribution of carabids (Coleoptera: Carabidae) in relation to a shelter belt. *Ent. Meddr.*, **48,** 133-140.

Mader, H-J (1990) Wildelife in cultivated landscapes. *Biological Conservation*, **54,** 167-173.

Mader, H.J.; Schell, C.; Kornacker, P. (1990) Linear barriers to arthropod movements in the landscape. *Biological Conservation*, **54,** 209-222.

Pollard, E. (1968a) The effect of removal of the bottom flora of a hawthorn hedgerow on the Carabidae of the hedge bottom. *Journal of Applied Ecology*, **5,** 125-139.

Pollard, E. (1968b) A comparison of the Carabidae of a hedge and field site and those of a woodland glade. *Journal of Applied Ecology*, **5**, 649-657.

Sherratt, T.N.; Jepson, P.C. (1993) A metapopulation approach to modelling the long-term impact of pesticides on invertebrates. *Journal of Applied Ecology,* **30,** 696-705.

Skellam, J.G. (1951) Random dispersal in theoretical populations. *Biometrika*, **38,** 196-218.

Sotherton, N.W. (1984) The distribution and abundance of predatory arthropods overwintering in farmland. *Annals of Applied Biology,* **105**, 423-429.

Sotherton, N.W. (1985) The distribution and abundance of predatory Coleoptera overwintering in field boundaries. *Annals of Applied Biology*, **106**, 17-21.

Speight, M.R.; Lawton, J.H. (1976) The influence of weed cover on the mortality imposed by artificial prey by predatory ground beetles in cereal fields. *Oecologia,* **23,** 211-223.

Thiele, H.U. (1977) *Carabid beetles in their environment: a study of habitat selection and adaptations in physiology and behaviour.* Berlin: Springer Verlag, 369pp.

Thomas, C.F.G. (1992) *The spatial dynamics of spiders in farmland.* PhD thesis, Southampton University.

Tischler, W. von (1958) Synokologische untersuchungen an der fauna der felder und feldgeholze. *Z. Morph. und Okol. Tiere,* **47,** 54-114.

Wallin, H. (1985) Spatial and temporal distribution of some abundant carabid beetles (Coleoptera: Carabidae) in cereal fields and adjascent habitats. *Pedobiologia,* **28**, 19-34.

Wallin, H; Ekbom, B.S. (1988) Movements of carabid beetles (Coleoptera: Carabidae) inhabiting cereal fields: a field tracing study. *Oecologia*, **77,** 39-43.

Williams, G. (1959) Seasonal and diurnal activity of Carabidae, with particular reference to *Nebria brevicollis* and *Feronia*. *Journal of Animal Ecology,* **28**, 309-330.

THE IMPORTANCE OF FIELD MARGIN ATTRIBUTES TO BIRDS

K.H. LAKHANI

NERC Institute of Terrestrial Ecology,
Monks Wood, Abbots Ripton, Huntingdon, Cambs, PE17 2LS

ABSTRACT

The research on the importance of field margin attributes to birds has particularly emphasised the importance of hedgerows. The current literature has been well reviewed and is outlined here. Also summarised here are some studies in progress. These have tended to emphasise the value of the structural dimensions of the different components of field margins to birds. Great care is required to make inferences from non-experimental field observations which might not be balanced over all covariates. The important topics of choice of sampling units, statistical independence and statistical modelling procedures are discussed.

INTRODUCTION

There is a considerable amount of literature on birds and field margins, particularly hedgerow, which suggests that on agricultural land field margins are important to birds (Moore *et al.*, 1967; Hooper, 1970a,b; Pollard *et al.*, 1974; Arnold, 1983; Osborne, 1984; O'Connor & Shrubb, 1986; O'Connor, 1987; Lack 1987, 1992; Parish *et al.*, 1993a,b, in press).

Brown (1969) and Fretwell (1972) suggested that the bird population of hedgerows may also provide immigrants to woodland populations. However, Krebs (1971) found that hedgerow nesting Great Tits (mostly yearlings) had a lower reproductive success and tended to abandon hedgerow territories for experimentally created vacancies in nearby woodland habitat, suggesting that they used hedgerows only as an "overflow" habitat. Pollard *et al.* (1974) argued that hedges provided refuges for woodland birds enabling them to breed in otherwise unsuitable areas. Murton and Westwood (1974) claimed that the removal of hedgerows resulted in only small losses in overall bird populations; but this view has been challenged by Osborne (1982) and O'Connor (1984). Bernstein, Krebs and Kacelnik (1991) concluded that the evidence of relatively stable woodland populations compared with those of field boundaries provided by Krebs and Perrins (1977) supported the earlier view put forward by Brown (1969) and Fretwell (1972).

CURRENT LITERATURE

In many parts of intensive agricultural areas, field margins provide the major habitat for many bird species (O'Connor & Shrubb, 1986). With considerable loss of hedgerows since 1945 (Pollard *et al.*, 1974; O'Connor & Shrubb, 1986; Lack, 1992) and the rate of hedgerow loss not appearing to have slowed down (Barr *et al.*, 1993), much of the research on the value of field margins to birds has concentrated on hedgerows.

The *volume* of the hedge was thought to be of particular interest to birds (Osborne, 1982; Arnold, 1983). Best & Stauffer (1980) showed that the volume of foliage around the nest site may reduce nest losses to avian predation. The ground layer cover promoted by tall and broad hedges may also protect against predation (Pollard *et al.*, 1974). Similarly, the number of herbs in a hedgerow base, and both the presence and variety of trees in a hedge had a positive association with the number of bird species and the number of individuals present in the hedge (Osborne, 1984, 1985; O'Connor, 1987). Shrub-rich hedges provide a greater variety of nesting locations as well as a greater variety of food (e.g. berries) for longer periods; and, different shrubs flower at different times, thus supporting a variety of invertebrates throughout the breeding season of the birds (da Prato & da Prato, 1977). Older hedges tended to be shrub-rich (Pollard *et al.*, 1974; Hooper, 1970b), and these may support a greater invertebrate fauna (O'Connor, 1987).

Field boundary features other than hedges have been studied less. Arnold (1983) found that in his sample plots of arable land in Cambridgeshire, plots containing ditches had twice as many bird species and nearly three times the density of birds, compared with the plots without ditches. He also found that, in winter, hedges with ditches alongside had nearly twice as many species (and if the ditch was large, there was a greater abundance of Blackbird, Song Thrush, Wren, Robin and Dunnock) compared with hedges without an adjoining ditch. A similar positive effect of ditches in the field margins was observed by da Prato (1985).

This paper is not intended to fully cover the literature on all field margin elements, but an important point is that the field margin components (field boundary, boundary strip and crop margin) provide further botanical and structural diversity enabling different species to co-exist (O'Connor, 1987). These components and their importance to game birds are reviewed in this volume (Aebischer *et al.*, 1994). O'Connor & Shrubb (1986) gave a detailed account of studies on farm structure and bird habitats, and the effects of hedges and hedgerow loss on farmland birds. For birds occurring naturally in lowland farms, the literature has been well summarised by Lack (1992), with informative chapters on hedgerows, other field boundaries and field margins; these chapters also describe management practices which are thought likely to benefit birds.

STUDIES IN PROGRESS

Present research on field margins as habitats for birds include studies by staff at the Wildlife Conservation Research Unit (WCRU), Oxford University, at the Royal Society for the Protection of Birds (RSPB), and at the Institute of Terrestrial Ecology (ITE). Also, the British Trust for Ornithology (BTO) has two current projects: The winter hedgerow survey carried out in 1987/88 and The organic farming project, with preliminary reports in BTO News numbers 164, 178 and 185 (P. Lack *pers comm*).

(a) The WCRU work (David Macdonald *pers comm*) is based on studying the association of bird populations with the attributes of 266 hedgerows in Buckinghamshire farms. Both botanical and structural variables were used to explain the observed bird distribution. For each hedgerow, the botanical attributes were the abundance of nearly twenty woody species recorded as absent, present, abundant or dominant. The structural characteristics included mean hedgerow height, hedgerow width at summit and base, the number of mature trees per m of hedgerow, the number of species of mature trees and of woody plants, and the proportion of length made up of gaps. Other categorised variables included ditches, gardens nearby, crop adjacent to the hedgerow, and presence/absence of trees, water and road. For each hedgerow, the birds were surveyed during April-July 1979, recording the nest locations and bird positions.

Multiple regression analysis relating bird variables to the botanical and structural variables showed that bird-rich hedges tended to be taller and had more species of shrub growing in them compared with hedgerows which had few bird species. Thus, the leading term in the final fitted model for total number of bird species was hedgerow height. The next important significant variable was the number of species of woody plant. Presence of dry ditches was also significant, as was the square of hedge height.

(b) The RSPB work (Green *et al.*, in press) is based on surveys of passerine birds during April-May and May-June 1988 and measurements of various attributes of hedgerows, field margin and adjacent land use. The study covered many types of farms in lowland England (46 farms), some with reduced spraying of herbicide and insecticide on the margin of cereal crops - Conservation Headlands. The study used 4760 sampling units of 50 m length of field margin.

For each 50 m section, the presence/absence of each bird species was recorded. The explanatory field margin variables included: number of trees, hedge height, hedge width, woody vegetation length, number of woody species, area of boundary strip, measures of geographical location and factors such as dominant shrub species, dominant plant growing under the hedge, dominant plant on the boundary strip, and adjacent land use.

Logistic regression models, based on those variables which significantly influenced the probability of occupancy of a 50 m section by the bird species, were developed for eighteen bird species. Most bird species preferred tall hedges with many trees, but Dunnock, Willow Warbler and Lesser Whitethroat preferred tall hedges with fewer trees; and, Linnet, Whitethroat and Yellowhammer preferred short hedges with few trees. The incidence of Robin, Song Thrush, Lesser Whitethroat, Whitethroat, Blue Tit and Yellowhammer was positively correlated with the number of woody species in the 50 m section; and the incidence of Lesser Whitethroat and Whitethroat was affected by the identity of the dominant woody plant species in the hedge. The adjacent land use (grass, tillage, roadside) had a significant effect on a number of bird species, with Willow Warbler, Blue Tit and Goldfinch preferring grass to tilled, but Greenfinch and Yellowhammer preferred tilled to grass. Both Goldfinch and Greenfinch preferred hedgerows bounded on one side by a road verge.

Most bird species occurred more often in hedges next to autumn than spring-sown cereals. Conservation Headlands appeared beneficial for bird occupancy in hedgerows adjacent to spring-sown cereals, but for hedgerows adjacent to autumn-sown cereals, the incidence of birds tended to be higher (particularly for Robin, Song Thrush and Greenfinch) when the spraying was not reduced. An explanation of such "negative" results might be along the lines suggested by Cracknell (1986) - dense weed growth in unsprayed areas hindering foraging activity and obscuring prey. Also, perhaps the Conservation Headlands need to be established for a longer time for its beneficial effects to be overriding and demonstrable.

(c) The ITE work (Parish *et al.*, 1993a,b, in press) covered winter and summer bird populations at two sites in East Anglia: Huntingdon (winters 1983, 1984; summer 1985) and Swavesey (winters 1985, 1986, 1991; summers 1986, 1987, 1991). At both sites the sampling units were 200 m long transects of field margin, extending 10 m into the crop each side.

The Huntingdon study was based on 79 transects for which the adjacent field type was small pasture (<20 ha), small arable or large arable. The Swavesey study used 131 transects in six sub-areas with a range of drainage regimes. The crop type was correlated with the drainage regimes; hence, the transects were categorised by the crop on either side: pasture/pasture, pasture/ley and ley/ley as "grass"; pasture/arable and ley/arable as "mixed", and the arable/arable transects as "arable".

The bird variables included various measures of species richness as well as the abundance of nearly thirty bird species. Further derived variables included the abundance of appropriate groups of similar species (FINCHES, WADERS, GAMEBIRDS, AQUATIC BIRDS, etc.) and also Simpson's index of diversity of birds.

The explanatory variables included, for each transect, the number and height of trees, hedge length (= 200 m, if there was no gap), hedge height, hedge crown width, hedge base width, verge width, ditch depth, ditch width. At Swavesey, the data included also the total number of woody, aquatic and herbaceous plant species, as well as the percentage of ground cover of woody, aquatic and herbaceous species and grass. Additional derived variables such as tree number x height, hedge length x height, hedge length x height x width (volume), ditch depth x width were also used.

Several hundred regression models were developed relating the bird variables to the field margin attributes and the adjacent land use. The factors LANDUSE (pasture, small arable, large arable) at Huntingdon and CROP (grass, mixed, arable) at Swavesey played a dominant part in the subsequent modelling. In the Huntingdon models, the significant terms were land use and the interaction of land use with variables reflecting the amount of "woody material" - the number and height of trees and the physical size of the hedge. The models explained a large proportion of the variation in winter, summer and breeding species richness variables at both sites. In the regression models for bird species abundance, land use and tree and hedgerow variables were significant for most woodland birds. Similarly, verge width was important for seed eating birds (most finches and buntings, Red-legged Partridge in summer, and for Carrion Crow in winter) and for insectivores e.g. Blue Tit. Some seed eaters (Linnet, Goldfinch, Reed Bunting) and insectivores (Blackbird, Great Tit, Skylark, Song Thrush and Wren) were associated positively with ditch dimensions; and, Kestrel and the groups RAPTORS, WADERS and AQUATICS with larger ditches. The number and height of hedgerow trees did not appear particularly beneficial to buntings, Skylark, Redwing, Goldfinch, Linnet and the groups GAMEBIRDS and WADERS. Most species favoured the field margins associated with pasture; but species such as Pheasant, Red-legged Partridge, Skylark, Carrion Crow and the group GAMEBIRDS showed greater abundance in field margins associated with large arable farms.

STATISTICAL CONSIDERATIONS

Sampling design

A marked difference between some of the major studies in the published literature (Arnold, 1983; Osborne, 1984) and the studies in progress is in the choice of sampling units. To be able to make inferences about the importance of the structure of the different elements of the field margin to birds, the sampling design should include measurements (number, height, width, depth, etc. as appropriate of trees, hedges, ditches, banks, verges, vegetation) which can be related directly and meaningfully to the bird diversity and abundance.

The sample units used by Osborne (1984) were 42 hedges of variable length, but one hedge was almost ten times as long as another. It is possible for a longer hedge to have more bird species as well as a large area (hedge length x hedge width). Thus, the result that the variable logarithm of hedge area explained 40.7% of the variation in the observed number of species per hedge tells us little about the bird population's needs for tall or wide hedges. Arnold (1983) surveyed bird species in 37 sites in eastern England using quadrats of area equal to 5 ha as the sampling units. The sites ranged from arable land without ditches, hedges and trees, to arable land and grassland with up to 200 m of ditch and/or hedge of various sizes. With such a sampling scheme it was impossible to directly assess the importance of various field margin attributes to birds. In contrast, the sampling units used by Green *et al.* (in press) and Parish *et al.* (in press) were fixed length transects of hedges and field margins respectively, and they recorded a large number of botanical and structural attributes of the transects. This was important for understanding the requirements of different bird species as far as the field margin attributes were concerned.

Statistical analysis and assumptions

Arnold (1983) found that, using data from those quadrats which contained hedges, the average number of species per plot clearly increased with the presence of increasingly larger hedgerows, but in his regression model the species numbers were significantly negatively correlated with hedge length. This is not surprising because it is difficult to obtain meaningful coefficients in modelling data from observational studies (James & McCulloch, 1990); and, this is particularly so if the choice of variables entering the model is based purely on statistical criteria (Parish *et al.*, in press). They found that the models for the same bird variable, based on data from successive years, might easily include different explanatory variables, by chance, particularly if the data were sparse. Such results conflict with ornithologically sensible expectation that the models for the same bird variable, in successive years, should contain the same or similar explanatory variables. They found that biologically consistent models were obtainable by relaxing the arbitrary significance level of $p<0.05$ to $p<0.1$ for a given explanatory variable in a particular year, if this variable was found to be highly significant in other years. An alternative approach is to pool the data of successive years. However, since birds are known to return to the same site year after year, the statistical requirement that the data from the different years should be independent will be violated for the pooled data.

The observations from the different sampling units should be statistically independent. This is not easy to achieve in practice because of the different territory size and mobility of different bird species. Care should nevertheless be taken to ensure that the sampling units are not contiguous, particularly if the units are small. However, in larger units the various structural variables (height, width etc.) might vary substantially, rendering the use of mean values (mean height, mean width etc.) unsatisfactory.

Interpretation of results

Most studies described here have been observational, and not experimental. In such studies, the different levels of the many factors of interest might not be represented in a balanced way in the data, requiring care in interpreting the results. For example, in Parish *et al.* (in press) the hedges in pasture fields tended to be larger than those in arable fields. Hence, though a simple plot of the total number of species against hedge height suggested a non-linear relationship, multiple regression analysis identified a combination of two linear relationships with a shallow slope for the data from arable transects, and a steeper slope for the data from pasture transects. Similarly, the significance of presence/absence of ditches might depend upon the species' daily need for water, its daily flight range and the distance to a water source, which might well be a ditch in a different sampling unit.

CONCLUSIONS

The synthesis of the large number of studies in the literature by O'Connor & Shrubb (1986) and Lack (1992) helps enormously to focus attention on the main findings. However, most studies have been observational, and only in summer and the breeding season - the exceptions being Arnold (1983) and the current work by ITE and BTO. Also, field margins consist of complex collectives of field boundaries with variable botanical and structural properties, as well as banks, verges and other strips of a variety of dimensions and attributes, and indeed crop margins varying in time and space. The true functional relationships between the bird variables and the numerous habitat variables may involve also a large number of other variables and processes (e.g. abundance of invertebrates, number and distribution of predators at different stages of life, climatic factors, competition between species, Markovian effects of species richness and density in previous years). A considerable amount of further research, based on *well-designed studies, is required for successful bird* conservation in coming decades.

ACKNOWLEDGEMENTS

I am grateful to my colleagues T. Parish and T.H. Sparks for their collaboration and to Dr R. Green and Dr D. Macdonald for communicating the findings of their ongoing research. Comments by them and by Professor T.M. Roberts and Drs N. Sotherton and P. Lack are gratefully acknowledged.

REFERENCES

Aebischer, N.J.; Blake, K.A.; Boatman, N.D. (1994) Field Margins as habitats for game. *This volume*.

Arnold, G.W. (1983) The influence of ditch and hedgerow structure, length of hedgerows, and area of woodland and garden on bird numbers on farmland. *Journal of Applied Ecology*, **20**, 731-750.

Barr, C.J.; Bunce, R.G.H.; Clarke, R.T.; Fuller, R.M.; Furse, M.T.; Gillespie, M.K.; Groom, G.B.; Hallam, C.J.; Hornung, M.; Howard, D.C.; Ness, M.J. (1993) Countryside Survey 1990 Main Report. *IFE/ITE Report to DoE, 174 pp.*

Bernstein, C.; Krebs, J.R.; Kacelnik, A. (1991) Distribution of birds amongst habitats: theory and relevance to conservation. In: *Bird Population Studies: Relevance to Conservation and Management*, C.M. Perrins, J.D. Lebreton and G.J.M. Hirons (Eds), Oxford: Oxford University Press, pp. 317-345.

Best, L.B.; Stauffer, D.F. (1980) Factors affecting nesting success in riparian bird communities. *Condor*, **82**, 149-158

Brown, J.L. (1969) The buffer effect and productivity in tit populations. *American Naturalist*, **103**, 347-354.

Cracknell, G.S. (1986) The effects on songbirds of leaving cereal crop headlands unsprayed. *A Report to The Game Conservancy. Research Report No. 18.* British Trust for Ornithology, Thetford.

da Prato, S.R.D. (1985) The breeding birds of agricultural land in south-east Scotland. *Scottish Birds*, **13**, 203-216.

da Prato, S.R.D; da Prato, E.S. (1977) The feeding ecology of Sedge Warbler, Whitethroat and Willow Warbler in a Midlothian scrub valley. *Edinburgh Ringing Group Report*, **5**, 31-39.

Fretwell, S.D. (1972) *Populations in a seasonal environment*. Princeton: Princeton University Press, 217 pp.

Green, R.E.; Osborne, P.E.; Sears, E.J. (in press) The distribution of passerine birds in hedgerows during the breeding season in relation to characteristics of the hedgerow and adjacent farmland. *Journal of Applied Ecology*.

Hooper, M.D. (1970a) Hedges and birds. *Birds*, **3**, 114-117.

Hooper, M.D. (1970b) Dating Hedges. *Area*, **4**, 63-65.

James, F.C.; McCulloch, C.E. (1990) Multivariate analysis in ecology and systematics: Panacea or Pandora's box? *Annual Review of Ecological Systematics*, **21**, 129-166.

Krebs, J.R.(1971) Territory and breeding density in the great tit (*Parus major* L.). *Ecology*, **52**, 2-22.

Krebs, J.R.; Perrins, C.M. (1977) Behaviour and population regulation in the great tit (*Parus major*). In: *Population control by social behaviour*, F. J. Ebling and D. M. Stoddart (Eds), London: Institute of Biology, 99, 23-47.

Lack, P.C. (1987) The effect of severe hedge cutting on a breeding bird population. *Bird Study*, 34, 139-146.

Lack, P.C. (1992) *Birds on Lowland Farms*. London: HMSO, 140 pp.

Moore, N.W.; Hooper, M.D.; Davis, B.N.K. (1967) Hedges I. Introduction and reconnaissance studies. *Journal of Applied Ecology*, 4, 201-220.

Murton, R.K.; Westwood, N.J. (1974) Some effects of agricultural change on the English avifauna. *British Birds*, 67, 41-69.

O'Connor, R.J. (1984) The importance of hedge to songbirds. In: *Agriculture and the Environment*, D. Jenkins (Ed.), *ITE Symposium 13*, Abbots Ripton: Institute of Terrestrial Ecology, pp. 117-123.

O'Connor, R.J. (1987) Environmental interests of field margins for birds. In: *Field Margins*, J. M. Way and P. W. Greig-Smith (Eds), *BCPC Monograph No. 35*, Thornton Heath: BCPC Publications, pp. 35-48.

O'Connor, R.J.; Shrubb, M. (1986) *Farming and Birds*, Cambridge: Cambridge University Press, 290 pp.

Osborne, P.J. (1982) The influence of Dutch elm disease on bird population trends. *Bird Study*, 29, 2-16.

Osborne, P.J. (1984) Bird numbers and habitat characteristics in farmland hedgerows. *Journal of Applied Ecology*, 21, 63-82.

Osborne, P.J. (1985) Some effects of Dutch elm disease on the birds of a Dorset dairy farm. *Journal of Applied Ecology*, 22, 681-692.

Parish, T.; Lakhani, K.H.; Sparks, T.H. (1993a) The assessment of the effects of field boundary attributes and of the adjacent land use upon bird population variables using multiple regression models: Huntingdon 1983-1985 and Swavesey 1985-1991. *ITE Report to MAFF*, 40 pp.

Parish, T.; Lakhani, K.H.; Sparks, T.H. (1993b) Field Margin Attributes and Bird Species Richness and Abundance. In: *The Proceedings of the First Meeting of the UK Region of the International Association of Landscape Ecology (IALE) meeting on LANDSCAPE ECOLOGY IN BRITAIN*, R. Haines-Young (Ed.), Nottingham: Nottingham University, pp. 116-119.

Parish, T.; Lakhani, K.H.; Sparks, T.H. (in press) Modelling the relation between bird population variables and hedgerow and other field margin attributes. I. Species richness of winter, summer and breeding birds. *Journal of Applied Ecology*.

Pollard, E.; Hooper, M.D.; Moore, N.W. (1974) Hedges. London: Collins, 256 pp.

FIELD MARGINS AND SMALL MAMMALS

T. E. TEW

Joint Nature Conservation Committee, Monkstone House, City Road, Peterborough. PE1 1JY.

I. A. TODD, D. W. MACDONALD

Wildlife Conservation Research Unit, Department of Zoology, University of Oxford, South Parks Road, Oxford. OX1 3PS.

ABSTRACT

Britain's small mammal species have colonised the agricultural ecosystem with varying degrees of success. The most adaptable species is the wood mouse which is found on open farmland throughout the year. However, the suitability of arable fields for small mammals is seasonally variable and hedgerows are valuable resources for wood mice through the winter, providing both food and cover, although the open field is still exploited by sections of the population. As the cover and food abundance in the fields increase during spring and summer, mice move out into the fields - nesting, mating and foraging entirely within the crop. Selective pesticide application onto field headlands significantly increases food availability for the small mammals. Harvesting drives most of the arable population back into the hedgerows, temporarily at least, in autumn.

Other small mammal species, such as the shrews and voles, are less adapted than wood mice to the arable ecosystem and live there almost entirely in the hedgerows and field margins. Bank voles were never recorded away from the field margins until May, when the high crop allowed them sufficient cover to venture into the fields. Similarly, the shrews, although common in hedgerows, rarely ventured into the field.

INTRODUCTION

The majority of Britain's small mammal species evolved in woodland and are most frequently, and abundantly, found there to this day. However, as with many other indigenous plant and animal species, small mammals have, in historical times, succeeded also in taking advantage of the opportunities afforded by the agricultural ecosystem.

Despite the more recent (post-war) 'intensification' of farming systems, which has almost certainly reduced its suitability for many species, small mammals still inhabit the agricultural ecosystem. This paper examines the importance of one facet of that ecosystem - the field margin (taken here to include hedgerow, boundary strip and field headland). The question asked was: What is the importance of field margins to small mammals, does it vary between species and/or seasonally, and can it be improved by management? As a useful paradigm for mammals on farmland (Macdonald *et al.* 1993), the wood mouse, *Apodemus sylvaticus* was the main study animal.

The importance of the hedgerow was also investigated, using both live-trapping and radio-tracking techniques. Radio-tracking was also used to determine the effects of 'conservation headlands' (Sotherton 1988) on small mammals. Since wood mice eat many of the species (Pelz 1989) that are known to increase in abundance on conservation headlands (Sotherton *et al.* 1989), it was hypothesized that the selective reduction of agricultural herbicides onto headlands of winter wheat fields would create localised food-rich patches for the wood mice. For convenience, conservation headlands and hedgerow use are considered separately below.

METHODS

Fieldwork was carried out on arable farmland adjacent to deciduous woodland at the Oxford University Farm, Wytham, Oxfordshire (OS Ref. SP4609) and on arable farmland at Sescut Farm, Woodeaton (OS Ref. SP5210), between May 1986 and October 1992.

Conservation headlands

The application of herbicides onto headlands of winter wheat fields was experimentally manipulated in 1986 and 1987 at Wytham, and in 1988 at Sescut. The application of all other chemicals (insecticides, fungicides, growth regulators, fertilisers) throughout this time was consistent both between plots and between plots and the rest of the field. During 1986 experimental plots, either normally sprayed ('sprayed') or completely unsprayed ('unsprayed') alternated along the field headland; each plot was 20m long and extended 10m into the field. During 1987 a pattern of 'sprayed', 'conservation' and 'unsprayed' plots (20m x 10m) alternated along the field headland. During 1988 the entire field was sprayed normally, receiving broad-spectrum herbicides in autumn 1987 and spring 1988, to provide a control to the previous two years. Both botanical and invertebrate sampling was conducted in each of the three years to determine the effects of the experimental spraying regimes on the floral and invertebrate communities. Full experimental details are given in Tew *et al.* (1992). Once the experimental plots were in place, resident wood mice were live-trapped (see below for methods) and radio-tracked (see below). Analysis of the radio-tracking data was either by calculation of a normalised index of preference (Duncan 1983) followed by a bootstrapping correction (Efron 1979), or by non-parametric multiple comparison analysis.

Hedgerow use - live trapping

Live-trapping data to investigate the use of hedgerows by small mammals was collected at Wytham between May 1990 and October 1992. Small mammals were captured in aluminium Longworth live traps which were deployed in two ways. Firstly, a regular grid was set across two adjacent cereal fields; 292 traps were used, with each trap separated from its neighbours by 24m; the trapping grid covered approximately 15 ha. Secondly, traps were set along approximately 1.2km of hedgerow, at 24m intervals, and set in the base of the hedgerow as far as possible. In this way, all the hedgerows surrounding the two cereal fields were trapped.

Traps were stocked with clean dry hay for bedding and were provisioned with whole grain winter wheat. Bedding and grain were replenished as necessary. Traps were checked thrice daily around the clock, to minimise trap mortality, for four consecutive days each month. Wood mice were weighed, sexed and individually marked with numbered metal ear tags. Other species were recorded and were weighed and sexed but not marked. Other species caught were harvest mouse, *Micromys minutus*, house mouse, *Mus domesticus*, bank vole, *Clethrionomys glareolus*, field vole, *Microtus agrestis*, common shrew, *Sorex araneus*, pygmy shrew, *Sorex minutus* and water shrew, *Neomys fodiens*. Of these, only wood mice, bank voles and common shrews were commonly trapped.

Hedgerow use - radio-tracking

Data on habitat use by individuals were collected by radio-tracking. Previous studies (Wolton 1985, Tew 1989) had demonstrated that wood mice were amenable to radio-tracking techniques and, where appropriate, wood mice were radio-collared. The radio-tags (SS1 transmitters: Biotrack U.K. Ltd., Wareham, Dorset, U.K.) weighed < 2g and were attached only to mice which weighed > 19g (Wolton 1985, Pouliquen *et al.* 1990). The transmitters were attached to the mice as radio-collars, using nylon cable-ties with a self-locking ratchet. Radio-collars were attached while the animals were lightly anaesthetised, for approximately one minute, with either pure diethyl-ether or methoxy fluorane ('Metofane'). The radio-collared mice were returned to the trap, allowed to recover full locomotor activity, and released at its site of capture. The mice were retrapped, usually after one month, and the radio-collars removed, also under light anaesthesia.

The mice were tracked on foot and could generally be located from 30m, using a Mariner receiver (Model M57: Mariner Radio, Lowestoft, Suffolk, UK.) and hand-held three-element Yagi aerial. Locational radio-fixes were taken continuously every 10 minutes throughout the night, from first to last activity, to an accuracy of 5m. Individual mice were radio-tracked for a minimum of three complete nights. The mice habituated quickly to the radio-trackers' presence and could be approached to within 5m. To facilitate accurate data collection a 50m grid was marked out across the study sites using fibre-glass canes marked with coloured reflective tape.

Habitat utilisation was analysed using the compositional analysis technique described by Aebischer *et al.* (1993). To facilitate this, the study site was divided up into 5m x 5m grid cells and each grid cell was assigned one of four habitat variables - wheat, barley, rape and hedge. Since crop characteristics are highly seasonal, and are likely to affect habitat utilisation in a highly seasonal manner, analysis of the data was divided into two broad time periods - winter and summer.

For each animal's home range the proportion of each of the four habitats - wheat, barley, rape and hedge - contained within it was calculated (the 'available' habitat). Following this, the location of every radio-fix was assigned to a habitat type and the proportions of each habitat calculated (the 'utilised' habitat). Following Aebischer *et al.* (1993), one of the habitats was chosen arbitrarily (in this case wheat) and ratios calculated for each of the other habitats by dividing its proportion by the proportion of wheat and then calculating the logarithm of this value, resulting in a logratio. These values were calculated for each habitat for both the available and utilised habitats. If there is no preference for a particular habitat, then the logratios for available and utilised habitat will not differ significantly from zero. A matrix is constructed first to test for an overall non-random use of habitats, followed by a final ranking of the habitats and analysis of significant differences between pairs of habitats.

RESULTS

Conservation headlands

Floral census

In 1986 unsprayed headland plots contained significantly ($P < 0.001$) more weeds than did sprayed headland plots for all three species sampled (blackgrass, *Alopecurus myosuroides*, sterile brome, *Bromus sterilis,* and wild oats, *Avena sp.*). In 1987 33 weed species were recorded from the headlands (with a maximum of fifteen from any one headland plot); the abundance of three species was significantly (single factor ANOVA - $P < 0.001$) influenced by the spray treatments and a Student-Newman-Keuls multiple range test indicated that black-grass and wild oats were more abundant ($P < 0.05$) in the unsprayed plots, whilst field forget-me-not was more abundant in the conservation plots ($P < 0.05$). In 1988 there were no significant differences in the abundance of any weed species between areas sampled 1, 5 and 8m from the hedge, except for cleavers, *Galium aparine*, and sterile brome, which were absent from the 5m and 8m quadrats. Neither was there any significant variation in the abundance of weeds within the field headland nor between headland and mid-field for any species.

Invertebrate census

In 1986 15 Orders of invertebrate were identified from the samples. There were significant ($P < 0.05$) differences in abundance between the sprayed and unsprayed treatments for four of the fifteen orders (Collembola, Hemiptera, Diptera and Parasitica). In 1987 8 Orders of invertebrate were identified (all of which also appeared in the 1986 sample); there were no significant differences in the abundance of any of the eight orders of invertebrate between the three spray treatments.

Spatial Use of the home-range by wood mice

In 1986 871 radio-fixes were recorded from two males and one female tracked over the experimentally sprayed areas. Preliminary analysis employing a normalized index of preference (PI) (Duncan 1983), indicated that the mice appeared to be selecting the unsprayed plots (Figure 1a).

Further analysis, to determine if this preference was significant for each of the mice, applied the bootstrap method (Efron 1979) to mean nightly values of PI to circumvent the problems caused by non-independence (*sensu* Swihart & Slade 1985) of radio-fixes. A value of PI = 0.3 indicates no habitat preference, so a percentile interval that does not include this value indicates preference (>0.3) or avoidance (<0.3). Full details are given in Tew *et al.* (1992). The results (Figure 1b) show a significant preference for the unsprayed plots for two of the three mice.

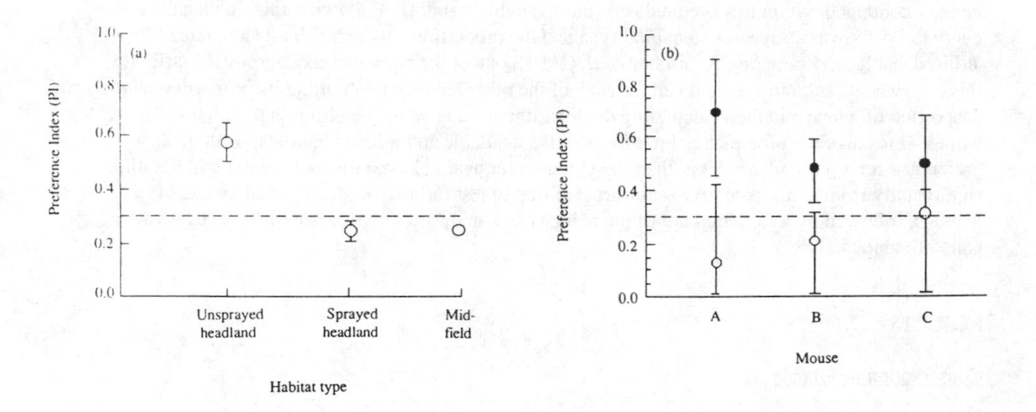

Figure 1. Radio-tracking data from Wytham 1986. a) Normalised preference indices (mean +/- SE of the mean) for different habitat types. Values > 0.3 indicate selection, values < 0.3 indicate avoidance. b) Bootstrap distribution, with percentile confidence limits, of the normalised preference indices for each of the mice in the two experimental treatments. Closed circles - unsprayed plots, open circles - sprayed plots. Bars represent 95% confidence limits.

The unsprayed plots, which the mice preferred, were characterised by abundant grass weeds and observation by torchlight revealed the mice to be feeding on these. Typically, the mouse was seen eating blackgrass seeds, having detached the seed head from the stem. Individual seed husks were ejected from the mouth and these, together with fallen seeds, left characteristic piles of debris on the ground. Thus, it was often possible to retrace the movements of the previous night's radio-tracking to find small piles of weed debris from the mouse's feeding activity.

In 1987 the home-ranges of twelve radio-tracked mice encompassed the experimentally-produced sprayed, conservation and unsprayed plots and 6427 radio-fixes were recorded; once again the mice appeared to select the headland plots that received reduced herbicide application. A Friedman two-way analysis of variance by ranks rejected the Null Hypothesis that the radio-fixes were equally distributed throughout the habitat types ($X_r^2 = 23.65$, $P < 0.001$); and non-parametric multiple comparison analysis revealed that the radio-fixes were more abundant in the unsprayed (1) and conservation headlands (2) than in the sprayed headlands (3) and mid-field areas (4) (S.E. = 4.47, $Q_{0.05,40,4} = 3.63$; 1 vs. 2: $Q = 0.45$, N.S.; 1 vs. 3: $Q = 4.03$, $P < 0.05$; 1 vs. 4: $Q = 5.59$, $P < 0.05$; 2 vs. 3: $Q = 4.47$, $P < 0.05$; 2 vs. 4: $Q = 6.04$, $P < 0.05$; 3 vs. 4: $Q = 1.57$, N.S.).

In 1988 17 mice were radio-tracked to test the null hypothesis that the mice would not prefer conventionally sprayed field headlands over mid-field areas. Of these, only 5 (29%) were ever observed to enter the field headland and, in fact, the mice showed a significant overall preference for the mid-field (Mann-Whitney one-tailed test, $P < 0.05$). Data from the five mice that did use both mid-field and headland showed no significant preference for either.

Hedgerow use - Live-trapping

Shrews and voles

Both common and pygmy shrews were caught entirely in the hedgerows during the winter months, but as crop cover increased over the summer were occasionally also caught in the cereal field itself (Figure 2). Bank voles showed a similar seasonal distribution and were only ever trapped (with one exception) away from the hedgerow during the summer (Figure 2).

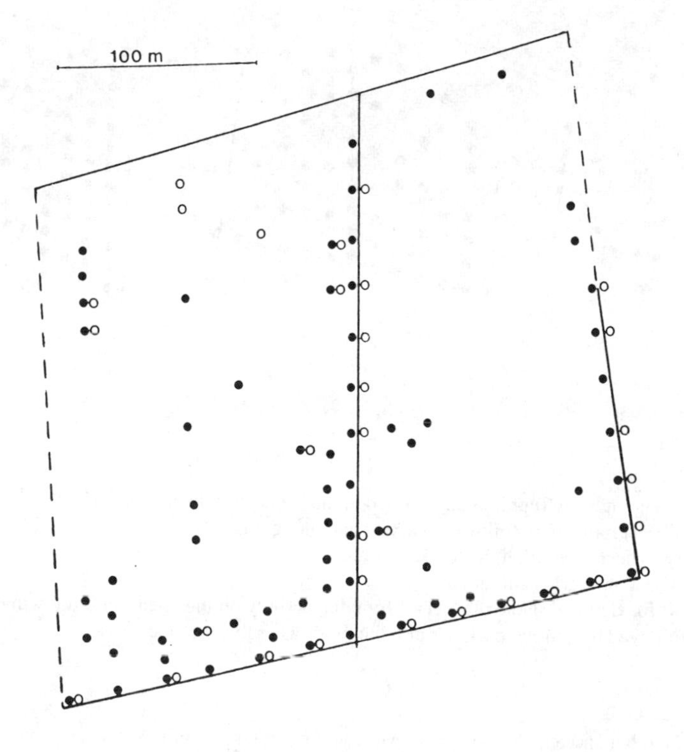

Figure 2. Diagrammatic map showing the capture points of bank voles (closed circles) and common shrews (open circles) between April and October 1991. Solid lines indicate hedgerows, broken lines indicate a field boundary other than a hedgerow. The left field was sown with winter wheat, that on the right with spring barley.

Because the bank voles were not individually marked it is not possible to say how many different individuals were captured. However, between May to August inclusively there were on average approximately 40 bank vole captures per monthly trapping round in the cereal fields, compared to zero for the period October to April inclusively. There was also a corresponding decrease in the number of bank vole captures in the hedgerow over the summer period.

Wood mice

Wood mice, on the other hand, were frequently caught away from the hedgerow throughout the year. Immediately following harvest there was a marked reduction in the number of captures on the open ploughed field, but this was temporary in nature and over the winter months mice were again caught away from the field margin (Figure 3).

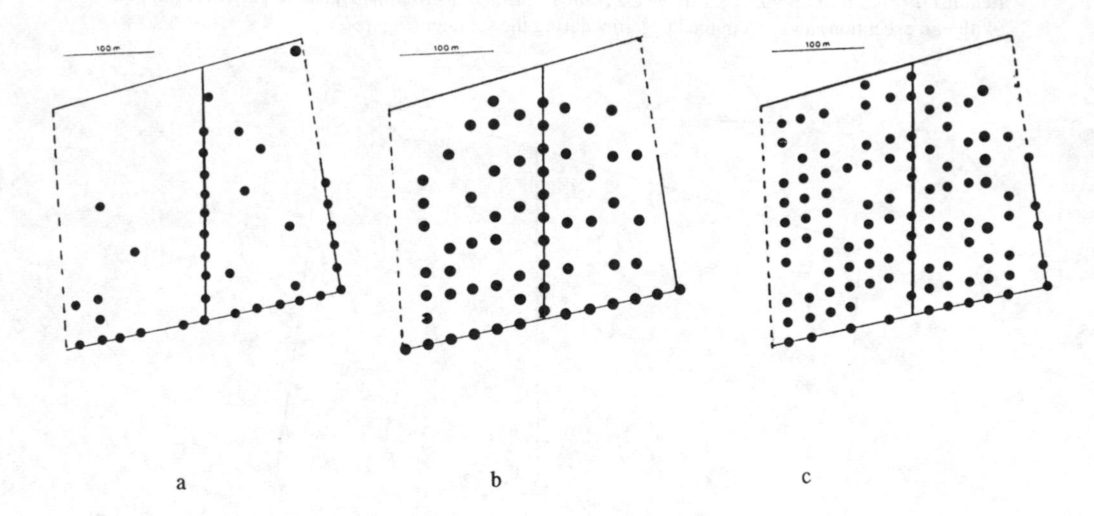

Figure 3. Diagrammatic representation of wood mouse capture locations during:
a - September, ground bare following harvest and ploughing
b - January, wheat 12cm high, barley 5cm high
c - June, wheat and barley full grown.
Symbols as for figure 2. Note the lack of foraging activity on the open field following harvest, followed by a move back out into the fields as crop cover increases.

As with the other species, the hedgerows were an important part of the arable ecosystem for the wood mice throughout the winter months. Between October and April up to 70% of the mice caught in the open field were also caught in the hedgerow, but between October and February only 30% of the mice caught in the hedgerow were also caught in the field. From May onwards, there was no interchange between the field and hedgerow and the two 'populations' were discrete.

Hedgerow use - radio-tracking.

Wood mice - Winter

During the winter wood mice occupied small home ranges that predominantly included hedgerows, although they were also observed to forage on the field surface (Figure 4). Ten animals provided data for at least three of the four habitats analysed during the winter period (November - March). The compositional analysis indicated that overall there was a significant deviation from random use of the habitats (X^2=15.36, v=3, P<0.01). The ranking of habitat use was hedgerow > rape > wheat > barley.

Wood mice - Summer

During the summer mouse home-ranges increased dramatically and were largely, and very often exclusively, in the field (Figure 4). However, nineteen animals provided data on at least three of the four habitats during the summer period prior to harvest (June - August). Overall, there was no significant deviation from random use (X^2=5.23, v=3, NS), although the ranking once again showed that hedgerows were the most preferred habitat type (hedgerow > rape > barley > wheat).

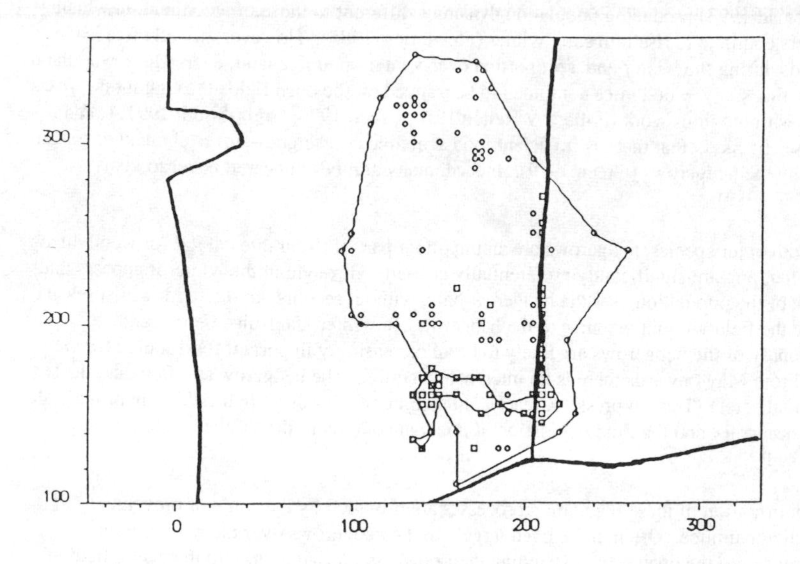

Figure 4. Diagrammatic representation, as an example, of an animal's radio-locations and home-range (calculated using a restricted polygon technique) over winter (open squares) and summer (open circles). Over winter the animal is largely dependant on the hedgerow, over summer the same animal nests and forages largely in the field. Bold lines denote hedgerows. Cropping as in figure 2. Axes scale in metres.

DISCUSSION

Shrews are widespread wherever there is ground cover and are frequently live-trapped in hedgerows. As cover in the cereal fields increases from May onwards, common shrews are also occasionally caught away from the hedgerow, although it is likely that most of these captures will be of animals nesting in the hedgerow but foraging in the field. Shrews take a wide variety of invertebrate prey (Churchfield 1991), many of which are seasonally available in the lowland cereal fields of southern England (Aebischer 1991). The restricted distribution of shrews away from hedgerows in the summer may be limited by their poor burrowing abilities (Churchfield 1991).

For bank voles, as with shrews, ground cover is an important habitat requirement (Gurnell 1985) and their distribution on arable land is severely restricted. However, whilst previous trapping studies on arable land have suggested that the bank vole is never caught away from hedgerows (Pollard & Relton 1970, Loman 1991, Greig-Smith 1991), in this study, as the cover afforded by the crop increased throughout the summer, the bank voles frequently foraged into the field. Food availability is likely to be low in the crop, since the preferred foods of bank voles are fleshy fruits and seeds and the leaves of woody plants (Watts 1968, Hansson 1985), but it is possible that the invertebrate fauna of the cereal field represents an alternative to the pre-fruiting hedgerow.

Many arable wood mouse populations are self-sustaining and do not require seasonal immigration from other habitats to maintain numbers (Green 1979, Tew 1989, Loman 1991). However, through either emigration or predation arable mouse numbers drop markedly at harvest (Greig-Smith 1991, Tew & Macdonald 1993) producing population dynamics different to those of woodland, in which animal numbers continue to rise into early winter (Flowerdew 1985). However, not all mice leave the arable fields during the winter and a proportion over-winter on arable land, even where woodland is close by. In this study, wood mice continued to be trapped on the open fields throughout the year, in accordance with previous work (Pollard & Relton 1970, Green 1979, Greig-Smith 1991, Loman 1991), and it seems likely that there is sufficient food available (waste grain, invertebrates) to sustain them away from the hedgerows (Green 1979). Indeed, many animals appeared never to visit hedgerows (Tew 1989).

Thus, as with other species, hedgerows are an important part of the arable habitat for wood mice during the winter, and are significantly preferentially utilised. Throughout the winter it appears that a large section of the population uses the hedgerow only, without recourse to the field, whilst fewer are able to use the field without recourse to the hedgerow. In winter, the fruits, berries and invertebrate supply of the hedgerows are likely to be an increasingly important food source for the arable mice. From May onwards there is no interchange between the hedgerow and field and the two populations are discrete (Tew, in press). Clearly, at this time of year the cereal fields can provide all the resources necessary and the field population of mice, in contrast to that of the hedgerow, increases markedly.

Further confirmation of these population processes are provided by the study of movement patterns of individual mice. Often, mice lived largely in the hedgerows over the winter, with occasional forays onto the open field, expanding their range and moving out into the cereal field in the summer as the crop grows. The social system of arable mice during the summer is one of female defence polygyny (Tew 1989, Tew & Macdonald, in press), with females defending small (0.4 ha) territories intra-sexually and males occupying large (1.5 ha) home-ranges that overlap both inter- and intra-sexually (Tew 1992). Seasonal territoriality by females may explain the seasonal nature of the interchange between the hedgerow and field populations.

For arable-dwelling wood mice cereal grains, weed seeds and arthropods are preferred food items in the summer (Green 1979; Pelz 1989; Plesner Jensen 1993). The floral and entomological data from this study and other related studies (e.g. Sotherton 1988), show that the reduced input of certain herbicides onto cereal headlands creates areas of high weed and insect abundance. Given the food preferences and diet of wood mice, we conclude that reduced herbicide input increased food availability. Small mammals in general, and wood mice in particular (Don 1979; Angelstam *et al.* 1987), are well able to recognise and exploit resources that are highly variable in both space and time. This study demonstrates that wood mice are able to recognise, and take advantage of, the localised conditions of high food abundance produced by the experimental spraying regimes. Wood mice spent significantly more time in the conservation and unsprayed headland plots, where food availability was high, than would be expected from the relative abundance of such plots within their home-range. Furthermore, direct observations suggested that the time spent in the food-rich patches was largely spent feeding. These conclusions are strengthened further by the inter-annual comparison since in the absence of experimentally-created areas of high food abundance, the mice showed no preference for the field headlands. This suggests that the mice were indeed reacting to the high food abundance in the earlier experiments and were not preferring the cereal field headlands for other reasons such as richer seed banks or increased cover from predators.

The analyses presented here have differentiated the hedgerow, headland and the body of the field. A finer level of resolution has considered the grassy margin between hedge and crop, which conservationists have viewed as a sanctuary for wildlife but which farmers have often viewed as a reservoir for weeds. Another experiment at Oxford University's farm at Wytham sought to discover whether grassy margins could be managed in ways which optimised the integration of nature conservation and farming. Ten management regimes were imposed on 2m margins in a complete randomised block design replicated eight times around arable fields (Smith & Macdonald 1989). Plesner Jensen examined the effects of eight of the management regimes on farmland rodents (Plesner Jensen 1993, Smith *et al.* 1993). Although wood mice preferred taller vegetation, margin management regimes were less important to small mammals than was the presence of boundary features such as hedges and ditches. Boundaries with hedgerows had relatively high numbers of wood mice, bank voles and common shrews, whereas bank voles were more numerous in margins with ditches than in those without.

ACKNOWLEDGEMENTS

Grateful thanks are extended to the many fieldworkers, too numerous to mention individually, of the Oxford 'JAEP team'. This work was funded by a NERC CASE studentship and by the Joint Agricultural and Environmental Programme (JAEP/NERC), grant reference number GST/02/479.

REFERENCES

Aebischer, N. J. (1991). Twenty years of monitoring invertebrates and weeds in cereal fields in Sussex. pp 305 - 332. In: *The ecology of temperate cereal fields.* Eds. Firbank, L. G., Carter, N., Darbyshire, J. F. & Potts, G. R. Blackwell Scientific Publications, Oxford.
Aebischer, N. J., Robertson, P. A. & Kenward, R. E. (1993). Compositional analysis of habitat use from animal radio-tracking data. *Ecology* **74,** 1313-1325.
Angelstam, P., Hansson, L. & Pehrsson, S. (1987). Distribution borders of field mice *Apodemus:* the importance of seed abundance and landscape composition. *Oikos,* **50,** 123-130.
Churchfield, S. (1991). The common shrew. pp 51-58. In: *Handbook of British Mammals.* Eds. Corbett, G. B. & Harris, S. Blackwell Scientific Publications, Oxford.
Don, B.A.C. (1979). Gut analysis of small mammals during a sawfly (*Cephalcia lariciphila*) outbreak. *Journal of Zoology, London,* **188,** 290-294.
Duncan, P. (1983). Determinants of the use of habitat by horses in a mediterranean wetland. *Journal of Animal Ecology,* **52,** 93-109.
Efron, B. (1979). Bootstrap methods: another look at the Jacknife. *Annals of Statistics,* **7,** 1-26
Flowerdew, J. R. (1985). The population dynamics of wood mice and yellow-necked mice. *Symposium of the Zoological Society of London.* **55,** 315 -338.
Green, R. E. (1979). The ecology of wood mice (*Apodemus sylvaticus*) on arable farmland. *Journal of Zoology, London.* **188,** 357 - 377.
Greig-Smith, P. W. (1991). The Boxworth experience: Effects of pesticides on the fauna and flora of cereal fields. pp 333 - 371. In: *The ecology of temperate cereal fields.* Eds. Firbank, L. G., Carter, N., Darbyshire, J. F. & Potts, G. R. Blackwell Scientific Publications, Oxford.

Gurnell, J. (1985). Woodland rodent communities. *Symposium of the Zoological Society of London.* **55**, 377 - 411.

Hansson, L. (1985). The food of bank voles, wood mice and yellow-necked mice. *Symposium of the Zoological Society of London.* **55**, 141 -168.

Loman, J. (1991). The small mammal fauna in an agricultural landscape in southern Sweden, with special reference to the wood mouse *Apodemus sylvaticus. Mammalia.* **55**, 91 - 96.

Macdonald, D. W., Tew, T. E. & Todd, I. A. (1993). The arable mouse: lessons from the wood mouse for wildlife on farmland. NERC News, **26**. Natural Environmental Research Council, Swindon.

Pelz, H-J. (1989). Ecological aspects of damage to sugar beet seeds by *Apodemus sylvaticus. Mammals as pests* (Ed. by R.J. Putman), pp. 34-48. Chapman and Hall, London.

Plesner-Jensen, S. (1993). Temporal changes in food preferences of wood mice, *Apodemus sylvaticus,* L. *Oecologia*, **94,** 76-82.

Pollard, E. & Relton, J. (1970). A study of small mammals in hedges and cultivated fields. *Journal of Applied Ecology.* **7**, 549 - 557.

Pouliquen, O., Leishman, M. & Redhead, T.D. (1990). Effects of radio collars on wild mice, *Mus domesticus. Canadian Journal of Zoology,* **68,** 1607-1609.

Smith, H. & Macdonald, D. W. (1989). Secondary succession on extended arable field margins: its manipulation for wildlife benefit and weed control. *Brighton Crop Protection Conference - Weeds.* pp 1063-1068. Brighton Crop Protection Council, farnham, Surrey.

Smith, H., Feber, R. E., Johnson, P., McCallum, K., Plesner Jensen, S., Younes, M. & Macdonald, D. W. (1993). The conservation management of arable field margins. *English Nature Science,* **18.** English Nature, Peterborough.

Sotherton, N.W. (1988). The Cereals and Gamebirds Research Project: Overcoming the indirect effects of pesticides. *Britain Since Silent Spring* (Ed. by D.J.L. Harding), pp. 64-72. Proceedings of a Symposium of the Institute of Biology, London.

Sotherton, N.W., Boatman, N.D. & Rands, M.R.W. (1989). The "Conservation Headland" experiment in cereal ecosystems. *The Entomologist,* **108,** 135-143.

Swihart, R.K. & Slade, N.A. (1985). Testing for independence of observations in animal movements. *Ecology,* **66,** 1176-1184.

Tew, T. E. (1989). The behavioural ecology of the wood mouse (*Apodemus sylvaticus*) in the cereal field ecosystem. Unpublished D. Phil Thesis, University of Oxford.

Tew, T. E. (1992). Radio-tracking arable wood mice. pp. 532 - 539. In: *Wildlife Telemetry*. Eds. Priede, I. G & Swift, S. M.

Tew, T. E. (In press). Farmland hrdgerows: habitat, corridors or irrelevant? A small mammal's perspective. In: *Hedgerow management and nature conservation.* Ed. Watt, T. A. Wye College Press, Wye, Kent.

Tew, T. E. & Macdonald, D. W. (1993). The effects of harvest on arable wood mice, *Apodemus sylvaticus. Biological Conservation,* **65**, 270-283.

Tew, T. E. & Macdonald, D. W. (In press). Dynamics of space use and male vigour amongst wood mice, *Apodemus sylvaticus,* in the cereal ecosystem. *Behavioural Ecology and Sociobiology.*

Tew, T. E., Macdonald, D. W. & Rands, M. R. W. (1992). Herbicide application affects microhabitat use by arable wood mice *Apodemus sylvaticus. Journal of Applied Ecology.* **29,** 532-539.

Watts, C. H. S. (1968). The foods eaten by wood mice (*Apodemus sylvaticus*) and bank voles (*Clethrionomys glareolus*) in Wytham woods, Berkshire. *Journal of Animal Ecology.* **37**, 25 - 41.

Wolton, R.J. (1985). The ranging and nesting behaviour of wood mice, *Apodemus sylvaticus*, as revealed by radio-tracking. *Journal of Zoology, London,* **206,** 203-224.

FIELD MARGINS AS HABITATS FOR GAME

N.J. AEBISCHER & K.A. BLAKE

The Game Conservancy Trust, Fordingbridge, Hants, SP6 1EF

N.D. BOATMAN

Allerton Research and Educational Trust, Loddington House, Loddington, Leics., LE7 9XE

ABSTRACT

Several game species make considerable use of field-margin habitats, in particular grey partridges, red-legged partridges, pheasants and hares. Generally speaking, field margins play a multiple role for these animals: they provide cover for nesting or shelter, they provide brood-rearing areas and they provide food. This paper briefly reviews current knowledge about such use, and presents results from recent work on nest-site selection by grey partridges. It highlights the features that the animals seek out preferentially, and discusses the implications in terms of integrating field-margin management with modern crop husbandry and EU regulations.

INTRODUCTION

Since the 1950s, many of the traditional game species often associated with agricultural land have undergone severe decline. The grey partridge *Perdix perdix* in particular has fallen to 20% of its pre-war abundance on average (Potts, 1986, Marchant *et al.*, 1991), and has disappeared completely from some areas of the British Isles (Gibbons *et al.*, 1993). Bag records suggest that numbers of brown hares *Lepus europaeus* have followed a similar pattern (Tapper, 1992), and numbers of red-legged partridges *Alectoris rufa* and pheasants *Phasianus colchicus* appear stable only because of large-scale releases of reared birds (Hill & Robertson, 1988, Gibbons *et al.,* 1993).

Directly or indirectly, the cause of the decline has been the steady intensification of agriculture: introduction of herbicides in the late 1950s, greater mechanisation leading to hedgerow removal and field enlargement, the abandonment of mixed farming in favour of all-arable agriculture increasingly geared towards winter crops, the rise in the use of fungicides, insecticides and other inputs in the 1980s (Jenkins, 1984, Potts, 1990). For game, the consequences have been a loss of the habitat types that are needed for food, cover or reproduction at one or more times of year. With the sanitisation of the crop itself, game has been squeezed into the intercrop zones of farmland. The relative importance of field margins is thus much greater now than in the past. We describe below the habitat requirements of game during the breeding period (spring-summer) and the non-breeding period (autumn-winter) based on recent work by The Game Conservancy Trust. The emphasis will be on the grey partridge, a bird currently in the UK Red Data Book (Batten *et al.,* 1990), but reference will also be made to the red-legged partridge, the pheasant and the hare. We also discuss the management implications of these habitat requirements.

FIELD-MARGIN REQUIREMENTS OF GAME

Requirements during the breeding season

For adult grey and red-legged partridges, food availability during this period is generally not a problem. The habitat requirements are two-fold: (1) availability of suitable nesting cover and (2) availability of suitable brood-rearing areas for birds that nested successfully. Early work, summarised in Potts (1986), found that partridges nested along hedgerows, fence lines and other linear features of arable landscapes. The physical structure and vegetation characteristics of hedgerows have been identified as a factor affecting the suitability of hedges for nesting (Blank *et al.*, 1967, Hunt, 1974). A more detailed study of nest-site selection by Rands (1986, 1988) showed that grey partridges sought out slightly elevated nesting locations such as ones on hedge banks, which were correspondingly well drained. The presence of dead grass at the bottom of a hedge was also a feature of preferred nesting locations, as it provided cover from predators, cryptic-coloured nesting material and shelter from the weather. Red-legged partridges too selected nest-sites in areas where the amount of dead grass was greater than in the surrounding vegetation and also where the amounts of leaf-litter, bramble *Rubus spp.* and common nettle *Urtica dioica* were higher (Rands, 1986, 1988).

More recent work on radio-tagged partridges at two sites in Wiltshire and Hampshire in 1991 provided an objective assessment of the nesting preferences of grey partridges (Table 1). Although more nests than expected from previous studies were found outside field margins, i.e. within crops or grass fields, two-thirds of nests were situated in marginal habitat such as hedge bottoms, verges and odd corners of uncultivated land, confirming the importance of field margins in the broad sense.

TABLE 1. Choice of habitats for nesting by radio-tagged grey partridges at two sites in Hampshire and Wiltshire in 1991.

	Field margins	Winter cereals	Peas/beans	Pasture/hay	Game crops	Total
Wiltshire	12	2	0	4	0	18
Hampshire	8	2	1	0	2	13
Frequency	65%	13%	3%	13%	6%	100%

This work studied the vegetation characteristics of the 31 nest-sites compared to randomly chosen locations at two levels: an "extensive", low-resolution level whereby 40 points were selected randomly within each of three habitats (cropped areas, field margins, other areas), and an "intensive", high-resolution level whereby each nest was paired with a point selected at random within the same patch of habitat containing the nest. The percentage cover of plant species within four 0.25 m^2 quadrats placed around each site was recorded and averaged. Boatman *et al.* (in press) carried out a preliminary regression analysis of these data. They found that nests were associated with dead grass, leaf litter and tall forbs of moderately disturbed ground, but recommended a canonical correspondence analysis of the plant communities; this is presented here using CANOCO (Ter Braak, 1988). The vertical structure of the vegetation was measured as the percentage of a graduated measuring board that was not visible from a fixed observation point at each of 11 height categories (0-5 cm, 5-10 cm, 10-20 cm, ... 90-100 cm), averaged over the four quadrats at each site.

After removing the effects of study area and habitat type, at the extensive level the vertical structure around nest sites was significantly denser at heights above 20 cm than around randomly chosen non-nest sites; the difference was most marked in the range 30-90 cm (Fig. 1). At the intensive level, a similar difference in density was observed at heights from 60 to 90 cm (Fig. 1).

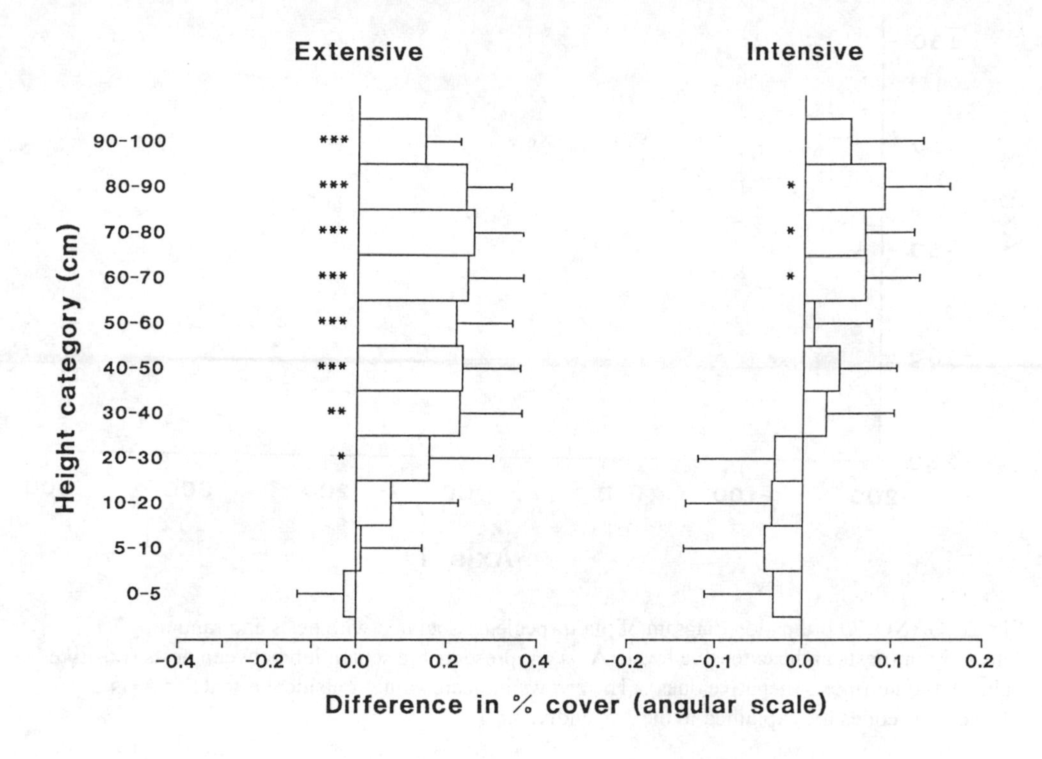

Fig. 1. Differences in vertical structure of vegetation around nests and randomly selected non-nests at two levels: extensive (left) and intensive (right). The percentage cover at each height category was normalised by angular transformation. Positive difference values indicate that cover was greater above nests than above non-nests, and vice versa; error bars represent 95% confidence limits. * $P<0.05$; ** $P<0.01$; *** $P<0.001$.

The CANOCO ordinations revealed that, at the intensive level, there were no detectable differences between the vegetation communities around nests and around non-nests. There was, however, a significant difference ($P<0.05$) in plant communities at the extensive level. In the ordination diagram (Fig. 2), the first axis represents the separation between nests and non-nests. The vegetation surrounding nests had features typical of mesotrophic rough grasslands, with species such as common nettle, cock's-foot *Dactylis glomerata* and upright hedge-parsley *Torilis japonica*. The non-nest plant community represented mainly vegetation of disturbed chalk, either low-lying species such as clovers *Trifolium* spp. and plantains *Plantago* spp. or taller clump-forming species colonising bare ground (e.g. mugwort, *Artemisia vulgaris*). Between the two extremes lay a group of species belonging to coarse chalk grassland.

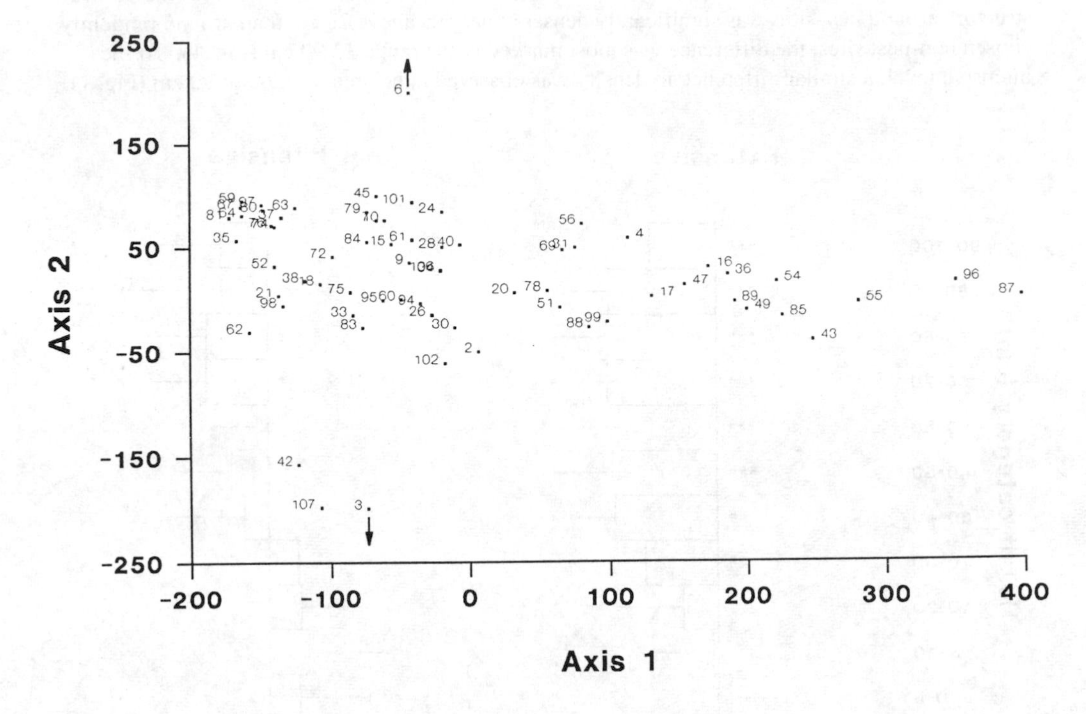

Fig. 2. CANOCO ordination diagram of plant species associated with nests and randomly selected non-nests at the extensive level. Axis 1 represents the separation between nests (positive values) and non-nests (negative ones). The arrows indicate values outside the scale of Axis 2. The species codes are explained in the Appendix.

The plant community characterising nest sites was therefore on average taller, and had a more continuous canopy, than that associated with non-nest sites. In the latter case, the vegetation cover was patchy, and the canopy tended to be lower and much more open. The vegetation classification agreed well with the results on vertical structure (Fig. 1). Taken together, the results implied that partridges preferred to nest in the type of plant community that provided them with the best cover around and over the nest; moreover within such a community they sought out those patches that best fitted those criteria.

For pheasants in most of Britain, field margins are important only if adjacent to woodland edges and shrubby cover (Woodburn & Robertson, 1990), and are used early in spring by males setting up their territories. Territory density was three times higher when such woodland edges were adjacent to cereal crops than to grass because the crop edges offered better feeding opportunities for the hens in each male's harem (Robertson, 1992, Robertson *et al.,* 1993). For nesting purposes, radio-tracking studies have shown that unlike the situation for partridges, grassy banks and hedgerows are not preferred habitats (Hill & Robertson, 1988). Work in the USA suggests however that grass strips are attractive as nesting cover provided that the vegetation height is sufficient to afford concealment from avian predators (Mankin & Warner, 1992).

Radio-tracking of hares (Tapper & Barnes, 1986) has shown that on intensive arable areas, hares suffer a food bottleneck during the summer. Frylestam (1980a) has shown that during this period grass strips are very attractive to hares as feeding areas, and that there is a link between nutrition and reproduction (Frylestam, 1980b). Field margins also offer shelter and resting places for hares in the form of hedgerows and grass banks (Tapper & Barnes, 1986).

Grey partridge and pheasant chicks both require an insect-rich diet in the first weeks of life (Hill, 1985, Potts, 1986), but a high-protein diet seems less critical for young red-legged partridges (Green, 1984, Rands, 1988). The structure of the vegetation in which the chicks forage is also important, as it must be tall enough for concealment from predators, yet sufficiently open to allow easy passage. In wet weather, the chicks must be able to avoid becoming soaked through contact with vegetation and, if wet, must be able to dry out. The structure of cereal crops are ideal from these points of view, and radio-tracking has demonstrated that hens lead their broods from the nesting site into adjacent cereal crops, where they spend most of their time (Green, 1984, Hill, 1985). Cereals can provide the insects that the chicks require as long as enough broad-leaved weeds are present as insect hosts; too often, however, this weedy understory is eliminated by the use of herbicides (Southwood & Cross, 1969, Potts, 1986). In Poland, where up to 70% of crop area was not sprayed, the mean brood size of grey partridges was 9.3 (Panek, 1992) compared to 4.8 on intensively farmed English land (Sotherton & Robertson, 1990); the latter increased to 7.4 in insect-rich, selectively-sprayed, weedy cereal headlands.

Pheasant brood sizes increased from 2.8 in fully-sprayed fields to 4.9 in ones with selectively-sprayed headlands (Sotherton & Robertson, 1990). In Austria, differential brood sizes of pheasants on conventionally farmed land and in untreated cereal mix on set-aside land were 4.9 ($n = 153$) and 6.8 ($n = 53$) respectively (P.A. Robertson, pers. comm.). Direct insect mortality through summer aphicide use may also negatively affect chick survival rates; Potts (1990) found that the chick survival rate of grey partridges and of pheasants was 50% lower in cereals sprayed with a broad-spectrum insecticide in June than in cereals that had not received the aphicide treatment.

Requirements during the non-breeding season

Field margins in winter are mostly useful to partridges as a source of cover in the form of hedgerows and rough grass. To a limited extent, they may supply some food items as well, such as green vegetable matter and weed seeds. Hedgerows and most grass strips play a more important role in late winter, as the birds start to form pairs and space themselves out. Potts (1980) found that the emigration rate of grey partridges was negatively related to the density of linear features (km/km^2) in the landscape. The importance of these features on a local scale was confirmed by Rands (1986), who observed that the length of potential nesting habitat (field boundaries including hedgerows) explained up to 81% of the variation in grey partridge pair density within farms, and up to 98% of that for red-legged partridges.

For hares in winter, field margins are less important as a food source as they are able to feed on winter crops. However, hedgerows and strips of long grass are still intensively used as shelter, as in the summer, particularly by day (Tapper & Barnes, 1986).

Pheasants in winter concentrate in areas of shrubby cover that provide shelter adjoining suitable feeding areas. As in the summer, densities are higher where cereal fields rather than grass abut the cover, by a factor of two (Robertson, 1992).

DISCUSSION AND IMPLICATIONS FOR MANAGEMENT

The relevance of different forms of field margins to farmland game is summarised in Table 2. The grey partridge, red-legged partridge, pheasant and hare all rely on field margins at some point in their life-cycle, be it for food, shelter or nesting.

TABLE 2. Summary of field margin types and their importance to game in spring/summer and autumn/winter.

Species	Hedgerow	Grass strip	Crop edge	Rough ground
Spring/summer				
Grey partridge	nesting	nesting	brood-rearing	nesting/brood-rearing
Red-legged partridge	nesting	nesting	brood-rearing	nesting/brood-rearing
Pheasant	-	nesting	brood-rearing	nesting/brood-rearing
Hare	shelter	feeding	-	shelter/feeding
Autumn/winter				
Grey partridge	shelter/feeding	pairing	feeding	shelter/feeding
Red-legged partridge	shelter/feeding	pairing	feeding	shelter/feeding
Pheasant	-	feeding	feeding	feeding
Hare	shelter	feeding	-	shelter/feeding

A number of management options have been developed and proposed as a means of integrating game conservation and modern farm management. Below, we review these for each of the different types of field margin that are relevant to the requirements of game.

Hedgerows

Between 1989 and 1990, 52000 km of hedgerows in Great Britain were removed while only 25400 km of new hedges were planted (Brown, 1992). This represents a continuation of the process of hedgerow removal described by Barr *et al.* (1986). The planting of new hedgerows is encouraged by government grants, now supplemented by a new Countryside Commission "Hedgerow Incentive Scheme" to encourage improved hedgerow management. Other schemes such as Countryside Stewardship or Environmentally Sensitive Areas also seek to favour a more sympathetic management of linear habitat features. Such management includes maintenance of the woody structure of the hedge at a height of approximately 2 m (Pollard *et al.,* 1974), rotational cutting of the hedge, bank and grass margin every two to three years to allow cover to develop (Rands, 1987), no spraying or selective spraying of the hedge bottom to control agricultural weed pests such as cleavers *Galium aparine* or brome *Bromus spp.* (Boatman, 1992), and establishment of a buffer strip between the crop and the hedge to protect the hedge from fertilizer and pesticide drift (Jepson *et al.,* in press) and the crop from weed encroachment (Boatman & Wilson, 1988).

Grass strips

Grass strips are valuable for game only if part of them at least is allowed to grow up and provide cover. Ideally, strips should be cut every two to three years, on a rotational basis around the farm (Rands, 1987). They can be planted alongside hedgerows, tracks or roads, or around crops. The Game Conservancy Trust, in conjunction with Southampton University, has developed "Beetle Banks", raised strips across fields that are planted with a mixture of tussocky grasses such as cock's-foot or Yorkshire fog (Thomas *et al.*, 1991, 1992). Besides harbouring high densities of natural aphid predators over winter, these banks may also provide alternative shelter and nesting cover for hares and partridges. Work in the USA has shown that wide strips are better for nesting gamebirds than narrow ones, as predation rates were much higher in narrow strips than in ones 10 m or more wide (Olsen, 1977).

Crop edges

Modern crops provide the right structure for gamebird chicks, but are deficient in the insect food that the chicks require (Southwood & Cross, 1969, Potts, 1986). Conservation Headlands are a tried and tested way of restoring the understory of weeds and their invertebrate fauna, with beneficial effects upon gamebird chick survival (Sotherton and Robertson, 1990, Sotherton, 1991, Chiverton, 1994). The idea is that the outer 6-m band of cereal crop receives reduced and selective pesticide inputs that control grass weeds and cleavers, while enabling most broad-leaved weed species and beneficial insects to survive. The guidelines are constantly being updated so as to maximise the benefit to game while minimising the agricultural disadvantages (Boatman & Sotherton, 1988, The Game Conservancy Trust, 1993).

Set-aside

The latest MAFF guidelines and EU regulations concerning management of set-aside land are much more favourable towards game than in previous years (MAFF, 1993). The option of mixing rotational and non-rotational set-aside on the same farm is particularly promising (The Game Conservancy Trust, 1994). Strips or blocks of non-rotational set-aside can be strategically placed to provide shelter, nesting or brood-rearing cover for all species of game. Strips across large fields can constitute islands of game habitat and make large cultivated areas more diverse. Rotational set-aside following cereals can be used to make up the required area; it provides food over winter and, if a strong growth of volunteer cereals develops, becomes ideal brood-rearing habitat if left undisturbed until mid-July.

Conclusion

As regards the fortunes of farmland game in Britain, set-aside holds the greatest potential because it affects the greatest surface area and is already being implemented. This and the other options mentioned above show that there is now more scope for sympathetic management and financial support for such management than there has been for several decades. Time will tell whether the opportunity will be grasped, and whether we are at a turning point in the declining fortunes of our wild lowland game.

ACKNOWLEDGEMENTS

We thank Hugh Oliver-Bellasis and the Ministry of Defence for granting access to their land.

REFERENCES

Barr, C.J.; Benefield, C.; Bunce, R.G.H.; Riddesdale, H.; Whittaker, M.A. (1986). *Landscape changes in Britain.* Huntingdon: Institute of Terrestrial Ecology, Monks Wood.

Batten, L.A.; Bibby C.J.; Clement, P.; Elliot, G.D.; Porter, R.F. (1990). *Red data birds in Britain.* London: Poyser.

Blank, T.H.; Southwood, T.R.E.; Cross, D.J. (1967). The ecology of the partridge I. Outline of population processes with particular reference to chick mortality and nest density. *Journal of Animal Ecology,* **36,** 549-556.

Boatman, N.D. (1992). Improvement of field margin habitat by selective control of annual weeds. *Aspects of Applied Biology,* **29,** 431-436.

Boatman, N.D.; Blake, K.A.; Aebischer, N.J.; Sotherton, N.W. (in press). Factors affecting the herbaceous flora of hedgerows on arable farms, and its value as wildlife habitat. In: *Hedgerow Management and Nature Conservation,* T.A. Watt and G.P. Buckley (Eds), Wye: Wye College Press.

Boatman, N.D.; Sotherton, N.W. (1988). The agronomic consequences and costs of managing field margins for game and wildlife conservation. *Aspects of Applied Biology,* **17,** 47-56.

Boatman, N.D.; Wilson, P.J. (1988). Field edge management for game and wildlife conservation. *Aspects of Applied Biology,* **16,** 53-61.

Brown, A. (Ed.) (1992). *The UK Environment.* London: H.M.S.O.

Chiverton, P.A. (1994). Large-scale field trials with Conservation Headlands in Sweden. In: *Field Margins - Integrating Agriculture and Conservation,* N.D. Boatman (Ed), BCPC Monograph No. 58, Farnham: BCPC Publications

Frylestam, B. (1980a). Utilization of farmland habitats by European hares, (*Lepus europaeus* Pallas) in southern Sweden. *Swedish Wildlife Research,* **11,** 271-284.

Frylestam, B. (1980b). Reproduction in the European hare in southern Sweden. *Holarctic Ecology,* **3,** 74-80.

Gibbons, D.W.; Reid, J.B.; Chapman, R.A. (1993). *The New Atlas of Breeding Birds in Britain and Ireland: 1988-1991.* London: T. & A.D. Poyser, 520 pp.

Green, R.E. (1984). The feeding ecology and survival of partridge chicks (*Alectoris rufa* and *Perdix perdix*) on arable farmland in East Anglia. *Journal of Applied Ecology,* **21,** 817-830.

Hill, D.A. (1985). The feeding ecology and survival of pheasant chicks on arable farmland.. *Journal of Applied Ecology,* **22,** 645-654.

Hill, D.A.; Robertson, P.A. (1988). *The Pheasant: Ecology, Management and Conservation.* Oxford: Blackwell Scientific Publications, 281 pp.

Hunt, H.M. (1974). Habitat relations and reproduction ecology of Hungarian partridge in a hedgerow complex in Saskatchewan. *Saskatchewan Department of Renewable Resources and Wildlife Report 3,* 1-51.

Jenkins, D. (Ed.) (1984). *Agriculture and the Environment.* Cambridge: Institute of Terrestrial Ecology, 196 pp.

Jepson, P.C.; Cilgi, T.; Cuthbertson, P.; Sotherton, N.W. (in press). Pesticide spray drift into field boundaries and hedgerows. Direct measurements of drift deposition. *Environmental Pollution.*

MAFF (1993). *Arable Area Payments 1993/94. Explanatory Guide: Parts I & II.* London: MAFF, 108 pp.

Mankin, P.C.; Warner, R.E. (1992). Vulnerability of ground nests to predation on an agricultural habitat island in east-central Illinois. *American Midland Naturalist,* **128,** 281-291.

Marchant, J.H.; Hudson, R.; Carter, S.P.; Whittington, P. (1990). *Population Trends in British Breeding Birds.* Tring: British Trust for Ornithology, 300 pp.

Olsen, D.W. (1977). *A literature review of pheasant habitat requirements and improvement methods.* Utah State Department of Natural Resources, Division of Wildlife Resources, Publication No. 77-7.

Panek, M. (1992). The effect of environmental factors on the survival of grey partridge (*Perdix perdix*) chicks in Poland during 1987-89. *Journal of Applied Ecology,* **29,** 745-750.

Pollard, E.; Hooper, M.D.; Moore, N.W. (1974). *Hedges.* London: Collins, 256 pp.

Potts, G.R. (1980). The effects of modern agriculture, nest predation and game management on the population ecology of partridges *Perdix perdix* and *Alectoris rufa. Advances in Ecological Research,* **11,** 2-79.

Potts, G.R. (1986). *The Partridge: Pesticides, Predation and Conservation.* London: Collins, 274 pp.

Potts, G.R. (1990). The causes of the decline in population of the partridge (*Perdix perdix*) and effect of the insecticide dimethoate on chick mortality. In: *The Future of Wild Galliformes in the Netherlands,* J.T. Lumeij, and Y.R. Hoogeveen (Eds) , The Hague: Gegevens Koninklijke Bibliotheek, 62-71.

Rands, M.R.W. (1986). Effect of hedgerow characteristics on partridge breeding densities. *Journal of Applied Ecology,* **23,** 479-487.

Rands, M.R.W. (1987). Hedgerow management for the conservation of partridges *Perdix perdix* and *Alectoris rufa. Biological Conservation,* **40,** 127-139.

Rands, M.R.W. (1988). The effect of nest-site selection on nest predation in Grey Partridge *Perdix perdix* and Red-legged Partridge *Alectoris rufa. Ornis Scandinavica,* **19,** 35-40.

Robertson, P.A. (1992). *Woodland Management for Pheasants.* London: Forestry Commission Bulletin 106, H.M.S.O. 18 pp.

Robertson, P.A.; Woodburn, M.I.A.; Neutel, W.; Bealey, C.E. (1993). Effects of land use on breeding pheasant density. *Journal of Applied Ecology,* **30,** 465-477.

Sotherton, N.W.; Robertson, P.A. (1990). Indirect impacts of pesticides on the production of wild gamebirds in Britain. In: *Perdix V: Grey Partridge and Ring-necked Pheasant Workshop,* K.E. Church, R.E. Warner and S.J. Brady (Eds), Emporia: Kansas Department of Wildlife and Parks, 84-102.

Sotherton, N.W. (1991). Conservation Headlands: a practical combination of intensive cereal farming and conservation. In: *The Ecology of Temperate Cereal Fields,* L.G. Firbank, N. Carter, J.F. Derbyshire, and G.R. Potts (Eds) , Oxford: Blackwell Scientific Publications, 193-197.

Southwood, T.R.E.; Cross, D.J. (1969). The ecology of the partridge III. Breeding success and the abundance of insects in natural habitats. *Journal of Animal Ecology,* **38,** 497-509.

Tapper, S.C. (1992). *Game Heritage: An Ecological Review from Shooting and Gamekeeping Records.* Fordingbridge: Game Conservancy Limited, 140 pp.

Tapper, S.C.; Barnes, R.F.W. (1986). Influence of farming practice on the ecology of the brown hare (*Lepus europaeus*). *Journal of Applied Ecology,* **23,** 39-52.

Ter Braak, C.J.F. (1988). CANOCO - an extension of DECORANA to analyze species-environment relationships. *Vegetatio,* **75,** 159-160.

The Game Conservancy Trust (1993). *The Game Conservancy's Field Margin: Guidelines for the Management of Cereal Field Margins 1992/93.* Fordingbridge: The Game Conservancy Trust, 10 pp.

The Game Conservancy Trust (1994). *Factsheet 3. Game, Set-aside and Match.* Fordingbridge: The Game Conservancy Trust, 12 pp.

Thomas, M.B.; Wratten, S.D.; Sotherton, N.W. (1991). Creation of 'island' habitats in farmland to manipulate populations of beneficial arthropods: predator densities and emigration. *Journal of Applied Ecology,* **28,** 906-917.

Thomas, M.B.; Wratten, S.D.; Sotherton, N.W. (1992). Creation of 'island' habitats in farmland to manipulate populations of beneficial arthropods: predator densities and species composition. *Journal of Applied Ecology,* **29,** 524-531.

Woodburn, M.I.A.; Robertson, P.A. (1990). Woodland management for pheasants: economics and conservation effects. In: *The Future of Wild Galliformes in the Netherlands.,* J.T. Lumeij, and Y.R. Hoogeveen (Eds) , The Hague: Gegevens Koninklijke Bibliotheek, 185-198.

APPENDIX

Plant species recorded in four 0.25 m^2 quadrats around each of 31 grey partridge nests and 120 randomly-selected non-nests. Numbers represent codes allocated by CANOCO in Figure 1.

1	Bare ground	59	*Linum catharticum*
2	Dead grass	60	*Lolium perenne*
3	Spring barley	61	*Lotus corniculatus*
4	Winter wheat	62	*Chamomilla suaveolens*
6	Peas	63	*Medicago lupulina*
9	Winter rape	64	*Melilotis officinalis*
15	*Achillea millefolium*	67	*Odontites verna*
16	*Agrimonia eupatoria*	69	*Papaver rhoeas*
17	*Elymus repens*	70	*Pastinaca sativa*
18	*Agrostis stolonifera*	72	*Phleum pratensis*
20	*Arrhenatherum elatius*	74	*Plantago lanceolata*
21	*Artemisia vulgaris*	75	*Plantago major*
24	*Bromus erectus*	76	*Plantago media*
26	*Bromus sterilis*	78	*Poa trivialis*
28	*Centaurea scabiosa*	79	*Potentilla anserina*
30	*Cirsium arvense*	80	*Potentilla reptans*
31	*Cirsium vulgare*	81	*Prunus vulgaris*
33	*Convulvulus arvensis*	83	*Ranunculus repens*
34	*Crataegus monogyna*	84	*Reseda lutea*
35	*Crepis capillaris*	85	*Rubus fruticosa*
36	*Dactylis glomerata*	87	*Rumex crispus*
37	*Daucus carota*	88	*Rumex obtusifolius*
38	*Epilobium hirsutum*	89	*Senecio jacobea*
40	*Festuca rubra*	94	*Stachys sylvatica*
42	*Fumaria officinalis*	95	*Stellaria media*
43	*Galium aparine*	96	*Torilis japonica*
45	*Galium verum*	97	*Trifolium pratense*
47	*Geranium dissectum*	98	*Trifolium repens*
49	*Glechoma hederacea*	99	*Urtica dioica*
51	*Heracleum sphondylium*	100	*Veronica persica*
52	*Holcus lanatus*	101	*Vicia cracca*
54	*Knautia arvensis*	102	*Viola arvensis*
55	*Lamium album*	106	*Taraxacum sp.*
56	*Lathyrus pratensis*	107	*Helianthemum sp.*

THE FLORA OF FIELD MARGINS IN RELATION TO LAND USE AND BOUNDARY FEATURES

J.O.MOUNTFORD, T.PARISH, T.H.SPARKS

NERC Institute of Terrestrial Ecology, Monks Wood, Abbots Ripton, Huntingdon, PE17 2LS

ABSTRACT

Within the Swavesey Project, the flora of field margins was recorded in 131 200m long transects in 1986. The exercise was repeated in 1991. For each species, the ground cover was recorded for each component of the field margin. 71 species were recorded in greater detail and the effect of adjacent land-use and type of boundary investigated. The impact of land-use change following drainage is discussed.

INTRODUCTION: THE SWAVESEY PROJECT

In 1985, a flood protection and pump drainage scheme at Cow Fen, Swavesey (Cambridgeshire) led to land-use changes following an altered water table (Harris *et al.*, 1991). 131 field boundary transects (each 200m long) formed the vegetation survey, covering the undrained Mare and Middle Fens, old pumped fen at Overcote, and Cow Fen itself. The survey included areas of low-input grass and intensive arable systems. This paper summarises the surveys and examines the importance of land- use and both the nature and size of the boundary features on a subset of the flora.

METHODS: DATA & ANALYSIS

Each field boundary consisted of one or more components: ditch, ditch bank, hedge, verge and grass track. A vegetation survey of each component for each of the 131 transects was carried out in 1986 and repeated in 1991. The flora was recorded as a list of species and a visual estimate of the ground cover of each. Information from each component was pooled, according to the contribution of that component to the transect, to form a weighted estimate of ground cover for the whole boundary. Transects were classified in three types depending upon the adjacent land-use: **i)** with pasture on both sides of the transect (*GRASS*); **ii)** with arable crops on both sides (*ARABLE*); and **iii)** with pasture on one side and arable on the other (*SEMI*). The present paper focuses on a subset of 71 species, identified either as present in abundance or of particular interest. Between 1986 and 1991 there were modifications to the cropping adjacent to some of the boundary transects. These modifications resulted mainly, but not exclusively, in the redefinition of land-uses from *SEMI* to *ARABLE* as hitherto marginal land was converted to arable.

Preliminary analysis of survey data adopted a rather broad-brush approach. Correlations between boundary features and species ground cover were carried out in order to identify major linear relationships rather than for formal significance testing. Changes in ground cover between 1986 and 1991 were analysed using ANOVA with land-use, fen or modified land-use as factors.

RESULTS AND DISCUSSION

The results are summarised in two tables. Table 1 lists those species that showed a correlation of >¦0.2¦ with one or more measure of boundary size or type. Table 2 shows species whose mean cover changed between 1986 and 1991: I. in particular land-uses or fens; II. where land-use itself changed from *SEMI* to *ARABLE*; and III. where the trend in Cow Fen (site of the 1985 pump drainage scheme) differed in type or degree from other fens.

Species showing correlations with hedge size and tree number

Eight measures of boundary type reflected the extent of hedge habitat (Table 1) and the occurrence of many species was correlated with all or most of these variables. It was not surprising that component shrubs (e.g. *Crataegus monogyna, Prunus spinosa* and *Rhamnus cathartica*), invasive shrubs (e.g. *Rubus* spp. and *Sambucus nigra*) and standard trees (e.g. *Fraxinus excelsior* and *Ulmus* spp) were positively correlated with hedge dimensions. Similar trends were shown in forbs typical of partial shade (e.g. *Glechoma hederacea* and *Torilis japonica*). Tall herbs that are eliminated by cultivation or intense grazing survived, protected by the larger hedges (*Conium maculatum* and *Urtica dioica*).

In contrast, species which were negatively correlated with hedge size were generally intolerant of woody shade. This group included aquatics (*Glyceria maxima* and *Phragmites australis*), arable weeds (*Convolvulus arvensis, Fallopia convolvulus, Papaver rhoeas* and *Sinapis arvensis*) and biennials of disturbed, fallow or neglected areas (*Carduus* spp, *Cirsium* spp, *Dipsacus fullonum* and *Lactuca serriola*).

Species showing correlations with size of drainage channel

Many of the field boundaries included a ditch component, which varied from wide and water-filled to very narrow, shaded and dry for all or most of the year. As expected, macrophytes were positively correlated with ditch dimensions (DD and DW) (*Alisma plantago-aquatica, G.maxima* and *P.australis*). There was also a positive correlation between ditch dimensions and both the occurrence of thistles and overall species richness. This result followed the disturbance associated with regrading of drains provided a suitable habitat for invasive biennials. The increased habitat diversity due to the presence of standing water also led to higher species totals. Many of these species showed a positive correlation with the overall width of the boundary (MW).

Those plants negatively correlated with ditch size (and overall boundary width) included arable weeds and hedgerow species. Water-filled ditches occurred primarily in grassland where they partially replaced hedges as the stock-proof barrier. Wide arterial drains occurred in all land-uses but were seldom associated with hedgerows since effective drain management demands ready access to the channel.

Species showing correlations with adjacent land-use

Eight species were positively correlated with transects in grassland. These included species tolerant of grazing (*Cirsium* spp, *D.fullonum, Potentilla anserina* and *U.dioica*), tall herbs protected from grazing by the stock-proof thorn hedges and a component shrub of the old hedges surviving in the unimproved grass areas of Middle Fen (*R.cathartica*).

Six of the species that were positively correlated with arable were annuals, intolerant of defoliation, but which produce abundant seed and can germinate, flower and fruit rapidly. The perennial *C.arvensis* is favoured by the fragmentation of its rhizome during ploughing. The remaining species may be able to tolerate grazing but are eliminated by the hay-cut that often precedes the introduction of stock (*Bryonia dioica, C.maculatum, Heracleum sphondylium, Lamium album* and *Picris echioides*).

Species changing in abundance between 1986 and 1991

Some species increased or decreased in all fens or all land-uses, but most showed some association with a particular management. The increased area of arable land, particularly following the pump drainage in Cow Fen, favoured eight species, among which *F.convolvulus* and *Galium aparine* were serious weeds in 1986 when much land was converted to arable. *Ranunculus repens* and *Taraxacum* agg. spread where *SEMI* transects were converted to *ARABLE*. Some of the species increasing in grassland may have spread as a result of works associated with regrading of the main drains for drainage or pond creation for conservation (*Cirsium* spp, *L.serriola, Rumex* spp and *U.dioica*). More extensive grassland management in all fens and gaps in the sward following drought may have been responsible for the observed increase in total species numbers in pasture transects.

It was thought that lowered water-tables in Cow Fen would affect the vegetation of field margins and a study of ditches supported this view (Mountford *et al.*, in press). However, few species recorded during the present study showed a significant response (Table 2, III). *Phalaris arundinacea* and *R.repens* increased in drained land along newly dug ditches, whilst *B.dioica* and *L.serriola* declined, as a result of hedge trimming and the use of herbicides. Herbicide use increased in Cow Fen, reducing the perennial vegetation of the boundary strip, opening up the boundary for annuals.

Many correlations were as predicted, but changes in Cow Fen were less dramatic than expected, since the drought and lack of flooding in the undrained fens produced changes similar to those in Cow Fen. The contrast between pumped and unpumped fens was much reduced.

REFERENCES

Harris, G.L.; Rose, S.C.; Parish, T.; Mountford, J.O. (1991) A case study observing changes in land drainage and management in relation to ecology. In: *Hydrological Basis of Ecologically Sound Management of Soil and Groundwater*, H.P. Nachtnebel & K. Kovar (Eds.). IAHS Pub. No. 202.

Mountford, J.O.; Lakhani, K.H.; Rose, S.C. (in press) Pump drainage impacts on the aquatic macrophytes and bank vegetation of a drainage network. In: *Nature Conservation and the Management of Drainage System Habitats*, J. Harpley & P.M.Wade (Eds). Wiley, Chichester.

Table 1: SWAVESEY: Correlations of species ground cover with boundary features in 1986 and 1991 (see notes for key).

Species	TN	TH	HL	HH	CW	BW	VW	DD	DW	MW	HV	BA	Crop
Acer campestre	++	..	..	..	..	..	..	..	..	..	..	..	..
Alisma plantago-aquatica	..	..	..	..	..	..	..	++	++	..	..	..	..
Anthriscus sylvestris	+.	..	..	++	++	++	.+	..	..	..	++	++	..
Arctium lappa	..	.+	..	.+	.+	..	..	.-	.-	..	.+	..	.G
Bryonia dioica	..	..	..	..	..	..	.-	-.	--	--	..	..	AA
Callitriche spp.	..	..	..	..	..	..	..	+.	..	..	..	..	..
Calystegia sepium	..	..	..	..	..	..	..	+.	++	..	..	..	..
Carduus crispus	..	..	--	..	..	..	++	+.	++	++	..	..	..
Cirsium arvense	..	..	--	..	..	..	+.	..	+.	+.	..	-.	GG
Cirsium vulgare	..	..	-.	..	..	..	++	+.	+.	++	..	..	.G
Conium maculatum	..	..	++	..	..	..	-.	..	..	-.	..	..	A.
Convolvulus arvensis	..	..	..	--	--	--	..	..	--	--	.-	.-	AA
Crataegus monogyna	..	..	++	++	++	++	--	--	--	--	++	++	..
Dipsacus fullonum	..	..	--	..	..	..	++	..	++	++	..	..	.G
Epilobium hirsutum	+.	..	..	..	..	..	..	..	..	..	..	..	..
Fallopia convolvulus	..	..	..	.-	.-	..	..	.-	.-	.-	..	..	.A
Filipendula ulmaria	..	..	..	..	..	+.	..	..	..	.+	+.	..	..
Fraxinus excelsior	++	++	..	++	..	..	..	..	..	..	++	..	..
Galium aparine	..	..	+.	..	..	..	--	-.	--	--	..	..	AA
Geranium dissectum	..	..	..	-.	-.	-.	..	..	..	..	..	..	..
Glechoma hederacea	..	..	..	++	++	..	..	-.	..	..	++	..	..
Glyceria maxima	..	+.	--	..	..	..	..	..	.+	..	..	.-	..
Hedera helix	++	++	..	++	+.	..	..	..	..	..	++	..	..
Heracleum sphondylium	..	..	..	..	..	..	..	..	..	..	..	..	AA
Lactuca serriola	..	..	--	..	..	..	++	..	.+	++	..	..	..
Lamium album	..	..	..	..	..	..	..	..	..	..	..	..	.A
Malva sylvestris	..	..	..	..	..	..	..	..	..	..	.+	..	..
Myosotis arvensis	..	..	..	..	..	..	..	..	..	..	..	..	A.
Papaver rhoeas	..	..	.-	..	..	..	+.	..	++	+.	..	..	..
Phragmites australis	..	..	.-	..	..	..	..	..	++	..	..	..	..
Picris echioides	..	..	..	..	..	..	.+	.+	..	.+	..	..	A.
Potentilla anserina	..	..	..	..	..	..	..	..	..	..	..	..	.G
Prunus spinosa	..	..	..	..	..	.+	..	..	..	..	..	++	..
Quercus robur	++	++	..	..	..	..	..	..	..	..	..	..	..
Ranunculus repens	..	++	+.	..	..	..	.+	..	..	..	..	..	..
Rhamnus cathartica	..	++	..	++	++	++	..	-.	--	..	++	++	.G
Rosa canina	..	..	++	..	..	..	.-	-.	--	--	..	..	..
Rubus fruticosus	..	..	++	++	++	++	..	-.	-.	--	++	++	..
Rumex conglomeratus	..	..	+.	+.	..	..	..	..	..	..	+.	..	..
Rumex crispus	..	..	.+	++	++	.+	..	..	..	..	++	..	.G

Table 1: SWAVESEY - Correlations (continued)

Species	TN	TH	HL	HH	CW	BW	VW	DD	DW	MW	HV	BA	Crop
Salix spp.	++	++	..	++	++	..	..	..	..	..	..	..	..
Sambucus nigra	..	++	..	++	++	..	..	-.	.-	..	++	..	..
Scrophularia auriculata	..	..	.-	..	..	..	..	+.	..	..	..	..	..
Senecio vulgaris	..	..	..	..	..	..	..	..	..	..	..	..	.A
Sinapis arvensis	..	..	-.	-.	..	..	++	..	..	.+	..	..	A.
Sonchus arvensis	..	..	..	..	..	..	..	.+	..	..	..	..	..
Sparganium erectum	..	..	-.	..	..	..	..	+.	++	..	..	..	..
Torilis japonica	..	..	.+	..	..	..	..	..	--	.-	..	..	..
Tripleurospermum inodorum	..	..	..	..	..	..	..	..	..	..	..	..	.A
Tussilago farfara	..	..	..	..	..	..	..	.+	..	..	..	..	..
Ulmus spp.	++	++	..	..	..	..	..	..	..	..	..	..	..
Urtica dioica	++	.+	..	.+	++	++	.+	..	.+	++	..	..	GG
SPECIES RICHNESS (examined subset)	..	..	..	..	..	++	++	++	++	++	..	++	..

Notes:

i) key to variables

TN number of trees
TH tree height
HL hedge length
HH hedge height
CW crown width
BW basal width
VW verge width
DD ditch depth
DW ditch width
MW margin width
HV hedge volume
BA basal area

ii) Key to Correlations (For each variable and each species two results are presented, for 1986 and 1991 respectively):

+ (>+0.2)
- (<-0.2)
. (-0.2<r<0.2)
A (positive with arable)
G (positive with grass)

iii) The following species showed no correlations >¦0.2¦ with any recorded boundary feature: *Centaurea nigra*, *Medicago lupulina*, *Phalaris arundinacea*, *Pastinaca sativa*, *Persicaria maculosa*, *Polygonum aviculare*, *Potentilla reptans*, *Rubus caesius*, *Rumex acetosa*, *Rumex obtusifolius*, *Senecio jacobaea*, *Sonchus asper*, *Sonchus oleraceus*, *Taraxacum* agg., *Trifolium pratense*, *Trifolium repens*, *Typha latifolia*, *Veronica chamaedrys* and *Veronica persica*.

iv) Significance testing of correlation coefficients requires assumptions about bivariate normality. Some data will not meet these requirements (particularly crop - a 3 point scale). The results of correlations presented here are intended for guidance only and hence no formal significance testing has been attempted

Table 2: SWAVESEY: Field boundary species changing in abundance between 1986 and 1991 (Significance in brackets).

I. Increasing in particular land-uses or fens:
A: Arable; C: Cow Fen; G: Grassland; Gen: Generally; Ma: Mare Fen; Mid: Middle Fen; and O: Overcote Fen.

II. Significantly changing in abundance as *SEMI* transects are converted to *ARABLE*: Dec: Decreasing; and Inc: Increasing.

III. Significantly changing in abundance in Cow Fen compared to other fens: DA: Decreasing in arable; DG: Decreasing in grassland; IA: Increasing in arable; and IG: Increasing in grassland.

Species	I	II	III
Bryonia dioica	A (***)	Inc (*)	DA (+)
Calystegia sepium	C (*)	.	.
Cirsium arvense	Mid (**)	.	.
Cirsium vulgare	Gen	.	.
Convolvulus arvensis	A (*)	.	IG (+)
Dipsacus fullonum	Gen	.	.
Epilobium hirsutum	.	.	DG (+)
Fallopia convolvulus	A (***)	.	.
Galium aparine	A (**)	Inc (**)	.
Geranium dissectum	Gen	.	DG (+)
Glechoma hederacea	G (+)	.	.
Heracleum sphondylium	Gen	.	.
Lactuca serriola	G (*)	Dec (+)	DA (*)
Lamium album	A	Inc (**)	.
Papaver rhoeas	.	Inc (+)	DA (+)
Pastinaca sativa	.	Dec (*)	.
Phalaris arundinacea	Gen	.	IG (+)
Phragmites australis	O (***)	.	.
Potentilla reptans	Gen	.	IA (+)
Prunus spinosa	.	.	IG (*)
Ranunculus repens	G (**)	Inc (*)	IA (+)
Rumex crispus	Ma (*)	.	.
Rumex obtusifolius	Ma (**)	.	.
Senecio vulgaris	A (*)	.	.
Sonchus asper	O (**)	Inc (*)	.
Sparganium erectum	Gen	.	.
Taraxacum agg.	Gen	Inc (*)	.
Torilis japonica	Gen	.	.
Urtica dioica	Ma (***)	.	.
SPECIES RICHNESS	Ma (***)	.	.

Note: Significance: + = <10%; * = <5%; ** = <1%; *** = <0.1%.

THE EFFECTS OF FIELD MARGINS ON BUTTERFLY MOVEMENT

G. L. A. FRY

Norwegian Institute of Nature Research, PO Box 5064, N-1432 Ås-NLH, Norway.

W. J. ROBSON

Department of Biology, The University, Southampton, SO9 5NH, England.

ABSTRACT

This study examines the effects of hedgerows on the inter-field movement of butterflies, using observation of behaviour both at field margin sites and at an artificial hedge. It was found that the number of butterflies leaving a meadow varied considerably depending on the structure of boundary vegetation. The percentage of butterflies crossing boundaries increased with the percentage of the boundary length where vegetation was less than 1.5m tall. When tested experimentally, it was found that even a 1m high hedge, without gaps, can significantly reduce the number of butterflies crossing a field boundary. Implications for the management of field margins are discussed.

INTRODUCTION

Over the last few decades, the European landscape has changed considerably. Modern intensive farming techniques have resulted in vast crop monocultures, whilst native vegetation has become increasingly fragmented, persisting as small patches against a backcloth of pasture and arable fields (Fry, 1991). On marginal land, the trend has been towards abandonment of farming with the result that many traditionally lightly cultivated or grazed meadows have succumbed to succession and reverted to forest (Erhardt, 1985). Again this results in increased isolation of habitat patches since remaining meadows are separated by forest and continuity of habitat type is lost.

Butterflies are just one of many groups which have suffered as a result of these changes in land use. In intensively farmed areas, pesticides affect butterflies directly (Davis *et al.,* 1991), and herbicides and fertilisers induce vegetation changes resulting in losses of important nectar sources and larval food plants. Recently, there have been attempts to enhance the wildlife value of farmland, and field margins have become a focus of attention. Whilst butterflies have benefitted from the change to selective spraying regimes and from increasing the area of low herbaceous vegetation in field margins (Rands & Sotherton, 1986; Dover, 1989; Dover *et al.,* 1990; Lagerlof *et al.*, 1992), the role of hedges and shelterbelts is less clear-cut.

Hedgerows provide shelter, larval food plants, protection from agrochemicals and good growth conditions for nectar resources for butterflies. Both woodland and grassland species are able to use the ecotone habitat provided by hedges. For example, the rare black hairstreak (*Strymonidia pruni*), a woodland species, finds its food plant, sloe (*Prunus spinosa*), in hedgerows, whilst the meadow brown (*Maniola jurtina*), a species of open grassland, makes use of the grasses growing in the undisturbed land at the hedge bottom (Dowdeswell, 1987). It has been suggested that when hedgerows link habitat patches, they act as movement corridors, allowing butterflies to disperse through alien habitats and colonise new patches (Dover, 1990).

This study considers the possibility that hedges may also have negative impacts on butterflies by acting as barriers to their movement. If hedgerows reduce movement across landscapes, the probability of dispersing individuals reaching new habitat patches will be reduced, resulting in isolation of populations and increasing the risks that local extinction rates will be higher than recolonisation rates.

METHODS

The study site was an abandoned hay meadow system in Telemark county, Southern Norway. The main meadow complex lies at about 450m a.s.l., is 1.5-2ha in area and is surrounded by several different land-use types, including bog, deciduous forest, conifer plantation, clear-felled area and other meadows.

Observation of butterfly behaviour at field boundaries

Eight boundaries, of various lengths, were recognised according to their type, structure and adjacent land use. Each boundary was divided up into sections 5m long and extending 3m into the meadow from the boundary line. The flight path of every butterfly entering a recording section, within a one minute period, was recorded. Map location was recorded to the nearest metre.

Three categories of flight path were recognised: entering the meadow; leaving the meadow; and "rebounds". The latter included cases where a butterfly approached the boundary but turned back into the meadow rather than crossing, or entered the recording section and moved parallel to the boundary.

Butterfly flight behaviour was recorded, following the same protocol, along a line transect through the open meadow to act as a control. Records were collected throughout the day on 18 suitable days, between late June and early August 1991. Recordings were not made when the mean air temperature was less than 17 °C, during strong wind or during precipitation. Species were identified according to Chinery (1989). Arcsin transformed percentage data were analysed by linear regression.

Experiment using an artificial hedge

To investigate the barrier effect of hedges, an artificial hedge was constructed, comprising 12 x 5m long sections, with three replicates of each treatment. The treatments were: A = 0m high, B = 1m high, C = 2m high and D = 3m high.

Wooden stakes were set into the ground at 5m intervals from each other and green tarpaulins stretched between the stakes. The tarpaulin of each section could be fixed at 1m, 2m or 3m high, independently of the height of the rest of the hedge, by means of hooks screwed into the stakes. This design meant that the hedge could be quickly dismantled during unsuitable weather conditions and at the end of each day. The hedge was orientated in an east-west direction in order to minimise the effects of shadow.

Butterfly movements were monitored by four observers, each recording butterfly activity over one 5m section for 30 minutes. This allowed simultaneous recording of each of the height treatments, thus standardising recording for each run. The experiment used a randomised block design in which the sequence of treatments was changed for each recording of all height replicates.

RESULTS

Simple observation indicated that butterfly responses to field boundaries were highly dependent on boundary structure (see Fig. 1).

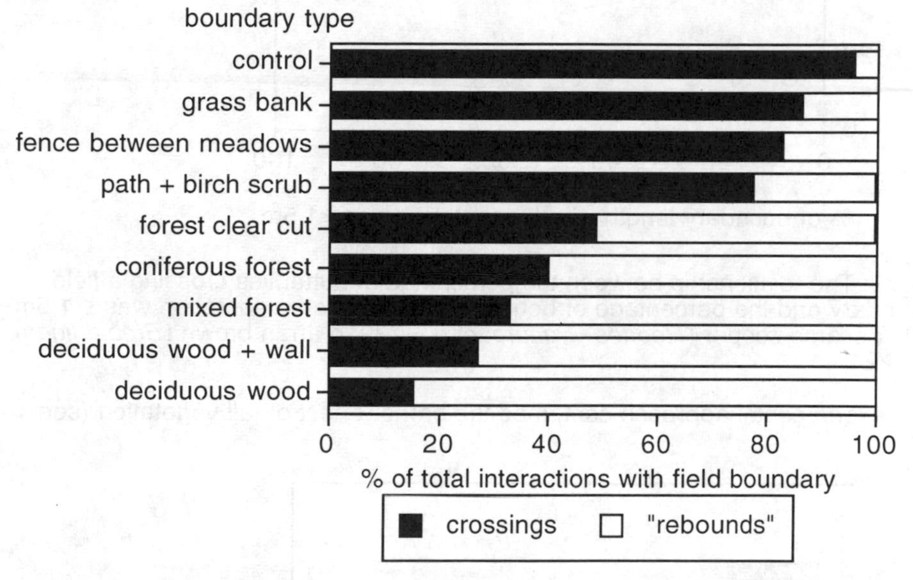

Fig. 1: The percentage of butterflies crossing the different field margins of the Sverveli meadow system.

Of all crossings observed over natural boundaries, 91% occurred through gaps in the boundary vegetation (total sample size, including all species = 1047 butterflies).

The number of butterflies crossing each natural boundary, as a proportion of the total number of approaches to the boundary, was positively correlated with the percentage of the boundary length were vegetation was less than 1.5m tall (r^2=0.9; p<0.001 for all species combined).There appeared to be an optimum degree of openness, about 30-40%, beyond which further increases in openness had little effect on movement rates across the boundary. This threshold was lower for Arran browns (*Erebia ligea*) than scarce coppers (*Heodes virgaureae*) (See Fig. 2).

There was no consistent relationship and no significant correlation between the number of butterflies crossing through individual gaps in the boundary vegetation and gap size (r^2=0.005; p>0.10).

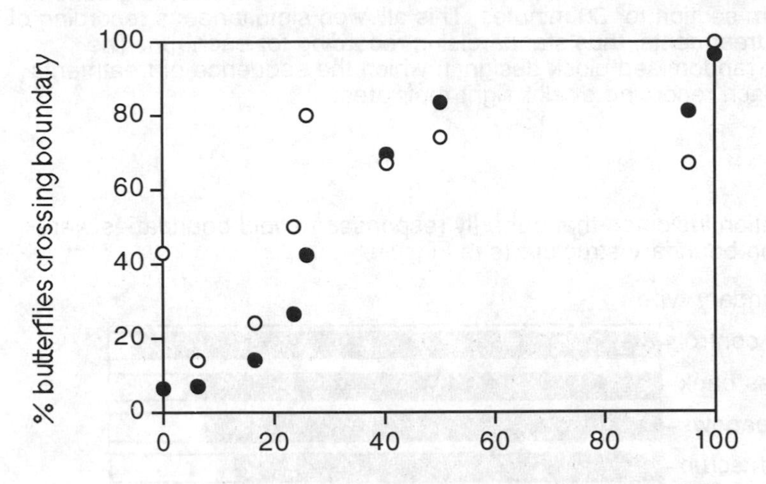

Fig. 2: The relationship between the percentage of butterflies crossing a field boundary and the percentage of boundary length where vegetation was ≤ 1.5m tall. ● scarce copper (*Heodes virgaureae*) n = 739; o arran brown (*Erebia ligea*) n = 224.

The experimental approach confirmed the barrier effect of tall vegetation (see Fig. 3).

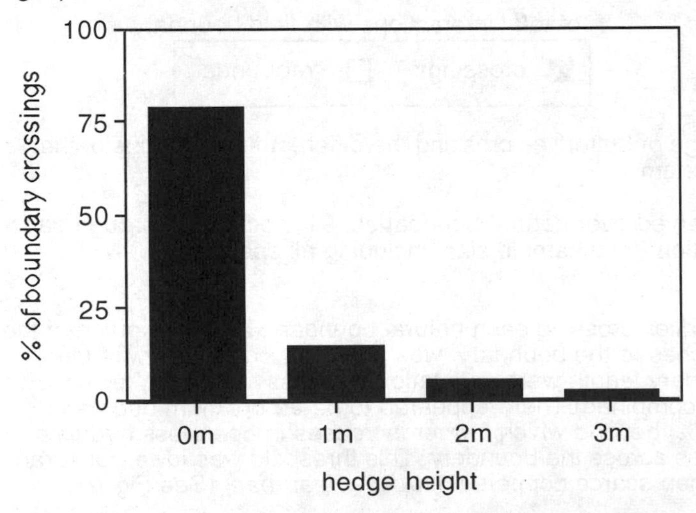

Fig. 3: The percentage of butterflies (all species) crossing the different experimental height treatments (n=1808).

Even a 1m high hedge significantly reduced landscape permeability to butterflies. Pairwise replicated G-tests comparing the number of butterflies (all species included) crossing different height treatments (45 replicates of each treatment) showed that each of the treatments differed significantly from all others ($p<0.001$ in all cases).

DISCUSSION

These results indicated that field margins can have a previously unrecognised negative role in butterfly dispersal at the landscape level; margins of tall vegetation form barriers to the movement of butterflies. The degree of barrier effect can vary for different species, for example, *E. ligea* , a species of open woodlands, accepts a lower threshold of openness than *H. virgaureae*, a meadow species. This type of threshold, where permeability shows no further increase after a certain percentage openness, was proposed theoretically by Stamps *et al.* (1987) who examined edge permeability using computer models. They found that a small increase in permeability of a 'hard' (relatively impermeable) edge, resulted in a dramatic increase in emigration, whilst for moderately permeable edges further increases in permeability had little effect on emigration rates.

Observations of butterflies, both at natural boundaries and at the artificial hedge, revealed that they often fly parallel to tall boundary structures and cross when they reach a gap. A few individuals, from a wide range of species, did cross tall structures suggesting that, while most of the species studied were capable of crossing tall hedges, they were inhibited by behavioural mechanisms. Whether the barrier effect is physical or behavioural, the result is the same: dispersal is reduced. This increases the isolation of fragmented populations, diminishing 'rescue effects', i.e. the chance that populations close to extinction will be revived by the arrival of immigrants, and reducing gene flow between populations (Descimon & Napolitano, 1993).

Hodgson (1993), in his life-strategy approach to factors influencing the abundance of British butterflies, states that butterflies of 'open' or migratory population structures tend to be commonest. We suggest that the permeability of field boundaries plays an important role in this mobility and thus in the abundance of butterflies. Further, we believe that more information is required on factors determining boundary permeability to improve modelling of population dynamics at a landscape scale.

Whilst the extension of field margins benefits butterflies by increasing nectar resources and food plants, the structure of constituent hedges should be taken into account so that inter-field movement is not restricted. Management of farm hedges for wildlife conservation requires planning, both in terms of the characteristics of the hedge and its position in relation to other hedges, woodlots, uncultivated land etc. The key characteristic of any hedge is variation; for butterflies this means, in particular, variation in height. Tall hedges provide sheltered habitat conditions, basking and perching sites (Rosenberg, 1984; Scott, 1974) whilst gaps in the hedge increase inter-field movement. If hedges link habitat patches they can act as corridors, facilitating dispersal of individuals. Therefore a combination of extended field margins and tall hedges could be used to direct butterfly movements over agricultural land, for example between patches of uncultivated meadow - perhaps as part of a future set-aside scheme. However, where tall hedges surround fields or have no links with other habitat fragments, they can act as barriers to butterfly movement across landscapes and isolate populations, increasing their vulnerability.

ACKNOWLEDGEMENTS

Many thanks to all those involved in collecting field data, to Jon Grøstad for allowing field work on his land and to Jan Heggenes for use of his farm. This study was made possible by a Norwegian Government Cultural Agreement Scholarship and funding from the Norwegian Institute Of Nature Research.

REFERENCES

Davis, B. N. K.; Lakhani, K. H.; Yates, T. J. (1991) The hazards of insecticides to butterflies of field margins. Agriculture, Ecosystems and Environment, 36, 151-161.

Descimon, H.; Napolitano, M. (1993) Enzyme polymorphism, wing pattern variability, and geographical isolation in an endangered butterfly species. Biological Conservation, 66, 117-123.

Dover, J. W. (1989) The use of flowers by butterflies foraging in cereal field margins. Entomologist's Gazette, 40, 283-291.

Dover, J. (1990) Butterflies and wildlife corridors. The Game Conservancy Review of 1989, 62-64. The Game Conservancy Trust.

Dover, J.; Sotherton, N.; Gobbett, K. (1990) Reduced pesticide inputs on cereal field margins: the effects on butterfly abundance. Ecological Entomology, 15, 17-24.

Dowdeswell, W. H. (1987) Hedgerows and verges. London: Allen and Unwin.

Erhardt, A. (1985) Diurnal Lepidoptera: sensitive indicators of cultivated and abandoned grassland. Journal of Applied Ecology, 22, 849-861.

Fry, G. (1991) Conservation in agricultural ecosystems. In I. Spellerberg, B. Goldsmith, G. Morris (Eds.), The Scientific Management of Temperate Communities for Nature Conservation. British Ecological Symposium Volume 31. London: Blackwell, pp. 415-443.

Hodgson, J. G. (1993) Commonness and rarity in British butterflies. Journal of Applied Ecology, 30, 407-427.

Lagerlöf, J.; Stark, J.; Svensson, B. (1992) Margins of agricultural fields as habitats for pollinating insects. Agriculture, Ecosystems and Environment, 40, 117-124.

Rands, M. R. W.; Sotherton, N. W. (1986) Pesticide use on cereal crops and changes in the abundance of butterflies on arable farmland in England. Biological Conservation, 36, 71-82.

Rosenberg, R. H. (1984) The effect of the landscape on the population structure of the admiral butterfly, Limenitis weidemeyerii. In J. Brandt & P. Agger (Ed.), The International Association for Landscape Ecology - "Methodology in Landscape Ecological Research and Planning", 5. Roskilde, Denmark: Roskilde Universitetsforlag, pp. 143-144.

Scott, J. A. (1974) Mate-locating behaviour of butterflies. American Midland Naturalist, 91, 103-117.

Stamps, J. A.; Buechner, M.; Krishnan, V. V. (1987) The effects of edge permeability and habitat geometry on emigration from patches of habitat. American Naturalist, 129 (4), 533-552.

THE INFLUENCE OF FIELD BOUNDARY STRUCTURE ON HETEROPTERAN DENSITIES WITHIN ADJACENT CEREAL FIELDS

S.J. MOREBY

The Farmland Ecology Unit, The Game Conservancy Trust, Burgate Manor, Fordingbridge, Hants, SP6 1EF

ABSTRACT

Densities of the true bugs (Hemiptera: Heteroptera) within cereal field headlands were studied with respect to headland aspect and structure of the adjacent boundary. The influence of these features on the dominant heteropteran species *Calocoris norvegicus* and two pooled Heteropteran groups, grass feeders and predatory species were studied. Boundary type affected the distribution of all groups with most individuals being found in headlands adjacent to hedgerows. The effect of aspect was less obvious, *Calocoris norvegicus* was most numerous in west-facing headlands, while the groups of predatory and grass-feeding species were more numerous in north-facing ones.

INTRODUCTION

Within the last five decades many previously common species of flora and fauna of the arable landscape have declined in numbers, with many once common species now becoming the subject of concern among conservationists, while others are now classified as endangered and are to be found in the British Red Data Book (Batten *et al.*, 1990). One such species, the grey partridge (*Perdix perdix L.)*, a once common sight in Britain's cereal fields, has declined dramatically. The major reason for this decline on arable land was found to be the indirect effects of pesticides acting to disrupt food chains (Potts, 1986). These products, especially herbicides and insecticides, caused a reduction in the densities of arthropod species that were essential food items in the diet of young partridge chicks. While the Heteroptera are one of the important groups of these so-called chick-food arthropods, little is known about their distribution and abundance within field headlands, the main brood-rearing area for partridges (Green, 1984), or about the effect of adjacent field boundaries. This study examined the heteropteran fauna of cereal field headlands adjacent to three common types of field boundary, post and wire fences, hedgerows and woodland edges, and examined the influence of boundary structure and aspect upon this important, non-pest insect group.

MATERIALS AND METHODS

Hetroptera were studied on a large Hampshire estate near Basingstoke during early July in 1988. Samples were collected using a Thornhill vacuum suction sampler (Thornhill, 1978) and five samples of five, 0.1m² sub-samples were collected at each headland site. To reduce possible variation between sites all sampling was carried out over a six-hour period (10 a.m.-4 p.m.) within spring barley crops, 3m from the field edge. Only selectively sprayed Conservation Headlands (Sotherton, 1991) were sampled to overcome the effects of pesticides. Within these headlands large distinct monoculture areas of arable weeds were avoided. Twenty sites per boundary type (hedgerow, woodland edge or post and wire fence), were sampled over nine fields. All field boundaries were sited on a slight bank at the field edge (about 50cm high) and covered with grass and dicotyledonous species. Post and wire fence boundaries were simply barbed-wire fences separating two fields, the hedgerow boundaries varied between 2-3 m in height and consisted of shrub and tree species cut every two-five years, while the wooded boundaries contained shrubs and trees 3-15m high, separating fields from wide shelter belts, small woods or copses. For every boundary type, five samples were taken for each aspect, north, south, east or west-facing. In the following sections post and wire fence boundaries will be referred to as grass.

All Heteroptera were identified to species level and were then either grouped for analysis or, in the case of the dominant species, *Calocoris norvegicus (Gmelin)*, analysed with respect to stage of development (nymph or adult). Three groupings were used, Total Heteroptera, grass-feeders (Stenodemini) and predatory species (Nabidae and *Anthocoris* spp.). A mean of the five samples from each site were calculated and these data were analysed using a two-way ANOVA to measure the effects of boundary structure and aspect.

RESULTS

All groups of Heteroptera were more numerous in crops adjacent to hedgerows compared to those adjacent to grass boundaries. These differences were significant for *Calocoris norvegicus* nymphs, total numbers of *C. norvegicus* and, because of the dominance of this species in the samples, for the total numbers of Heteroptera (Table 1). The presence of a hedgerow was also associated with more Heteroptera in the crop compared to wooded boundaries for all groups except the pooled group of predatory species. However none of these differences were significant. Only *C. norvegicus* nymphs were significantly more numerous adjacent to wooded boundaries compared to grass. One group, the grass-feeders, exhibited a significant boundary/aspect interaction, however none of the individual boundary or aspect comparisons were significant.

When comparing *C. norvegicus* numbers in relation to boundary aspect, lowest numbers of *C. norvegicus* occurred in north-facing headlands, with highest numbers of both nymphs and adults occurring in west-facing ones. None of the differences between south-, east- or west-facing headlands were significant but significantly more adults occurred in south- and west-facing headlands compared to north-facing ones (Table 2). While no significant differences were found between the headlands in respect to numbers of predatory and grass-feeding heteropteran species, highest densities of both groups occurred in north-facing headlands.

DISCUSSION

The dominant heteropteran species collected during this study was *Calocoris norvegicus*, a species which made up 60-80% of all Heteroptera found in headlands regardless of boundary or aspect type. While woody species are preferred for oviposition (Southwood & Leston, 1959), cereals appeared to be the major food plant for *C. norvegicus* within the field. The influence of the boundary structure on this species, which was found in significantly higher numbers in headlands adjacent to hedges or woods rather than grass, was more likely to be due to the suitability of the field boundary vegetation as oviposition and overwintering sites for eggs. While headland aspect seemed to influence adult feeding sites, (twice the number of individuals occurred in the warmer east-, west- and south-facing headlands), the effect of aspect may have been indirect, with possibly a greater diversity of flowering plant species occurring in these areas. If aspect had a similar effect on adults in the previous year, no noticeable effect on oviposition was observed, as the numbers of nymphs collected were similar in all headlands.

Predatory Heteroptera may have been directly influenced by the crop microclimate or indirectly affected by the abundance of suitable prey items, such as Hemiptera and other soft bodied prey groups. These prey may have themselves been more numerous in the microclimatic conditions offered by headland crops shaded by hedgerows and trees. The distribution of the grass-feeding species, (twice the number of individuals occurring in headlands adjacent to hedges compared to grass), was perhaps unexpected. However, while this group may be found in grass-weed free crops, their abundance and distribution is primarily effected by the presence of tall grass weed species in the field such as black-grass (*Alopecurus myosuroides* Huds.).and rough meadow-grass (*Poa trivialis* L.)

REFERENCES

Batten, L.A.;Bibby, C.J.;Clement,P.;Elliot, G.D; Porter R.F. (1990). Red Data Birds in Britain. London, Poyser.

Green, R.E. (1984) The feeding ecology and survival of partridge chicks (*Alectoris rufa* & *Perdix perdix*) on arable farmland in East Anglia. *Journal of Applied Ecology*, **21**, 817-830.

Potts, G.R. (1986) The Partridge: Pesticides, Predation & Conservation. Collins, London.

Sotherton, N.W. (1991). Conservation Headlands: a practical combination of intensive cereal farming and conservation. In: The Ecology of Temperate Cereal Fields. Eds. L.G. Firbank, N. Carter, J.F.

Darbyshire, & G.R. Potts, Blackwell Scientific Publications. Oxford , 373-397.

Southwood, T.R.E.; Leston, D. (1959) Land and Water Bugs of the British Isles. Frederick Warne & Co. London.

Thornhill, E.W. (1978) A motorised insect sampler. *Pans*, **24**, 205-207.

TABLE 1. Mean number of Heteroptera per $0.5m^2$ (± S.E.) occurring in three boundary types

	Boundary Structure			
	Hedge	Wood	Grass / wire fence	
Total Heteroptera	18.02 ± 1.63	11.82 ± 1.27	10.32 ± 1.63	H-G **
Calocoris norvegicus - Total	14.87 ± 1.79	7.20 ± 1.12	7.31 ± 1.56	H-G **
Calocoris norvegicus - Adults	4.91 ± 0.89	1.10 ± 0.23	2.76 ± 0.71	ns.
Calocoris norvegicus - Nymphs	9.95 ± 1.33	6.09 ± 1.00	4.55 ± 1.07	H-G *** W-G *
Predators	1.76 ± 0.58	3.80 ± 0.56	1.44 ± 0.44	ns.
Grass-feeders	0.96 ± 0.22	0.45 ± 0.11	0.47 ± 0.14	ns.

Boundary - Hedgerows = H
- Wood = W
- Grass/fence = G

*P <0.05 ** P <0.01 *** P <0.001

TABLE 2. Effect of headland aspect on mean number of Heteroptera per $0.5m^2$ (± S.E.)

	Aspect				
	North	South	East	West	
Total Heteroptera	11.47 ± 1.49	12.40 ± 2.06	13.65 ± 1.86	15.22 ± 2.29	ns.
Calocoris norvegicus - Total	7.23 ± 1.37	9.81 ± 2.20	9.72 ± 1.77	12.29 ± 2.29	N-W*
Calocoris norvegicus - Adults	1.17 ± 0.21	3.67 ± 1.07	3.34 ± 0.96	3.39 ± 0.80	N-S* N-W**
Calocoris norvegicus - Nymphs	6.06 ± 1.36	6.13 ± 1.38	6.37 ± 1.16	8.87 ± 1.79	ns.
Predators	3.20 ± 0.89	1.51 ± 0.65	2.89 ± 0.64	1.65 ± 0.29	ns.
Grass-feeders	0.83 ± 0.19	0.39 ± 0.11	0.46 ± 0.10	0.43 ± 0.12	ns.

Aspect - North = N
- South = S
- East = E
- West = W

* $P < 0.05$ ** $P < 0.01$ *** $P < 0.001$

THE DISTRIBUTION OF *EMPIDIDAE* (*DIPTERA*) IN HEDGEROW NETWORK LANDSCAPES

N. MORVAN, Y.R. DELETTRE, P. TREHEN

Université de Rennes I, URA 696 (CNRS),Station Biologique de Paimpont, F-35380 Plélan-le-Grand, France.

F. BUREL

Université de Rennes I, URA 696 (CNRS), Laboratoire d'Evolution des Systèmes Naturels et Modifiés, Campus de Beaulieu, F-35042 Rennes cedex, France.

J. BAUDRY

Institut National de la Recherche en Agronomie, SAD Armorique, 65 route de Saint-Brieuc, F-35042 Rennes Cedex, France.

ABSTRACT

Hedgerows are important uncultivated elements in some agricultural landscapes. They are temporary refuge or constant habitats for different groups of insects. Considering Empididae (Diptera), hedgerows are shelters for adults and development sites for larvae. These uncultivated elements are included in a farming mosaic both defining landscape structure. We are studying how hedgerows structure, hedgerow network and at coarser scale "bocage" landscapes can influence diversity and distribution of Empididae.

INTRODUCTION

Brittany landscapes are characterized by a mosaic of relatively small fields (1 to 5 ha) surrounded by hedgerows: this is the "bocage". Since the 1950's, modification of agricultural practices (helped by subsidies and reallotement programmes) has led to the removal of many hedgerows and dramatic changes in many places. As planning programmes are on a municipality basis, continuous changes across the countryside can be seen.

The ecology of hedgerows and hedgerow networks has been widely studied (Pollard *et al.*, 1974; Forman & Baudry, 1984; Burel & Baudry, 1990 a; 1990 b). These researchs have barely considered the land use mosaic, having focused either on hedgerow structure and species composition or on the role of interconnected hedgerows as a possible route for forest species to move through farmland. This later point, the corridor effect, is one of the major research topics in landscape ecology.

A NEW RESEARCH PROGRAMME ON "BOCAGES"

Our current research has a different perspective: all the landscape components (mosaic and networks) and their spatial relationships are taken into account at a variety of spatial and time scales (Baudry *et al.*, 1993). This programme aims at understanding the factors driving landscape dynamics and how different groups of species "perceive" the landscape. The landscape is no longer seen as a mere framework, but as a place where farming activities and vegetation dynamics continuously change the spatial structure either at short time scale (annual crop production techniques and associated field margin management) or medium scale (hedgerow removal, plantation, enlargement of fields, etc.). The core of this programme is carried out south of the Mont-Saint-Michel Bay, north of Rennes (Brittany). Three study areas of 500 ha each have been chosen because of their differences in landscape structure (hedgerow density and field size).

<u>The *BOCAGE* and *AGRICULTURE* databases</u>

All the hedgerows of the study areas are described individually, recording their structure and type of management. The observations are gathered in various databases (structure, adjacent field use, bank, woody vegetation..) and managed using *BOCAGE* software (Denis *et al.*, unpubl.). The databases are linked to a GIS (raster format) to allow the study of landscape structure and changes in scales (by aggregating cells). As far as time is concerned, we consider that the different variables have different time scales. For example, tree species management (pollarding, shredding..) is a slow variable, constant over 10-20 years, while pruning is intermediate (done every 9 years) and bank mowing is fast (yearly).

AGRICULTURE is a set of databases related to farm structure and farming practices. From an ecological point of view, it allows one to integrate such variables as pesticide inputs or dates of mowing and ploughing as variables driving species presence or abundance.

Therefore, field boudaries (the major component of biodiversity in agricultural landscapes) can be seen as part of both the landscape ecological system and the farming system. This allows a better understanding of the factors regulating biodiversity, and definition of management rules. The later may apply to landscape planning as well as to agricultural practicies or boundaries management. With this perspective, the multiple scale approach of species distribution in landscapes should be profitable.

<u>Insects in bocages</u>

Among insects, forest carabids (i.e. species found mostly in woodland) have been widely studied and exibit a sensitivity to hedgerow vegetation structure as well as to landscape structure, such as hedgerows intersection, connectivity of the network, presence of lanes bordered by parallel hedgerows (Burel, 1991). Studies of the relationships between landscape structure and flying insects are few; this is certainly due to problems in sampling, identification and to their dispersal ability.

STUDY OF EMPIDIDAE

Empids (Diptera) require different sites during their life cycle. Edaphic larvae of the genus *Hilara* develop in the soil of grassland and uncultivated elements such as hedgerows. Adults form two different types of swarms above or close to water: 1. hunting swarms where males search for small prey which are embedded in silk balloons and offered to females prior to copulation and 2. mating swarms, the location of which is different from the former. The environmental conditions of swarm development and location are strictly determined (temperature, hygrometry, wind, sunshine, vegetation structure and pattern)(Tréhen,

1971; Grootaert *et al.*, 1990). Individuals need to find both types of sites within a given radius in the landscape to complete their life cycle (Fig.1). Their flight from the emergence site to the mating site may be facilitated or inhibited by some landscape elements that act as corridor or barriers.

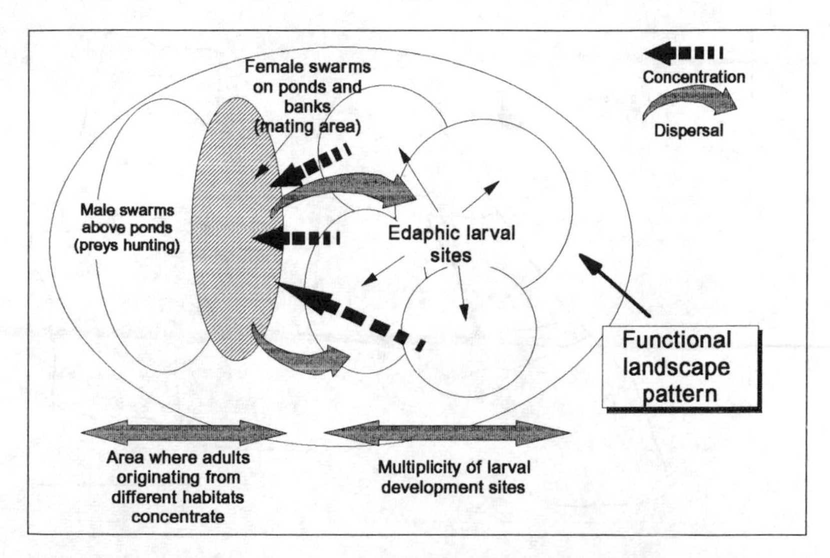

Fig.1. Functional landscape pattern needed for the life-cycle completion of an *Hilara* species.

We hypothesize that the "grain" of landscape structure relating the density and arrangement of the different habitats will influence the species assemblage. Agricultural practices,in particular pesticide sprays, influence the survival of individuals. Our second hypothesis is that species assemblages depend on agricultural practices. This study has begun recently, so this paper only presents the general context and describes the methods.

Methods

Empid sampling

The sampling methods have been selected according to previous researchs on different groups of Diptera. Coloured traps may be used for all groups, but their efficiency differs according to the family sampled (Baillot *et al.*, 1976). Empid flies are not trapped when swarming but only when searching for food or egg-laying (Baillot & Tréhen, 1974). Traps are set on hedgerow ground along a transect from a brook (the water source) up to the middle of each study area (Fig.2). Hedgerows have been selected according to the type of landuse on both sides and to their degree of connectivity to other uncultivated elements. Two traps are set at each site and sampled for two days every week from April to August.

A second method of trapping is useful for our approach because of the mating behaviour of adults.Thus,walking along brooks, we have used hand nets to sample the swarms (5 back and forth net sweeps) during the reproduction period.

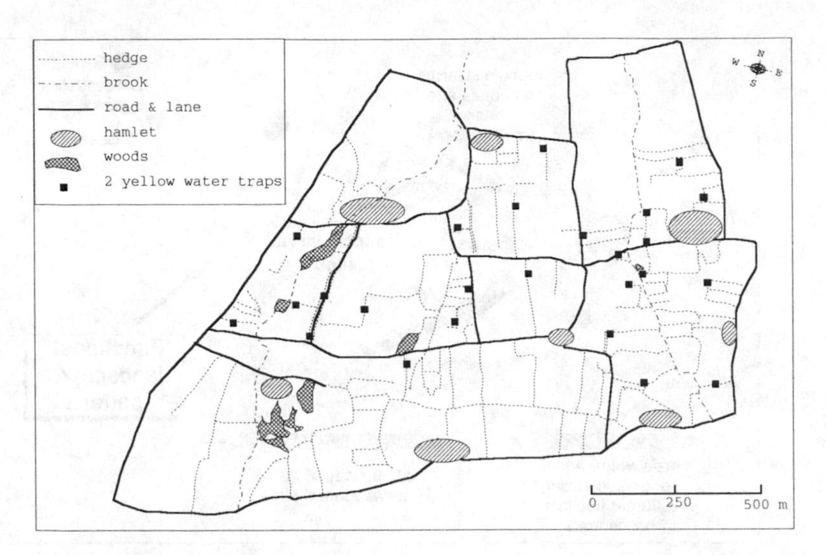

Fig.2. Distribution of traps in a bocage landscape, characterized by a median density of hedgerows.

Landscape data

Landscape structure has been analyzed in order to characterize its grain size and heterogeneity. Based on field work, data on hedgerows have been collected and stored in the BOCAGE database. Only parameters that have been previously suspected to be pertinent for Empididae will be extracted for used in the present work. They concern hedgerow structure (height, width, permeability, ...) and hedgerow flora on which many Empid species feed.

The most complex parameter is heterogeneity. The measure of heterogeneity depends not only on diversity of elements in landscape , but also on their spatial arrangement (Baudry, 1985). For Empididae, mosaic and networks must be included in the measure of heterogeneity. Distance between mating sites and emergence sites is important, and these complementary habitats, which can be defined as a functional spatial unit, are scale dependant, in relation to dispersal ability (Delettre *et al.*, 1992) (Fig.1).

Data on landscape pattern are spatially explicit and allow us to compute parameters such as heterogeneity, connectedness, contiguity, mean distance between elements. The measure of heterogeneity considers the spatial arrangement, and the diversity of landscape elements (Baudry, 1985). In our study we must integrate heterogeneity of the hedgerow network as well as heterogeneity of the landscape mosaic. Distance between potential emergence and mating sites is also assessed.

CONCLUSION

At the moment, no more details can be given because this study has just begun. However we hope that the consideration of both agricultural practices and landscape structure will provide a powerful tool to explain Empid species assemblages and conservation guidelines.

ACKNOWLEDGEMENTS

This study is included in the multidisciplinary programme "Organisation paysagère, agricole, écologique, sociale des bocages armoricains", financially supported by the "Ministère de l'Environnement".

REFERENCES

Baillot, S.; Tréhen, P. (1974) Variations de l'attractivité des pièges colorés de Moericke en fonction de la localisation spatio-temporelle de l'émergence, des comportements sexuels et des phases de dispersion de quelques espèces de Diptères. *Ann. Zool. - Ecol. anim.*, **6**, *575-584*.

Baillot, S.; Brunel, E.; Tréhen, P. (1976) Comparaison des Diptères produits et capturés dans une parcelle bocagère au moyen de nasses d'émergence et de pièges jaunes: signification dans le processus de colonisation de l'espace. In: *Les Bocages: Histoire, Ecologie, Economie,* INRA, CNRS, ENSA and Université de Rennes, pp.359-365.

Baudry, J. (1985) *Utilisation des concepts de Landscape Ecology pour l'analyse de l'espace rural. Utilisation du sol et bocages.* Thèse d'Etat Université de Rennes I (France), 472 pp.

Baudry, J. (1993) Continuing research on the ecology of hedgerow network landscapes from the 40's to the 90's. In: *Abstracts volume, IALE Congress Agricultural Landscapes in Europe*, Rennes France, pp. 185-185.

Burel, F. (1991) *Dynamique d'un paysage, `réseaux et flux biologiques.* Thèse d'Etat Université de Rennes I (France), 235 pp.

Burel, F.; Baudry, J. (1990a) Hedgerow networks as habitats for forest species. In: *Dispersal of Species in Agricultural Land*, R.G.H. Bunce and D.C. Howard, Belhaven Press

Burel, F.; Baudry, J. (1990b) Structural dynamic of a herdgerow network landscape in Brittany France. *Landscape Ecology*, **4**, 197-210.

Delettre, Y.; Tréhen, P.; Grootaert, P. (1992) Sapce heterogeneity, space use and short-range dispersal in Diptera: a case study. *Landscape Ecology*, **6**, 175-181.

Grootaert, P.; Tréhen, P., Brunel, E. (1990) Reproductive isolation of Hilara psecies: choice of habitat and swarming behaviour, *Abstract Volume of Second International Congress of Dipterology*, Bratislava, August 27 - September 1, 80.

Forman, R.T.T.; Baudry, J. (1984) Hedgerows and hedgerw networks in Landscape Ecology. *Environ. Manage.*, **8**, 499-510.

Pollard, E.; Hooper, M.D.; Moore, N.W. (1974) *Hedges, W.collins, and Sons,* London, 256 pp.

Tréhen, P. (1971) *Recherches sur les Empidides à larves édaphiques*, Thèse d'Etat, Université de Rennes I (France), 280 pp.

NEST SITE SELECTION AND TERRITORY DISTRIBUTION OF YELLOWHAMMER (*EMBERIZA CITRINELLA*) AND WHITETHROAT (*SYLVIA COMMUNIS*) IN FIELD MARGINS

C. STOATE, J. SZCZUR

The Game Conservancy Trust, Fordingbridge, Hampshire. SP6 1EF

ABSTRACT

Territory maps and nest locations were used to assess habitat selection by breeding yellowhammers and whitethroats in relation to vegetation in field boundaries. Both species selected herbaceous vegetation for nest sites and occurred at higher densities where extensive, perennial, herbaceous vegetation was present in field boundaries.

INTRODUCTION

One of the consequences of agricultural intensification in arable areas has been the loss of perennial, herbaceous vegetation from field boundaries. Grazing, ploughing, spraying and fertilizer drift into the hedge base have resulted in destruction and degredation of the perennial vegetation and its replacement by annual arable weeds (Boatman, 1988). Perennial herbaceous vegetation in field boundaries has been shown to provide a nesting habitat for grey partridges (*Perdix perdix*) (Rands, 1986), hibernating areas for coleopteran aphid predators (Sotherton, 1984) and feeding areas for Lepidoptera (Dover, 1991) and hoverflies (Cowgill *et al.*, 1993).

Hedge length and structure are known to influence the abundance of farmland passerines (Morgan & O'Connor, 1980, Arnold, 1983, Lakhani, this volume). In this paper we assess the influence of field boundary vegetation on yellowhammer (*Emberiza citrinella*) and whitethroat (*Sylvia communis*) territory distribution and nest site selection.

METHOD

The study was conducted in 1992 and 1993 at Loddington, Leicestershire, a 300 hectare mixed arable and livestock farm managed by the Allerton Research and Educational Trust. Both hedges with and hedges without vegetated bases are present at Loddington. Where field boundary vegetation has been lost it is mainly due to over-grazing by sheep, while field margins with extensive perennial vegetation are represented by hedgerow ditches. Field boundaries were classified into those with and those without vegetated bases. Double (ie roadside) hedges were not used in the study. Yellowhammer and whitethroat breeding territories were mapped over the entire farm using the method described by Marchant (1983). Territories per kilometre of field margin weighted for hedgerow length were calculated from the territory maps.

Systematic searches combined with observations of territorial birds were used to locate active nests of both species. Nest height was recorded as well as clutch and brood size and the vegetation type selected. Data were collected for Chaffinch (*Fringilla coelebs*) nests (n=26), as well as for Yellowhammer (n=42) and Whitethroat (n=35). Four nest site categories were recognized: herbaceous vegetation in ditch, herbaceous vegetation against hedge, shrubs overhanging ditch and the hedge itself. Nests were visited at two to three day intervals to monitor nest survival as part of a separate study. Calculation of nest survival to fledging was based on an assessment of daily nest survival (Hensler & Nichols, 1981).

RESULTS

Breeding territories of yellowhammers and whitethroats were widely distributed with yellowhammers, in particular, being associated with open unwooded areas of the farm. Ninety seven percent of yellowhammer and 84% of whitethroat nests were in field margins. Both yellowhammer and whitethroat territories were significantly more numerous in hedges with vegetated ditches than in hedges without them (yellowhammer: $F= 3.78$, $df= 9,9$, $P<0.05$, whitethroat: $F= 4.50$, $df= 9,9$, $P<0.05$) (Figure 1).

Nest heights of yellowhammer, whitethroat, and chaffinch were compared. Mean nest heights ($\pm$ 2SE) of yellowhammer (0.28m $\pm$ 0.09) and whitethroat (0.48m $\pm$ 0.11) were significantly lower in the hedge than those of chaffinch (1.26m $\pm$ 0.20) (yellowhammer: $F= 67.85$, $df\ 41,25$, $P< 0.001$, whitethroat: $F= 9.92$, $df=34,25$, $P<0.001$) with yellowhammers often building nests directly on the ground (Figure 2). Whitethroats selected nest sites both in herbs and in brambles (*Rubus fruticosus*) over a ditch while yellowhammer nests were strongly associated with rank perennial grasses and herbs ($\chi^2_3=13.81$, $P<0.01$).

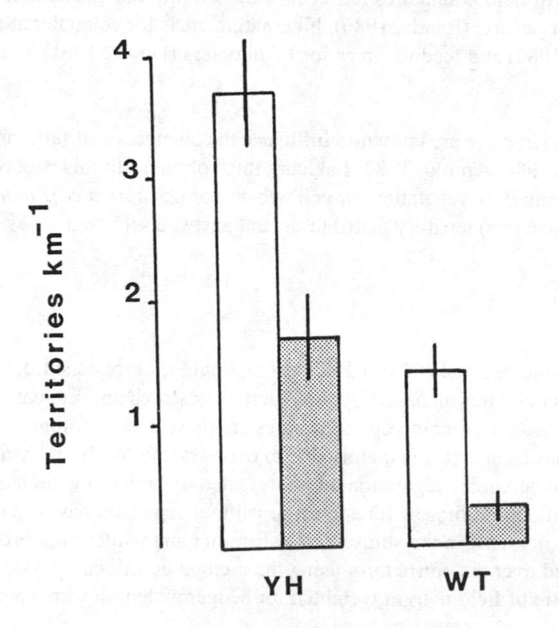

Figure 1. Yellowhammer and whitethroat territory density (means km^{-1}) weighted hedge length in field boundaries with (open) and without (shaded) extensive herbaceous vegetation. Vertical lines represent two standard errors.

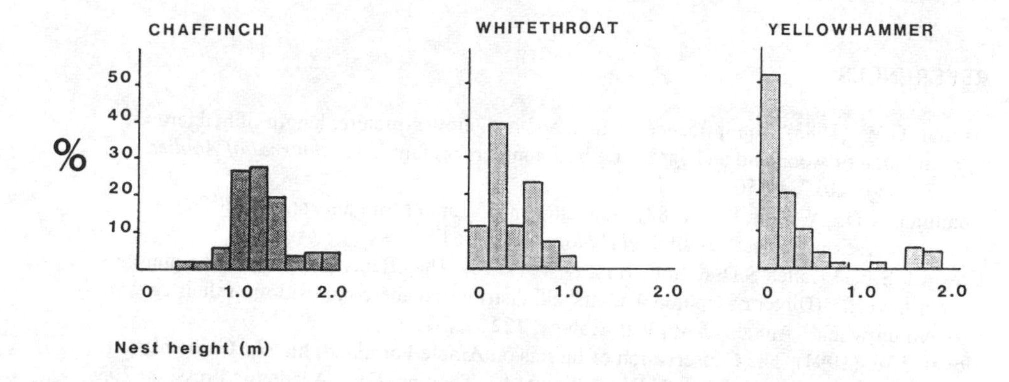

Figure 2. Frequency distribution of chaffinch, whitethroat and yellowhammer nest height.

DISCUSSION

The national yellowhammer population has remained relatively stable while that of whitethroats is at a low level following a considerable decline in 1969 (Marchant *et al.* 1990). Whitethroat nesting success may therefore be expected to be uniformly high because, at low densities, suboptimal territories remain unoccupied. Where there was a significant difference in the variability of clutch size (1992: $F=3.96$, $df=13,22$ $P<0.01$) or fledging success (1993: $F= 3.38$, $df=10,25$ $P<0.01$) yellowhammers were indeed more variable than whitethroats, suggesting that whitethroats occupy fewer suboptimal territories than yellowhammers.

Mason (1976) recorded significantly higher nest survival rates in whitethroat nests built within 60cm of the ground than in nests built above this height. There were insufficient data for a similar comparison to be made in the present study but, between species, nest survival rates to fledging (± 2SE) were higher for yellowhammer (0.74 ± 0.03) and whitethroat (0.84 ± 0.04) than for chaffinch (nesting higher in the hedge) (0.24 ± 0.04). However, yellowhammer and whitethroat nests may be more susceptible to predation in territories where low, herbaceous nest sites are not available.

Until now, data on nest site selection in relation to herbaceous field margin vegetation were only available for partridges (*Perdix perdix* and *Alectoris rufa*) (Rands, 1986). Our results strongly suggest that hedges lacking extensive herbaceous vegetation in their bases represent less suitable breeding territories for yellowhammer and whitethroat. Within individual farms management of this field boundary vegetation for game or integrated pest control may therefore improve the breeding habitat for these two species.

ACKNOWLEDGEMENTS

This study was funded by the Allerton Research and Educational Trust.

REFERENCES

Arnold, G.W. (1983) The influence of ditch and hedgerow structure, length of hedgerows, and area of woodland and garden on bird numbers on farmland. *Journal of Applied Ecology,* **20**, 731-750.

Boatman, N.D. ; Wilson, P.J. (1988) Field edge management for game and wildlife conservation. *Aspects of Applied Biology,* **16**, 53-61

Cowgill, S.E., Wratten,S.D. & Sotherton, N.W. (1993) The effects of weeds on the numbers of hoverfly (Diptera:Syrphidae) adults and distribution and composition of their eggs in winter wheat. Annals of Applied Biology, **122,** 223-231.

Dover, J.W. (1991) The Conservation of Insects on Arable Farmland. In: *The Conservation of Insects in their Habitats*, N.W. Collins and J. Thomas (Eds), Academic Press, pp 293-318.

Hensler, G.L. ; Nichols, J.D. (1981) The Mayfield method of estimating nesting success: a model, estimators and simulation results. *Wilson Bulletin,* **93,** 42-53.

Marchant, J.H. (1983) *Common Bird Census Instructions*. British Trust for Ornithology. Tring.

Marchant, J.H.; Hudson, R.; Carter, S.P. ; Whittington, P.A. (1990) *Population Trends in British Breeding Birds*. British Trust for Ornithology. Tring.

Mason, C.F. (1976) Breeding biology of the *Sylvia* warblers. *Bird Study,* **23**,213-232.

Morgan, R.A. ; O'Connor, R.J. (1980) Farmland habitat and yellowhammer distribution in Britain. *Bird Study,* **27**, 155-162.

Rands, M.R.W. (1986) Effects of hedgerow characteristics on partridge breeding densities. *Journal of Applied Ecology*, **23,** 479-487.

Sotherton, N.W. (1984) The distribution and abundance of predatory arthropods overwintering on farmland. *Annals of Applied Biology,* **105**, 423-429.

SMALL MAMMAL POPULATIONS IN HEDGEROWS: THE RELATIONSHIP WITH SEED AND BERRY PRODUCTION

S. M. C. POULTON

Wildlife & Ecology Development Team, ADAS Wolverhampton, Woodthorne, Wolverhampton, WV6 8TQ

ABSTRACT

This paper presents an analysis of one aspect of a ten-year small mammal trapping program undertaken in twelve sites in England and Wales. It examines the small-scale distribution within hedgerows of small mammal captures. A total of 476 wood mice (*Apodemus sylvaticus*) and 132 bank voles (*Clethrionomys glareolus*) were caught during the autumn of 1990. The distribution of these captures between 696 trap-groups was derived. In addition a survey of the distribution and abundance of berry-bearing species in the hedgerows was also undertaken. Mantel tests were used to i) investigate the spatial component of the distribution of captures, and ii) test the association between captures and the berry abundance. Wood mice were shown to have a very highly significant spatial component to their captures, but also a very strong association with berry abundance. In contrast, bank voles had no detectable spatial component, but also had a very strong association with berries. There is some evidence that this association is largely a result of the captures of adult male voles.

INTRODUCTION

In the spring of 1983, the Agricultural Development and Advisory Service of MAFF began a long-term survey of small mammal populations in hedgerows. Twelve sites, each consisting of a 650m length of hedgerow, were established around the country; their approximate locations are shown in Fig. 1. The work finished in the spring of 1992 with nearly ten years of mammal trapping data. In addition, several surveys were made of the botanical and structural composition of the hedges, the adjacent cropping regimes and the hedgerow management practices.

This paper reports on one aspect of the current survey. During the autumn 1990 trapping session the hedgerows were surveyed for berry and seed production. From the trapping results for this session, the distribution of captures within hedges were derived for two species of small mammal; wood mice (*Apodemus sylvaticus*) and bank voles (*Clethrionomys glareolus*). One problem with data derived from sequential points along a line or transect, is that they do not form an independent sample, and so they violate one of the principle conditions for parametric statistical testing. So the aims of the current analysis are two-fold. Rather than ignore this problem, the first aim is to investigate the spatial component of the capture index. Secondly, having accounted for any spatial effect, the true relationship between the mammal capture rates and berry production can be analysed.

METHODS

At each site, trapping sessions lasted for four nights during spring and autumn. Longworth live-catch traps were used, provided with bedding and suitable food for mice, voles and shrews. Within each site, 199 traps were laid in 58 trap-groups in two contiguous lines labelled "A" and "B". The A-line consisted of 33 groups of three traps at 5m spacing and the B-line, 25 groups at 20m spacing (see Fig. 2). Polynomial functions were used to equilibrate capture-rates between lines, resulting in a non-integer capture index. Separate indices were derived for i) total captures of each species, ii) adults only, iii) adult males only and iv) adult females only. This gave a total of eight indices.

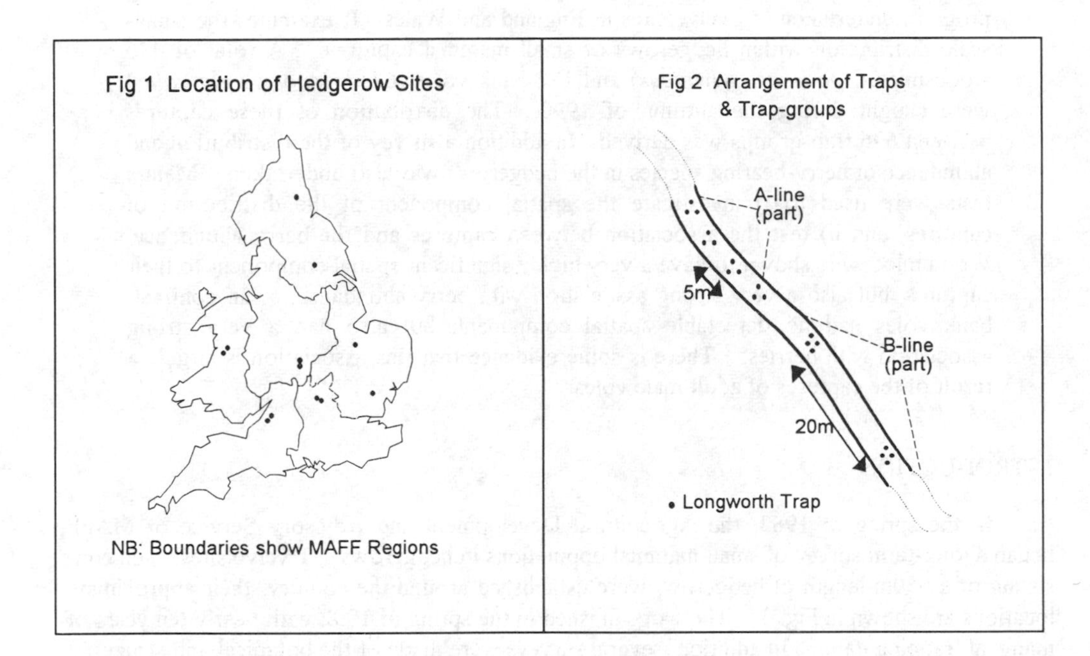

Hedgerows were divided into 5m sections, centred around the trap-groups. Within each section the abundance of seeds and berries (hereafter generically termed "berries") of all shrub and climbing species were recorded on a $\log_{10}$ scale (*e.g.* 1 = 1 to 9, 2 = 10 to 99, *etc.*). Within the B-line, this resulted in three intermediate sections between adjacent trap-groups. To make use of the data recorded in these sections, a weighted rolling average was used throughout to give an index of berry abundance.

RESULTS

A total of 608 mice and voles were caught during the trapping period. Full details of age and sex sub-groups are given in Table 1. Overall, about three times as many mice were caught as voles ($\chi^2_{(1)} = 196$, $P \approx 0$) Approximately half the mice were adults with more males than females ($\chi^2_{(1)} = 9.44$, $P < 0.01$). In contrast, only a third of the voles caught were adults, with more females than males ($\chi^2_{(1)} = 0.364$, $P > 0.05$). A total of 1442 captures were made during this period. The picture for captures was even more pronounced, with over five times the number of captures of mice than voles ($\chi^2_{(1)} = 658$, $P \approx 0$). These differences are shown by the capture ratios in Table 1.

Table 1 also gives a summary of the number of trap-groups containing animals. Mice were recorded in over half of the 696 trap-groups, whilst voles were found in less than 20% of the groups. Clearly, adult male and adult female voles were not widely caught. This table also gives the capture indices, derived from the trapping data, which are used in the subsequent analyses.

TABLE 1. Details of mammal captures by species and sub-groups

	ANIMALS			GROUPS[1]	CAPTURE INDEX[2]	
SPECIES	Individuals	Captures	Ratio		Max	Mean[3]
A. sylvaticus	476	1208	2.54	397	13.0	2.47
- Adult	244	628	2.57	239	12.6	1.42
- Adult male	146	385	2.64	167	8.5	0.71
- Adult female	98	243	2.48	112	8.3	0.51
C. glareolus	132	234	1.77	133	7.5	0.41
- Adult	44	87	1.98	55	5.0	0.16
- Adult male	20	47	2.35	25	5.0	0.04
- Adult female	24	40	1.67	28	3.0	0.07

NB: [1] Numbers of trap-groups containing animals
[2] Captures per trap-group
[3] Total captures / total trap-groups (696)

The results of the berry survey are given in Table 2 showing that 12 species were recorded with berries. In addition an aggregate "Other" category was derived to hold the very scarce species. This category was found in only 55 trap-groups and has been treated as a pseudo-species for the sake of the analysis. The vast majority of seed records consisted of bramble, hawthorn or rose, all three being found in at least ten sites. Oak and spindle were the rarest species, being found in only 31 and 22 sections in two sites respectively. Species richness per site also varied from only two species to ten or more in two sites. Note that this table only summarises presence of species in trap-groups and does not give the abundance of berries used in the subsequent analyses.

The methodology used for the analysis was based on the Mantel test, described in Manly (1986) and partial Mantel tests described in Leduc *et al* (1992). These test for associations between distance matrices. For the purposes of this analysis, Mantel tests have been used in the following way. Firstly, spatial distance matrices were derived from the distances between trap-groups. Berry distance matrices were then calculated using euclidean distances of the berry indices in multi-dimensional space. Finally, capture distance matrices were calculated individually for the eight capture variables. Note that all these matrices have been calculated separately within each of the twelve sites.

TABLE 2. Number of trap-groups and sites containing berries

	Total		Sites											
	Sections	Sites	1	2	3	4	5	6	7	8	9	10	11	12
Ash	25	6	0	0	0	0	12	1	0	2	4	4	0	2
Bramble	347	11	6	21	38	37	45	14	29	51	37	0	45	24
Blackthorn	105	7	0	0	1	0	15	1	42	0	12	3	31	0
Maple	81	4	0	0	0	0	39	0	0	0	39	2	1	0
Hawthorn	569	12	57	53	31	38	48	44	40	58	58	57	55	30
Oak	31	2	0	0	0	0	13	0	18	0	0	0	0	0
Privet	75	4	0	0	27	37	0	0	0	2	9	0	0	0
Rose	361	10	0	21	29	25	58	26	46	14	49	0	44	49
Spindle	22	2	0	0	0	0	20	0	0	0	2	0	0	0
Black bryony	160	7	0	0	30	24	41	47	0	4	5	9	0	0
Ivy	83	6	0	0	0	0	9	0	0	2	45	5	2	20
Clematis	47	3	0	0	17	0	29	1	0	0	0	0	0	0
Other	55	4	0	0	4	0	12	0	33	0	0	0	0	6
No of species.			2	3	8	5	12	7	6	7	10	6	6	6

Tests were carried out independently within sites, and also on an "aggregate matrix". Two test statistics were used. Firstly, the correlation coefficient between a capture matrix and the spatial matrix, and secondly, the partial correlation coefficient between a capture matrix and the berry matrix, holding the spatial matrix constant. 999 randomisations of the capture matrix were made within sites to give approximate one-tailed probability levels for the statistics. The first test statistic describes the spatial component of the capture index; a significant result shows that captures are clustered in space. The second statistic shows whether there is a significant association between the capture index and the berry index, having taken into account any spatial component of captures. The results from this suite of tests are given in Table 3. (Note that these significance levels have not been adjusted for the effects of multiple testing.)

Apodemus sylvaticus

Taking the aggregate data first, it was clear that all the mouse capture indices had a very highly significant spatial component. Capture rates were not constant or randomly varying along the trap lines but were clustered in groups of similar values. However, despite this spatial effect there was also a very significant association between mouse capture indices and berry indices. In other words, capture rates were clustered in some way in "berry space".

The individual site data provides evidence of the generality of the aggregate results. For total mouse captures, eight out of the twelve sites showed a spatial effect, with four of these also showing a highly significant association with the berry indices. In addition, one site revealed no spatial effect but did have a berry association. The most consistent capture index was adult males, where seven sites showed a significant association with the berry index.

Clethrionomys glareolus

The aggregate data here showed a quite different pattern to the mice capture indices. Firstly, for all captures, there was no spatial component to capture rates. Thus vole capture rates were either constant or, more likely, varied randomly between trap-groups. In contrast, there was a very highly significant association with the berry index. The three other capture indices must be interpreted with caution as they are based on a relatively small number of captures. Nevertheless, it is interesting that only male voles showed a spatial component to capture rates, but also a very highly significant association with berry indices. The individual site results were quite consistent, with seven out of eleven sites showing a very highly significant association with berry indices. Adult male and female voles were only caught in half the sites so the other capture indices can provide little evidence of individual site associations.

Table 3. Results of Mantel tests for aggregate data and individually by site

		Aggregate		Sites											
		r^2	P	1	2	3	4	5	6	7	8	9	10	11	12
A. sylvaticus	Spatial	0.102	***	**	-	**	-	-	-	*	***	***	*	***	***
	Berry	0.082	***	-	-	***	-	***	-	-	-	-	***	***	***
-Adult	Spatial	0.058	***	**	-	***	**	-	-	-	***	-	**	***	***
	Berry	0.093	***	-	**	***	-	***	-	-	-	-	***	-	***
-Male	Spatial	0.103	***	***	-	***	*	-	-	-	***	*	-	***	***
	Berry	0.072	***	-	***	***	-	**	-	*	**	-	***	-	***
-Female	Spatial	0.044	***	-	-	*	***	-	-	-	-	***	***	***	*
	Berry	0.069	***	-	-	*	-	***	-	-	-	***	***	*	***
C. glareolus	Spatial	-0.003	-	-	**	-	-	-		-	-	***	-	*	-
	-Berry	0.043	***	-	***	-	***	-		***	***	***	-	***	***
-Adult	Spatial	0.015	-	-		-	***	-				***	-	*	-
	Berry	0.003	-	-		-	-	-				*	-	-	**
-Male	Spatial	0.027	**	***		-		-				-		***	-
	Berry	0.053	***	-		-		-				***		***	**
-Female	Spatial	0.009	-	-		-		-				***		-	-
	Berry	0.003	-	-		-		-				-		-	***

NB: Probability levels: - NS, * < 0.05, ** < 0.01, *** < 0.001 Space = no captures

DISCUSSION

A number of workers have investigated the food preferences of mice and voles, both from field and laboratory studies. A review by Hansson (1985) showed that seeds, fruits and berries form a high proportion (>70% by vol.) of the stomach contents of wood mice. Furthermore, this proportion increases to over 90% in the winter months. In contrast, he showed that these components often constitute less than 40% of the stomach contents of bank voles, leaves and forbs usually formed the largest component. Clearly these food preferences will be reflected in foraging behaviour, but the question still remains whether this can be detected at a population level. In other words, the behaviour patterns of individual animals may be influenced by food availability (see e.g. Tew *et al*; 1992), but is this manifest as a population response?

One strength of this extensive trap-based survey was the large number of animals (and locations) which contributed simultaneously to the data-set; in this case over 600 animals and nearly 1500 captures. The results from this analysis indicated a number of population effects. Firstly, wood mice capture rates were very strongly associated with the abundance (and possibly species) of berries found in hedgerows. This is consistent with the food preferences indicated above. Secondly, a similar , though less pronounced association existed for bank voles, although interestingly it occurred in more of the study sites. This too is consistent with food preferences. These associations may not, of course, reflect any causative relationship between berry abundance and capture rates, or even any linear correlations, but the relationships were very highly significant. Furthermore, the inclusion of a spatial matrix in the analysis allowed for the effects of any other independent variables which may have had a spatial component. In other words, the berry associations were likely to be real and were probably not influenced by the effects of unrecorded variables.

Future analyses will attempt to identify what type of relationships exist between small mammal captures and berry production and with which hedgerow species. In addition, the association of captures with ground flora is currently being analysed, along with the overall species composition and structure of hedgerows and field boundaries. Finally, the influence of adjacent cropping patterns and agricultural inputs on capture rates is being analysed.

ACKNOWLEDGEMENTS

This work was carried out under a research contract with the LUCC division of MAFF. I am very grateful to many dedicated ADAS colleagues who contributed to the fieldwork and organisation of this project. I should like to thank Drs. T. Milsom and S. Langton of the Central Science Laboratory for their involvement with this work and Dr. R. Sanderson of Newcastle University for his advice on the analytical procedure.

REFERENCES

Hansson, L (1985) The food of bank voles, wood mice and yellow-necked mice. *Symposia of the Zoological Society of London,* **55,** 141-168.

Leduc, A., Drapeau, P., Bergeron, Y, & Legendre, P. (1992) Study of spatial components of forest cover using partial Mantel tests and path analysis. *Journal of Vegetation Science,* **3,** 69-78.

Manly, B.F.J. (1986) *Multivariate statistical methods: A Primer.* Chapman & Hall, London: 281 + xiii

Tew, T.E.; Macdonald, D.W.; Rands, M.R.W. (1992) Herbicide application affects micro-habitat use by arable wood mice *Apodemus sylvaticus. Journal of Applied Ecology,* **29,** 352-359

MODIFICATIONS OF FIELD MARGINS AND OTHER HABITATS IN AGRICULTURAL AREAS OF QUÉBEC, CANADA, AND EFFECTS ON PLANTS AND BIRDS

C. BOUTIN

National Wildlife Research Centre, Canadian Wildlife Service, Hull, Québec, Canada.

B. JOBIN

Gauthier & Guillemette Consultants Inc., St-Romuald, Québec, Canada.

J-L. DESGRANGES

Canadian Wildlife Service, Environment Canada, Sainte-Foy, Québec, Canada.

ABSTRACT

A study was undertaken in southern Québec, Canada, to investigate how wildlife habitat in the vicinity of farmland has been modified in the last 25 years and how this has affected native plants and bird populations. Results show that the landscape has been markedly modified in the last 25 years: many hedgerows and trees have disappeared, and fragmentation has increased due to an increasing number of roads and urban settlements. The herb layer vegetation in wooded habitats is composed of a large number of species typical of the original forested vegetation but many introduced species are also found. Herbicide use has had an adverse effect on the vegetation of field margins. The diversity of native plants and birds is greater in farmlands where some wooded areas remain.

INTRODUCTION

Major change in land use has been occurring in Canada since European settlement started in the 17th century. Large segments of the once dominant deciduous forest have vanished and the forest is still receding due to agriculture, timber production and urbanization. Only 7.5% of the land is under a cropping or pasture regime in Canada (constituting 67.7 million hectares) (Statistics Canada, 1991). While this appears very limited when compared to the UK with 75% agricultural land (Cobham & Rowe, 1986), most Canadian agricultural land is confined to the southern part of the country.

In the last 50 years or so, agriculture has been undergoing dramatic changes in Canada, as in many other countries (Freemark & Boutin, in press). The average farm size gradually increased from about 100 hectares to an average of 200 hectares (Statistics Canada, 1986). Accordingly, field sizes doubled in some parts of the country (Baldwin & Johnston, 1986), and monocultures were favoured with a corresponding reduced variability within and between fields. Woodlots are increasingly managed for wood production or are used as shelter or grazing areas for cattle, and hedgerows and streambanks are sprayed for control of noxious weeds. There is mounting evidence that wildlife (including plants) living in agricultural areas has been adversely affected by these modifications.

Herbicides are the most widely used pesticides in cultivated land of southeastern Canada, and their use has escalated sharply during the last 30 years (Statistics Canada,

1991). Direct acute toxicity to birds and mammals is not a major issue with herbicides; adverse impacts on resources through plants, soil organisms and other invertebrates is the main concern. Use of agricultural herbicides affects not only plants but also organisms at higher trophic levels (e.g. Sotherton *et al.*, 1988). This study was aimed at investigating the relationships between landscape modifications, farming practices, wildlife and wildlife habitats in the farm land of southern Québec.

MATERIAL AND METHODS

The present study is part of a broader project undertaken using Breeding Bird Survey (BBS) data obtained from 1966 to 1992 to assess trends of bird populations associated with agriculture (Jobin *et al.* 1994). The BBS data, gathered all over North America during the breeding season, consist of a series of routes, 40 km long, with a stop (0.5 km^2) every 800 m for a total of 50 stops per route (Falardeau & DesGranges, 1991). Our choice of sites was confined to the BBS routes situated in the St. Lawrence Valley. Originally 172 stops were selected within ten BBS routes; only stops having at least 50% or more of their area cultivated in the 1980s were selected for landscape inventory and bird census. Ninety-five of these stops were retained for a detailed study of vegetation, farming practices, and bird utilization. For 20 minutes in June 1992 birds seen and/or heard were recorded. Trends in birds using the BBS data are presented in Jobin *et al.* (1994). Change in farming areas and noncultivated habitats were determined for the 172 stops using interpretation of aerial photographs (1:15000) taken from 1959 to 1966 and 1981 to 1989. Variables measured were: 1) average crop area (cash crops, vegetables, forage crops, pastures could not be differentiated) and noncrop areas including old fields, forested, urban and aquatic areas, and other noncultivated habitats, 2) linear features such as hedgerows and length of woodland edges, and 3) features which could be counted such as woodlots, trees and buildings. The floristic composition and structure of 66 wooded hedgerows (minimum 5 m wide), 35 woodlots (less than two hectares), 57 woodland edges (woodland greater than two hectares with edges adjacent to cropfields), 50 old fields, and 47 ditches were described. At each stop habitats were selected so as to survey at least one habitat type present per stop. The point-intercept method (Bonham, 1989) was used for the ground vegetation; 3 X 10 m transects were used, 20 points per transect, except for woodlots where transects were 20 m long. For edge habitats the 10 m transects were positioned 5 m inside the noncrop area and 5 m outside into the crop. Only the 5 m transects inside the noncrop habitat are presented. For the woody vegetation, the line-intercept method was used. Information on farming practices and pesticide use was gathered by means of a questionnaire and interviews with farmers who owned the land at the stops. Responses were obtained from 183 farmers with land on the 95 stops.

RESULTS

Table 1 shows the trend in land use as interpreted from the aerial photographs. Cultivated areas have decreased between the 1960s and 1980s as have aquatic areas (mainly lentic systems) which were reduced by 12 ha. Forested areas (+83 ha), old fields (+52 ha) and urban settlements (+53 ha) have increased. In rural Québec many family farms have been abandoned hence the reduction of total cultivated areas which were gradually replaced by old fields and forested areas such as woodlots. In the remaining farmed land, features such as hedgerows, and trees have been removed and are thus present in a lower number of stops in the 1980s than in the 1960s (Table 1). Concomitant with an increase in urban areas, more roads and buildings are also present.

TABLE 1. Changes in the habitat types between 1960s and 1980s in the 172 stops

	Areas (ha)		n of stops		
	1960s	1980s	1960s	1980s	Prob.[a]
Total	8600	8600			
Cultivated[b]	7650	7468	172	172	.0001
Forested	509	592	94	96	.0052
Old field	229	281	72	68	.5012
Aquatic[c]	191	179	40	40	.0494
Urban	22	75	6	17	.0016
Others	0	5	0	2	-
Linear features (average m /stop)					
Hedgerows	247	218	109	96	.0272
Edges					
agri-forest	402	447	94	96	.3887
agri-urban	18	67	6	17	.0003
Roads	961	988	172	172	.0117
Element features (average # /stop)					
Woodlots	0.19	0.34	27	42	.0001
Trees	13	10	165	157	.0001
Buildings	11	17	160	163	.0001

[a] Wilcoxon signed rank test on change in area
[b] includes cash crops, forage crops, pastures, vegetables, summerfallow
[c] includes streams, lakes, ponds, wetlands

TABLE 2. Characteristics of the herb layer vegetation (%)

	Woodlot	Woodland edge	Hedgerow	Old field	Ditch
Status					
Native species	71.7	56.1	51.2	51.9	39.4
Introduced	11.1	24.2	31.5	30.1	46.7
Unknown	17.2	19.7	17.4	18,1	13.9
Lifespan					
Annual	2.9	2.2	5.4	4.6	16.7
Biannual	4.5	0.5	3.9	4.5	3.9
Perennial	76.1	80.2	73.6	73.8	65.9
Unknown	16.5	17.1	17.1	17.0	13.4
Habitat					
Shaded	45.1	16.5	5.1	1.4	0.7
Open	13.3	25.0	36.0	38.4	44.9
Ubiquitous	12.0	13.7	18.7	15.2	5.9
Wet	8.5	16.0	11.9	18.7	24.0
Crop	3.9	8.9	10.9	8.0	10.6
Unknown	17.2	19.9	17.4	18.3	13.9

Characteristics of the vegetation inventoried in the different uncultivated habitats are presented in Table 2. The vegetation of the herb layer was composed of a large number of introduced species; e.g. 47% of the vegetation in ditches. A large proportion of the species in the ditches were annuals typical of open areas. In contrast, woodland edges and small woodlots contained the highest percentage of perennial species typical of shaded areas. Crop species have penetrated the different habitats to comprise between 8 to 11% of the species in most habitats except woodlots. The shrub/tree layers of the different habitats were represented mostly by native species (not shown).

TABLE 3. Effects of herbicide use on vegetation in margins (probability values (ANOVA) for difference between unsprayed and sprayed with herbicide)

Variables	Hedgerow	Woodland edge	Ditch
# species	0.10 (↓)	0.12 (↓)	0.48 (NS)
Shannon index	0.01 (↓)	0.08 (↓)	0.69 (NS)
Cover	0.04 (↓)	0.69 (NS)	0.02 (↑)
Height	0.22 (NS)	0.82 (NS)	0.01 (↑)

TABLE 4. Importance of different farmland habitats for bird and plants

a) Plants (herb layer)

Habitats	Number surveyed	Total # of species	# of unique species	# (%) of typical spp.[a] MH	MBY
Woodlots	35	102	20	14 (35)	34 (69)
Woodland edges	57	99	13	10 (25)	24 (49)
Hedgerows	66	110	10	7 (18)	17 (35)
Old fields	50	120	25	5 (13)	13 (27)
Ditches	47	144	38	4 (10)	9 (18)

b) Birds

Stops with:	n	Cumulative # of bird species
1- Cropland only or with old fields, no wooded areas	10	38
2- 1 + wooded hedgerows, no wooded areas	13	39
3- 1 & 2 + woodlots <2 ha, <10% forested areas	16	45 (41.5)[b]
4- 1 & 2 & 3 + woodland >2 ha, <10% forest	25	56 (49.75)
5- As 4, >10% forested areas	31	64 (52.25)

[a] From Grantner (1966), 2 types of vegetation associations: in 3 routes, MH= sugar maple (*Acer saccharum*/) hickory (*Carya cordiformis*) +40 herb spp; in 6 routes, MBY= sugar maple/basswood (*Tilia americana*) /yellow birch (*Betula lutea*) +49 herb spp.
[b] Numbers within brackets refer to the average # of species counted when 13 stops were randomly selected 4 times in order to test for the effect of different sample size.

To assess the effect of herbicide use on the vegetation of hedgerows, woodland edges and ditches, stops where herbicide had been used (n =79) were compared to stops where no herbicide had been sprayed (n =16) for at least 6 years (Table 3). There were significant reductions in species diversity and cover where herbicide had been used in hedgerows and woodland edges. In contrast, for vegetation in ditches, cover and plant height was enhanced where herbicides had been used. Ditches are highly managed and subjected to agricultural runoff containing both nutrients and pesticides.

The importance of different farmland habitats for plant and bird diversity was assessed (Table 4). The number of unique species relates to species recorded in one habitat but not found in any other habitats inventoried during the study. Ditches have the highest number of distinct plant species, mostly introduced or typical of wet areas (Tables 2 & 4a). Two slightly different vegetation associations are found in the studied area both dominated by sugar maple (see Table 4a). The associations differ in the proportion of the main species present and in the composition of secondary species (Grantner, 1966). We compared these different recognised associations to the inventory performed during the study. Results indicated that woodlots harbour more plants that are representative of their traditional associations (Table 4a). The number of bird species was notably higher in stops with wooded habitats (Table 4b). Stops with large forested habitats (5 in Table 4b) provided shelter for a number of warblers and thrushes, among others, that were not present in stops with only old fields, hedgerows and small wooded areas.

DISCUSSION

The aerial photographs revealed a great diversity of habitats. As expected, however, some habitat features such as hedgerows were reduced and trees have been removed. Hedgerows are important habitats in agricultural landscape especially if they are connected with forests, woodlots or old fields (Wegner & Merriam, 1979). In the present study, although the total number of bird species did not increase with the presence of hedgerows, species such as the Great Crested flycatcher (*Myiarchus crinitus*) and Eastern Phoebe (*Sayornis phoebe*) were not seen in areas devoid of trees. In order to increase the area under cultivation or to facilitate the passage of farm equipment, wooded and vegetative margins are frequently reduced or eliminated. More recently, however, the establishment of shelterbelts has been advocated to replace vanished hedgerows to prevent soil erosion and to enhance soil moisture. Unfortunately, the recommendation given to farmers promotes the establishment of a few rows of trees together with eradication of the herb layer to prevent the intrusion of noxious weeds into the crop (Baldwin & Johnston, 1986). The width and structural diversity of hedgerow habitats are crucial in supporting a diversity of wildlife and protecting it against predation (Best *et al.*, 1990). Thus a simple shift in the management of shelterbelt/hedgerow features could enhance and protect biological diversity in agriculture.

The area under cultivation has slightly declined in the past 25 years in the study sites while old fields and forested areas have progressively expanded. Forested areas, primarily in small woodlots (<2 ha), have increased mainly due to regrowth of trees in abandoned fields. Buildings have been erected and the road system has extended contributing to the sectioning of the land. Clearly, the regrowth of forest due to abandonment of farms can only be beneficial to some bird species if fundamental requirements are satisfied, and, in some cases, these can only be met in large wooded areas (e.g. low predation and parasitism, food and nesting requirements, minimal disturbance). In the present study eight bird species were counted exclusively in stops with large forested areas, including three thrush species, three warblers and other forest species (Table 4b).

Uncultivated habitats, especially woodlots, appear to be important refuges for remnant plant species typical of the once dominant vegetation of southeastern Canada (Tables 2 & 4a). In addition three species rare for Québec (Bouchard *et al.*, 1983) were found - *Aster ontarionis* in a ditch, *Viola affinis* in a woodlot and *Waldsteinia fragarioides* in a hedgerow and a woodlot - illustrating the conservation value of such habitats associated with agriculture. In this context the effect of herbicides on plants situated at the margins of croplands (Table 3) is of concern. The occurrence of small amounts of herbicide drifting into field margins and affecting plants has been documented (Marrs *et al.*, 1989). The establishment of an unsprayed buffer zone at the margin of cropfields would reduce the drift of herbicide into wildlife habitats. However before this can be implemented properly in Canada more research on the agronomic impact and cost to farmers is needed.

REFERENCES

Baldwin, C.S.; Johnston, E.F. (1986) Windbreaks on the farm. *Ministry of Agriculture and Food,* Ontario, Canada, 20 pp.

Best, L.B.; Whitmore, R.C.; Booth, G.M. (1990) Use of cornfields by birds during the breeding season: The importance of edge habitat. *American Midland Naturalist*, **124**, 84-99.

Bonham, C.D. (1989) *Measurements for terrestrial vegetation*. John Wiley and Sons Ltd. New York, U.S.A., 338 pp.

Bouchard, A.; Barabé, D.; Dumais, M.; Hay, S. (1983) The rare vascular plants of Québec. The National Museum of Natural Sciences, Ottawa, *Syllogeus*, **48**, 75 pp.

Cobham, R; Rowe, J. (1986) Evaluating the wildlife of agricultural environments: an aid to conservation. In: *Wildlife Conservation Evaluation*, M.B. Usher (Ed.), Chapman and Hall Ltd., London, pp. 223-246.

Falardeau, G.; DesGranges, J.L. (1991) Sélection de l'habitat et fluctuations récentes des populations d'oiseaux des milieux agricoles du Québec. *Canadian Field Naturalist*, **105**, 469-482.

Freemark, K.E.; Boutin, C. (in press) Impacts of agricultural herbicide use on terrestrial wildlife in temperate landscapes: a review with special reference to North America. *Agriculture, Ecosystems and Environment*.

Grantner, M.M. (1966) *La végétation forestière du Québec méridional*. Les Presses de l'Université Laval, Québec, 216 pp.

Jobin, B.; Desgranges, J.-L.; Plante, N; Boutin, C. (1994) Relations entre la modification du paysage rural, les changements de pratique agricole et les fluctuations des populations d'oiseaux champêtres du sud du Québec (Plaine du Saint-Laurent). *Gauthier & Guillemette Inc. pour Environnement Canada. Série de rapports techniques* No **191**. Service canadien de la faune, Région du Québec.

Marrs, R.H.; Williams, C.T.; Frost, A.J.; Plant, R.A. (1989) Assessment of the effects of herbicides spray drift on a range of plant species of conservation interest. *Environmental Pollution*, **59**, 71-86.

Sotherton, N.W.; Dover, J.W.; Rands, N.R.W. (1988) The effects of pesticide exclusion strips on faunal populations in Great Britain. *Ecological Bulletin*, **39**, 97-199.

Statistics Canada (1986) Census of Canada. Agriculture. Ottawa, Canada.

Statistics Canada (1991) Agricultural Profile of Canada. Part 1. Ottawa, Canada.

Wegner, J.F.; Merriam, G. (1979) Movements by birds and small mammals between a wood and adjoining farmland habitats. *Journal of Applied Ecology*, **16**, 349-357.

Session 3

Management of Field Margins

Session Organisers & Chairmen	DR R J FROUD-WILLIAMS & DR H M CARNEGIE

MANAGING FIELD MARGINS FOR HOVERFLIES

R.W.J. HARWOOD, J.M. HICKMAN, A. MACLEOD, T.N. SHERRATT

Department of Biology, University of Southampton, Bassett Crescent East, SO9 3TU.

S.D. WRATTEN

Department of Entomology, PO Box 84, Lincoln University, Canterbury 8150, New Zealand.

ABSTRACT

In this paper we consider the effectiveness of providing flowering strips along the border of fields to augment the density of aphidophagous hoverflies. It is now fairly well established that additional floral resources can increase the local density of adult hoverflies. The evidence that higher densities of adult hoverflies actually promote significant control of aphid populations is however rather equivocal and the possible reasons for these incongruous results are discussed. Finally, we address some of the wider implications of this management programme including a discussion of its economic justification and the possibility of undesirable side effects.

INTRODUCTION

One reason for managing field margins is to enhance the local densities of predators of agricultural pests and thereby improve the degree of biological control. The larvae of a number of species of hoverfly (Syrphidae) feed on aphids and in this paper we consider:
i) whether the local density of adult hoverflies can be increased by providing strips of wild flowers along field margins
ii) whether higher densities of adult hoverflies results in less aphids, and
iii) some other implications of this management practice.

CAN FLOWERING STRIPS BE USED TO INCREASE THE DENSITY OF HOVERFLIES ?

A considerable amount of work has already demonstrated that the provision of floral resources on farmland may lead to an increase in both the local density of Syrphidae and their species diversity (e.g. Ruppert & Molthan 1991, Sengonca & Frings 1988, Weiss & Stettmer 1991). Work at Southampton has augmented this work by examining the effectiveness of a variety of modern field margin management regimes in promoting hoverflies.

Harwood, Wratten & Nowakowski (1992) drilled three field margins with mixtures of indigenous British wild flowers and grasses and compared hoverfly captures from transects of yellow pan traps perpendicular to these margins with hoverfly captures from transects of traps extending from three unmanaged field margins. The results suggested that the numbers of the total aphidophagous Syrphidae within an unmanaged arable field margin and within the adjacent field (up to 100m) can be increased by planting the field margin with wild flowers. This increase was not however shown for the most abundant aphidophagous syrphid, *Episyrphus balteatus*. The most significant treatment effect was seen in the Eristalinae (e.g. *Eristalis arbustorum*), whose larvae do not feed on aphids. Further studies have compared the number of adult syrphids observed in plots with (1) wild flower mixtures and grasses (managed by mowing and selective graminicide application), (2) grasses only, (3) bare ground, and (4) natural regeneration after cultivation, (Harwood, in prep.). Wild flower drilled margins were found to contain significantly more aphidophagous Syrphidae than bare ground, grass and natural regeneration treatments (Figure 1).

Figure 1. Number of aphidophagous syrphids observed in field-margin plots on 2nd August 1992 +/- S.E. Treatment 1 was drilled with a wild flower and grass mixture in September 1991 and mown in April and May 1992, Treatment 2 was cultivated in September 1991 and left to regenerate natural vegetation, Treatment 3 was a bare ground treatment (control) sprayed with a broad spectrum herbicide in September 1991, Treatment 4 was drilled with a grass mixture only in September 1991.

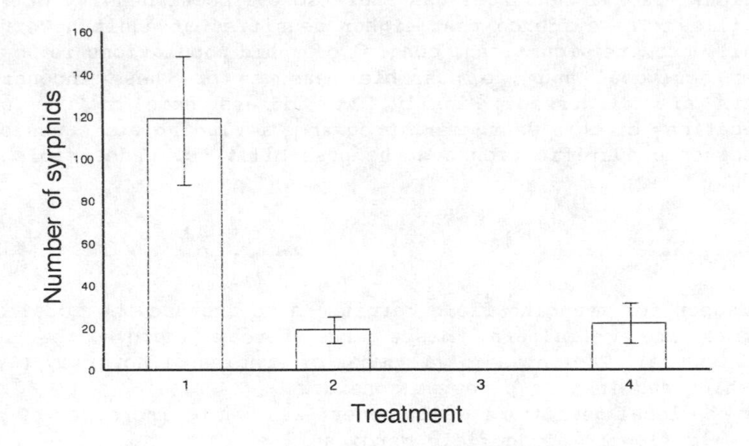

In a separate experiment, MacLeod (1992) looked at the role of alternative and novel crops as a source of pollen and nectar for foraging hoverflies. Results suggested that of all the flowering crops tested, species such as *Coriandum sativum* (Coriander) and *Fagopyrum esculentum* (Buckwheat) were the most attractive to hoverflies. However a field trial using strips of coriander along two edges of a cereal field did not produce significantly higher densities of hoverflies in the coriander bordered field than in an unmanaged control field.

Hickman & Wratten (in prep.) planted field margins with *Phacelia tanacetifolia*. This species is a member of the family Hydrophyllaceae, a family of plant which does not occur naturally in the U.K., and as such has a characteristic pollen shape. By using gut dissections of *Episyrphus balteatus* caught in transects of traps extending from the field margin into the crop it was possible to establish that hoverflies which had fed in the *Phacelia* strips moved up to 250 metres into the crop. Comparison of the numbers of hoverflies caught in transects extending from *Phacelia* margins and from control margins showed that hoverfly numbers (including *E. balteatus*) were increased in the treatment transects, and that this increase was greatest at a time when the *Phacelia* was in full bloom.

HOVERFLY OVIPOSITION AND APHID NUMBERS

The control of aphids is the basic rationale behind a number of research programmes concerned with managing field margins for hoverflies. Hoverfly adults are highly mobile and while one might expect them to lay more of their eggs close to where they have fed, this is not necessarily the case. Evidence for an increase in oviposition and reduction of the local density of aphids, as a result of this field margin management, is sparse. Studies by van Emden (1965), investigating syrphid oviposition on Brussels sprouts at different distances from flowering strips, and Pollard (1971), who compared oviposition in arable and woodland sites, did show higher rates of hoverfly oviposition near flowers. However the first study was unreplicated and the author of the second concluded that the difference between his sites could be explained by the fact that some species of hoverfly were restricted to woodland. Chandler (1968a) found no difference in oviposition in small replicated plots of Brussels sprouts between those with buckets of flowers added and those without. Sengonca & Frings (1988) recorded higher aphid density in control plots of sugar beet than in those with *Phacelia tanacetifolia* patches or borders but ironically the density of syrphid eggs was also higher in the controls. This highlights an important confounding effect: aphidophagous hoverflies tend to lay more eggs on stems that contain more aphids (Chandler, 1968b), so we should not treat evidence for high hoverfly egg densities alone as evidence for control of aphids!

Hickman (in prep.) conducted a large scale experiment on a farm in North Hampshire U.K. during 1992 and 1993. In this study she compared hoverfly oviposition and aphid numbers between three winter wheat fields with *P. tanacetifolia* borders along two of the four sides, and three control fields (different fields were used in the two years). In both years, and on several different dates from the time when the *P. tanacetifolia* flowered, a number of stems at eight distances from the field borders (up to 180m in 1992; 100m in 1993) were taken and examined for the presence of syrphid eggs and aphids. Although syrphid larvae were recorded when seen, they were not specifically searched for since they conceal themselves in the crop during the day and are most active at night when most of their predation occurs.

During the 1992 season very few syrphid eggs were found in the crop and no significant differences in the mean density of aphid populations were detected between treatments. One reason for this may have been that the wheat crop ripened about two weeks earlier than usual. The early emergence of wheat ears enabled them to be colonised by the grain aphid *Sitobion avenae* while numbers of gravid hoverflies remained low. Syrphid oviposition may also have been deterred by the early yellowing senescent condition of the wheat. The mean number of syrphid eggs found on aphid infested baits in the *P. tanacetifolia* - bordered fields (3.03 per bait) was higher than the mean number of eggs found on baits in the control fields (2.00 per bait) although the difference was not statistically significant ($P > 0.05$).

In 1993, 320 stems were checked in each experimental and control field during each week that aphids remained in the crop. A total of 61 syrphid eggs were found in *P. tanacetifolia* - bordered fields and 21 in control fields, with the majority being found between 11.6.93 and 27.6.93. No clear patterns could be seen between treatments for the percentage of stems infested with one or more aphids, until the week beginning 5.7.93.; from this point percentages of infested stems were lower in all *P. tanacetifolia* - bordered fields than any control field. This period coincided with the main appearance of third instar syrphid larvae (the most voracious instar) in the crop. Levels of parasitoid activity (as assessed by the number of aphid mummies seen) appeared similar between treatments and there was very little evidence of aphid death from pathogens. We therefore consider it likely that the differences in aphid levels between treatments were the result of increased predation by syrphid larvae in the *P. tanacetifolia* - bordered fields.

Figure 2 Percentage of wheat stems with one or more aphid in *P. tanacetifolia* - bordered and control fields on different dates. Each line represents a separate field. P = *P. tanacetifolia* borders; C = control.

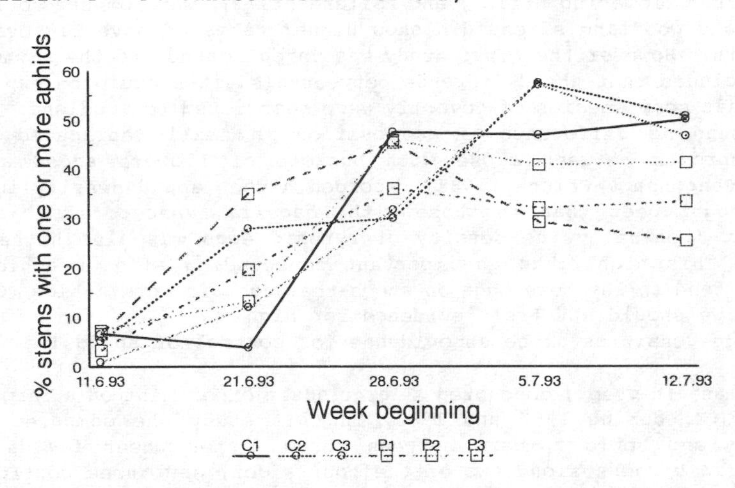

From the above it does appear that the provision of suitable flowering strips for hoverflies can sometimes, under relatively large scale conditions, promote measurable local control of aphids.

MISCELLANEOUS CONSIDERATIONS

The underlying philosophy behind managing field margins to promote control agents of agricultural pests is rarely made explicit. Since our intention is to manage field margins in order to promote hoverflies, then the following basic questions come to mind:

1) Does the provision of floral resources actually increase the regional total population size of hoverflies, or does it simply influence their spatial distribution?

2) Where are the floral resources best positioned for maximum control of aphids?

3) Is the provision of flowering strips for aphidophagous hoverflies likely to make economic sense?

4) Are there any potentially deleterious effects of providing additional floral resources along field margins?

These questions are now considered in turn;

1) Clearly from the time additional floral resources are first provided, then

initially any increase in the number of hoverfly adults comes about from the spatial re-distribution of adults. It has been known since 1948 that female syrphids require pollen to develop their ovaries (Schneider, 1948), but it is still unknown whether pollen availability is a limiting factor in the life of some species of hoverfly. Studies have been published on the key factors affecting survivorship of syrphids in the egg, larval and pupal stages, but factors affecting the adult stage have not been recorded (Verma & Makhmoor,1989). Furthermore it is still unclear as to whether or not the quality of the pollen and nectar resources influences total fecundity and egg fertility. Work is currently being undertaken at Southampton to address this possibility.

2) This question has never been considered directly. Clearly flowering strips should be in large enough blocks to attract hoverflies, but they should also be spread out over fields to ensure that the adult hoverflies penetrate into all sides over the fields before laying their eggs. Any impediments to movement of beneficial syrphids around an arable system ought to be considered before deciding upon the positioning of flowering strips in the landscape. Figure 3 shows how a higher proportion of syrphids containing *P. tanacetifolia* pollen are caught in yellow traps 10m from a strip of *Phacelia* when there is no physical barrier between the strip and the traps, than in traps also 10m from a *Phacelia* strip, when there are barriers between the pollen source and the traps (Hickman & MacLeod, in prep). So far our work suggests that hoverflies may be reluctant to cross such features which cause a break in vegetational ground cover.

Figure 3. Mean distribution of the percentage of all hoverflies containing *P. tanacetifolia* pollen caught on either side of linear features on a farm.

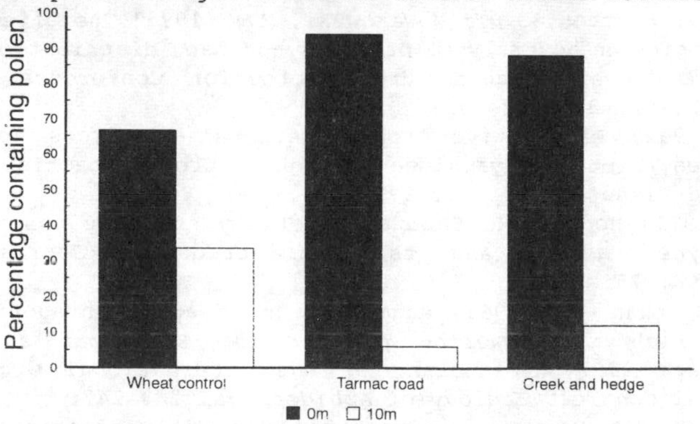

3) An examination of the costs involved in this management option indicates that even if *P. tanacetifolia* borders were provided all around a field the economic burden on the farmer would not be a great one. In the Southampton experiment, the seeds were drilled in what would otherwise have been a sterile strip round the crop, thus there was no reduction in crop area. *P. tanacetifolia*, drilled at the recommended rate of 5kg/ha costs approximately 0.5p/m^2; the cost of seeds for a 1m strip round the largest field would be little more than £10. Once drilled or hand broadcast the strips require no further maintenance, and can be ploughed in with the stubble after harvest. Thus any reduction in insecticide use should result in increased profit for

the farmer in addition to the other benefits of reducing inputs.

4) We can see two potentially deleterious effects of providing flowering strips along field margins. One concern is that by providing a "sink" for hoverflies, one may achieve a **local** reduction in aphid density but actually augment it over a **regional** scale. Thus, while a crop in one field benefits from an increased number of syrphid larvae predating upon aphids, another nearby crop suffers and could allow aphid numbers to build up with winged aphids emigrating to surrounding fields, making the pest worse in the long term. We are currently assessing the likelihood of this scenario using a simulation model which is parameterised from field data. Another concern is that by providing additional floral resources, we could potentially reduce the amount of pollination in native wild plants. Our feeling is that the amount of additional floral resources provided would have to be on a massive scale before these more subtle side effects became important.

REFERENCES

Chandler, A.E.F. (1968a) Some factors influencing the occurrence and site of oviposition by aphidophagous Syrphidae (Diptera). *Annals of Applied Biology*, **61**, 435-446.

Chandler, A.E.F. (1968b) The relationship between aphid infestations and oviposition by aphidophagous Syrphidae(Diptera). *Annals of Applied Biology* **61**, 425-434.

van Emden, H.F. (1965) The effect of uncultivated land on the distribution of cabbage aphid (*Brevicorne brassicae*) on an adjacent crop. *Journal of Applied Ecology*, **2**, 171-196.

Harwood, R.W.J.; Wratten, S.D.; Nowakowski, M.; (1992) The effect of managed field margins on hoverfly (Diptera: Syrphidae) distribution and within-field abundance. Brighton Crop Protection Conference - Pests and Diseases 1992, **3**, 1033-1037.

MacLeod, A. (1992) Alternative crops as floral resources for beneficial hoverflies (Diptera: Syrphidae). Brighton Crop Protection Conference - Pests and Diseases 1992, **3**, 997-1002.

Pollard, E. (1971) Hedges VI. Habitat diversity and crop pests: a study of *Brevicoryne brassicae* and its syrphid predators. *Journal of Applied Ecology*, **8**, 751-780.

Ruppert, V.; Molthan, J. (1991) Augmentation of aphid antagonists by field margins rich in flowering plants. In: *Behaviour and impact of aphidophaga 1991 4th meeting of the International Organization of Biological Control: Ecology of Aphidophaga*, 243-247.

Schneider, F. (1948) Beitrag zur kenntnis der Generationsverhaltnisse and Diapause rauberischer Schwebfliiegen. *Mitteilungen der Schweizerischen Entomlogischen Gesellschaft*, **21**, 249-285.

Sengonca, C.; Frings, B. (1988) Einfluss von *Phacelia tanacetifolia* auf Schadlings und Nutzlingpopulationen in Zuckerrubenfeldern. *Pedobiologia*, **32**, 311-316.

Verma, A.K. & Makhmoor, H.D. (1989) Development of life tables for *Metasyrphus confrater* (Wiedemann) Diptera: Syrphidae, a predator of the cabbage aphid (Homoptera, Aphididae) in cauliflower crop ecosystem. *Entomon*, **14**, 227-232.

Weiss, E.; Stettmer, C. (1991) *Agrikologie I: Unkrauter in deer agrarlandschaft locken blutenbesuchende nutzinsekten an*. Paul Haupt, Berne.

FIELD MARGIN FLORA AND FAUNA CHANGES IN RESPONSE TO GRASSLAND MANAGEMENT PRACTICES

J H McADAM, A C BELL

Department of Agriculture for N Ireland, Newforge Lane, Belfast BT9 5PX

T HENRY

Greenmount College of Agriculture and Horticulture, DANI, Antrim BT41 4PU

ABSTRACT

Agricultural land use in Northern Ireland is dominated by high stock numbers and grass-based enterprises. An experiment was established in 1990 to examine the impact of grass field boundary management strategies on flora and fauna species diversity. Three well-managed hawthorn hedges separating paired grass fields formed the sites for the research. Four treatments were imposed, each 30 m long and extending 10 m into the field either side of the hedge. The treatments were as follows: (i) fertilised and rotationally grazed with sheep; (ii) 2 m strip ploughed adjacent to the hedge and sown with a game cover crop, the remaining 8 m being taken for silage; (iii) as (ii) but with the ploughed strip left unmanaged; (iv) unmanaged control.

Plant species presence and percentage cover have been recorded annually in permanent quadrats located in the hedge, hedge base, and at 0.5, 2, 6, and 9 m into the field. Carabid beetles were trapped using pitfall traps placed 1-2 m from the hedge and 8-10 m into the field. Plant species diversity was greatest within 0.5 m from the hedge base. Grazing and fertiliser use significantly reduced species diversity of carabids and plants relative to all other management treatments. If wildlife is to be conserved, hedges and field margins must be protected from grazing and other intensive grassland management operations.

INTRODUCTION

In Northern Ireland agriculture is largely pastoral. Approximately 1.5 x 10^6 cattle and 2.3 x 10^6 sheep utilise 1.1 x 10^6 hectares of grassland and approximately 83% of gross margins are derived from livestock enterprises (Department of Agriculture for N Ireland, 1993). Farms are small (mean size 35.1 ha) with a relatively large number of small fields (mean size 1.8 ha) and there are estimated to be 152,000 km of field boundaries. In such a rural scenario, field boundaries form an important component of the visual landscape. It has been estimated that between 1976 and 1982 the rate of loss of hedges in the Republic of Ireland was 14% (An Foras Forbartha, 1985) though this estimate was based on a sample of only 12 km^2.

The rate of hedgerow loss in N Ireland is poorly documented. From a landscape ecological survey of the Mourne Mountains Area of Outstanding Natural Beauty, Cooper & Murray (1987a) estimated an annual loss rate of 0.5%. The rate of removal was greater in the lowland area. In the Antrim Glens and Causeway Coast Areas of Outstanding Natural Beauty 12.3% of field boundaries were removed in the lowland area between 1975 and 1987 (Cooper & Murray, 1987b). The overall loss rate was 5.2 %. In the Fermanagh district (Murray *et al.*, 1991) 13.9% of field boundaries were removed since 1962, with the greatest loss in the lowland areas where grassland management was most intensive.

It is well documented that an increase in soil nutrient status, disturbance by cultivation and spray drift has resulted in many field boundaries having a species-poor flora. Sheep densities have increased substantially on N Ireland farms over the past 10 years (DANI, 1992). Sheep graze close to the ground and this selects against the survival of non-rosette plants with apical meristems borne aloft. Hence, sheep grazing can reduce associated hedge flora diversity and can restrict regrowth of managed hedges by eating regenerating shoots. Agricultural practices such as dereliction, increased use of fertiliser, use of slurry, zero grazing, conservation for silage instead of hay have resulted in loss of botanical diversity in hedgerows and grassland in favour of species which respond positively to soils with a high nutrient status.

Much research has been conducted on the impact of cropping practices and conservation value of arable field margins (eg Way & Greig-Smith, 1987; Thomas *et al.*, 1991). However, research into the impact of agricultural practices on the grass field margins is much sparser. In view of the importance of hedges to an intensive, grassland farming country such as Northern Ireland, the decline in hedgerows and the damage which grassland management can have on hedges, an investigation was initiated to investigate the effects of four grassland management practices on the flora and fauna of grass field boundaries. The aim of this long term project was to recommend the best strategy for maximising wildlife value of field boundaries.

MATERIALS AND METHODS

Treatments

Three well-maintained, mature hedges separating paired grass fields formed the blocks for the study. The predominant hedgerow species was hawthorn (*Crataegus monogyna*). Within a block, 4 treatments, each 30 m long, were randomly arranged across the hedge extending 10 m into the fields. The treatments first imposed in 1990 were as follows:

(1) Fertilise/graze. Plots were fertilised (100 kg N/ha) and rotationally grazed with sheep down to a target sward height of 2-3 cm.

(2) Plough/game cover. A 2 m strip adjacent to the hedge was ploughed and sown initially with a game cover crop of kale, mustard and quinoa; the remaining 8 m was fertilised (150 kg N/ha) and two cuts of silage taken. In March 1993 Jerusalem artichokes were planted as the game cover crop. These subsequently failed to grow and a change in the cover species will be made in 1994.

(3) Plough/unmanaged. This was similar to the previous treatment except that the 2 m strip was left unseeded and allowed to colonise naturally.

(4) Unmanaged control. No fertiliser or management treatments.

Flora and fauna recording

Plant species presence and percentage cover were recorded in July 1991 and August 1993 in 1 m x 1 m permanent quadrats in the hedge and hedge base, and at 0.5, 2, 6 and 9 m into the field. Quadrats were placed along three randomly arranged line transects on both sides of the hedge. Carabid beetle species were trapped annually in each plot using three pitfall traps placed 1-2 m ("margin" sample) and 8-10 m ("field" sample) either side of the hedge. Monthly catches were taken in March, May, July and September of each year.

Data Analysis

The species recorded in each set of three quadrats at a particular distance from the hedge were added together giving a total of 144 samples for each sampling session. The resultant data matrix was subjected to classification and ordination using TWINSPAN (Hill, 1979a) and DECORANA (Hill, 1979b) respectively (Bell *et al.*,1994). An analysis of variance of the mean total number of species per treatment was carried out for treatments and distance from the hedge base. The catches for each set of three pitfall traps were combined to produce a total of 48 samples. The number of Carabid species occurring in each sample was counted and a Modified Simpson's Diversity Index (Usher, 1986) calculated. The modification is that the calculated Simpson's index figure is subtracted from 1 to produce an index, the magnitude of which is directly proportional to the degree of species diversity, (a species diversity of 0 would be a monoculture).

RESULTS

There were highly significant differences (P<0.001) in plant species number per quadrat between treatments and distance from the hedge (Table 1). Fertiliser application and grazing resulted in significantly (P<0.05) fewer plant species than all other treatments and although there were more species in the two ploughed strip treatments than the unmanaged control (8.3 vs 7.7) the difference was not significant (P>0.05). More species tended to occur close to the hedge than in the field and there were significantly more species in the quadrats placed 0.5 m from the hedge than in the hedge itself or further out into the field.

There were significant differences among treatments in species diversity of carabids in July and September (Field margin) and in September (Field). On these occasions there were significantly fewer carabids trapped in the fertilised/grazed plots than any of the managed treatments (Fig 1). From previous TWINSPAN analysis (Bell *et al*., 1994), eight species were selected as indicator species. *Abax parallelepipedus* a species normally found in hedgerows was found in significantly fewer numbers in the fertilised/grazed margin traps in May and July. *Agonum muelleri* was more abundant in field traps in all treatments except the unmanaged (especially in July). *Bembidion aeneum* was found in greater numbers in the fertilised/grazed traps in the field in July than any other month. *Clivina fossor* was found in significantly lower numbers in field traps outside the ploughed treatments than the fertilised/grazed or unmanaged treatments. *Leistus fulvibarbes* was not trapped in the fertilised/grazed plots in the field margin in September. There were significantly ($P<0.05$) greater numbers of *Loricera pilicornis* in field traps adjacent to the ploughed treatments and of *Pterostichus strenuus* in the fertilised/grazed margin in July. There were significantly fewer *Pterostichus melanarius* in margin traps in the fertilised/grazed plots in May compared with other treatments.

TABLE 1. The effect of hedge management techniques imposed in 1990 on the mean number of plant species at varying distances from the hedge in 1993.

TREATMENT	HEDGE	HEDGE BASE	0.5 m	2 m	6 m	9 m
Fertilised/grazed	5.7	7.0	7.0	6.0	5.0	5.3
Unmanaged	8.3	8.7	8.7	7.0	7.0	6.67
Plough/Unmanaged	8.7	9.3	9.3	8.3	6.3	7.7
Plough/Game cover	9.0	8.7	11.0	9.0	5.3	7.0
Mean	7.9	8.4	9.0	7.6	5.9	6.0

SEM Treatment = 0.32 ($P<0.001$); Distance from base = 0.39 ($P<0.001$)

DISCUSSION

The results of this experiment clearly illustrate the effect that sheep grazing and fertilisation have on the flora and carabid fauna of field margins. Two years after imposition of the alternative management treatments there are significantly more wild plant species and carabid beetles than in the fertilised/grazed treatment. Populations of four key (indicator) species - *Abax parallelepipedus*, *Clivina fossor*, *Leistus fulvibarbes* and *Pterostichus melanarius* were found to be reduced by grazing/fertilising, particularly in late summer. The game cover crop produced no effect on plant or carabid species diversity. The importance of the field margin as a source of biodiversity in grassland is clearly shown. Most species were found in the hedge, hedge base and 0.5 m out from the hedge, compared to the field.

If wildlife is to be conserved in field margins, protection from intensive grazing and fertiliser application is necessary. From a wildlife perspective, there is little advantage to be gained in fencing further out than approximately 1 m from the hedge base though each hedge and associated swards will have historic and management backgrounds which make prediction of the likely course of species colonisation very difficult. This is a long-term trial and monitoring will continue for a further 6 years.

ACKNOWLEDGEMENTS

The authors are grateful to Jenny Thompson, Jill Forsythe, John Anderson and Graham McCollum for assistance with sampling and recording and to Eugene McBride and the farm staff at Greenmount College for practical assistance.

REFERENCES

An Foras Forbartha (1985) *The state of the Environment*. Minister for the Environment, Dublin. An Foras Forbartha.

Bell, A. C.; McAdam, J. H.; Henry, T. (1994 - in press) Grassland Management and its effect on the wildlife value of field margins. In Haggar R. (ed) Grassland Management for Nature Conservation. BGS/BES symposium, Leeds.

Cooper, A. & Murray, R. (1987a) *A Landscape ecological study of the Mourne and Slieve Croob, Areas of Outstanding Natural Beauty*. Report to the Countryside and Wildlife Branch, Department of the Environment (DOENI), Department of Environmental Studies, University of Ulster, Coleraine.

Cooper, A. & Murray, R. (1987b) *A Landscape ecological study of the Antrim Coast and Glens and Causeway coast Areas of Outstanding Natural Beauty*. Report to the Countryside and Wildlife Branch, Department of the Environment (DOENI), Department of Environmental Studies, University of Ulster.

Department of Agriculture for N Ireland (1993) Statistical review of N Ireland Agriculture 1992. *Economics and Statistics Division DANI*, HMSO, Belfast.

Hill, M. O. (1979a) TWINSPAN: a FORTRAN programme for arranging multivariate data in an ordered two-way table by classification of the individuals and attributes. Cornell University, Ithaca, New York.

Hill, M. O. (1979b) DECORANA: a FORTRAN programme for detrended correspondence analysis and reciprocal averaging. Cornell University, Ithaca, New York.

Murray, R.; Cooper, A.; McCann, T. (1991) *A landscape ecological study of Fermanagh District*. Report to Countryside and Wildlife Branch, DOENI, Department of Environmental Studies, University of Ulster, Coleraine.

Thomas, M. B.; Wratten, S. D.; Sotherton, N. W. (1991) Creation of 'island' habitats in farmland to manipulate population of beneficial arthropods: predator densities and emigration. *Journal of Applied Ecology*, **28**, 906-917.

Usher, M. B. (1986) *Wildlife Conservation Evaluation*. London: Chapman and Hall.

Way, J. M.; Greig-Smith, P. W. (1987) *Field margins*. BCPC Monograph No. 35. London.: British Crop Protection Council.

Fig. 1 Histograms showing variation in carabid species diversity with season and treatment (modified Simpson's indices of diversity).

MARGIN

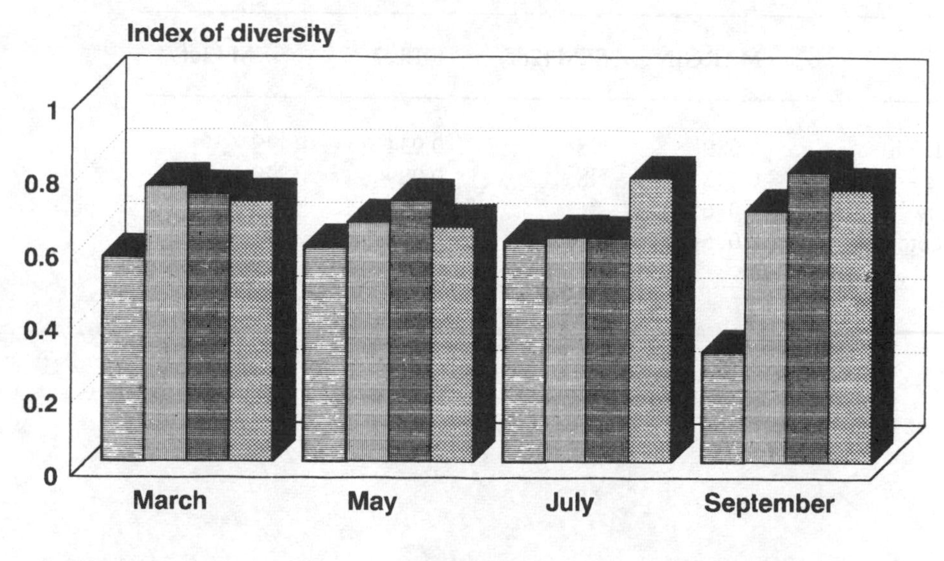

FIELD

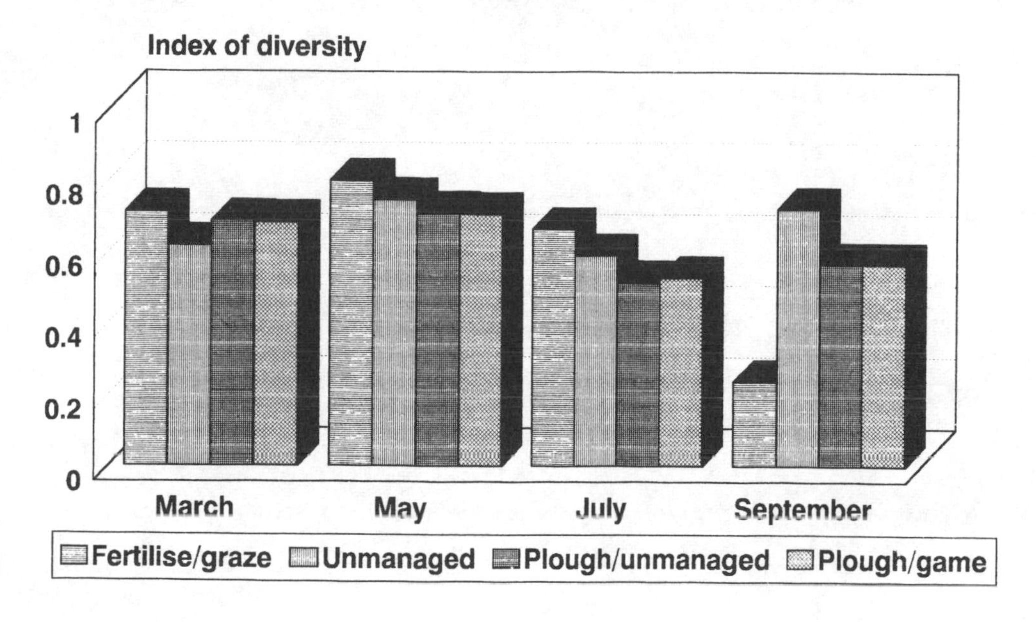

TABLE 2. Standard errors of the treatment means for the modified Simpson's index of carabid species diversity for the margin and field samples taken between March and September 1993.

	MARGIN	SEM (2df)	FIELD	SEM (2df)
March	0.089	NS	0.034	NS
May	0.043	NS	0.029	NS
July	0.036	*	0.054	NS
September	0.075	*	0.058	*

AN ECONOMIC ANALYSIS OF FARM HEDGEROW MANAGEMENT.

D.A. SEMPLE, E.C. BISHOP & J. MORRIS

Silsoe College, Cranfield University, Silsoe, Beds, MK45 4DT.

ABSTRACT

The paper assesses the effect to farm incomes attributable to changes in hedgerow management associated with environmental enhancement. Hedgerow features and characteristics with potential financial implications for farmers were identified. Data was obtained from standard sources and from farm surveys in various parts of England and Wales. The impact of different hedgerow management practices on cropped areas, yields, farm labour, machinery requirements, and on farm revenues and costs were assessed. Analysis compared the costs and benefits of existing practices to alternative hedgerow systems and fences. The analysis was conducted using three farm models: a 200 ha arable farm; a 75 ha mixed farm; and a 75 ha dairy farm. A review of existing grants for hedgerows was carried out. The need for specific incentives to encourage farmers to adopt preferred hedgerow management practices was considered.

INTRODUCTION

The study of economic impacts of hedgerow management was part of a larger investigation of preferred hedgerow management sponsored by the Directorate of Rural Affairs, Department of the Environment. The study aimed to establish the financial cost which farmers would incur by changing their hedgerow management practices to increase the environmental benefit.

Using information gained from the farm survey carried out by the Institute of Terrestrial Ecology and Cobham Resource Consultants, and other relevant literature, the hedgerow features that would affect farm income and expenditure were identified. The costs of existing hedgerow management practices were identified; covering labour, machinery, supplies and services and including fencing where relevant. The impact of different hedgerow management practices were investigated with regard to the effect on cropped areas, yields, regular farm labour and machinery requirements. Analysis compared the costs and benefits of existing practices to alternative hedgerow systems and fences. The analysis was conducted using three farm models: a 200 ha arable farm; a 75 ha mixed farm; and a 75 ha dairy farm. The data for these models was taken from the aforementioned survey and published sources (Nix, 1990). A review of existing grants for hedgerows was carried out. The additional cost borne by farmers due to environmentally beneficial hedgerow management was identified. Consideration was given to the type and scale of incentives required to encourage farmers to adopt preferred hedgerow management practices.

HEDGEROW FEATURES

The main features of hedgerows relate to the size and the configuration of the hedge itself, and the adjacent verge.

Hedge types

A matrix of hedgerow features was drawn up by the Institute of Terrestrial Ecology (Institute of Terrestrial Ecology, 1992) and four main types of hedge identified. These hedge types may be ranked according to their relative merits in terms of environmental enhancement.

Type 1 is an unmanaged hedge, approximately 5 m high by 4 m wide and is considered to be the best habitat for birds, invertebrates and small mammals, as well as having the greatest landscape, amenity and sport values.

Type 2 is a side trimmed hedge, 4 m high by 2 m wide and has similar merits to type 1. The hedge is cut every second year, preferably cutting half the length of hedgerow on the farm each year.

Type 3 is a fully trimmed hedge 2 m high by 1.5 m wide. It has fewer environmental benefits than those above and is cut every second year.

Type 4 is a fully trimmed hedge 1 m high by 0.75 m wide. It is the most common type of managed hedge that exists on farms, is cut annually and has minimal environmental benefits.

The type 4 hedge is taken as the baseline for the subsequent calculations.

Verges

A verge extending from the outer extremity of the hedgerow for at least 1m has been shown to be highly desirable for wildlife, particularly if it is left uncut. This not only creates an additional area of habitat but also helps reduce any overspill of agricultural field management, for example agrochemicals or mechanical operations, into the hedgerow.

ECONOMIC IMPLICATIONS OF HEDGEROW CHARACTERISTICS

Within the two main hedgerow features, namely hedges and verges, a number of hedgerow characteristics were identified as having potential economic cost and benefit implications for farmers. A change in hedge or verge dimension has implications for land loss, labour and machinery costs and shelter and shade. These characteristics are discussed below, along with fencing, which is the most common field boundary alternative.

Land Loss

The land lost to agricultural production due to a change in hedgerow management is a function of any increase in the width of the hedge or verge. It is affected by the size and shape of the field, and the proportion of the field boundary that is enclosed by a hedgerow. The value of the land loss due to an increase in hedge or verge size was taken as the area lost multiplied by the gross margin of the crop, where the latter comprises the value of the crop output less the direct variable costs such as seed and fertilizer.

Farm labour and machinery costs

Labour and machinery requirements for field operations change in proportion to the area of cropped land. For the purpose of this analysis it has been assumed that any change in crop area due to hedgerow management will not affect the size of the permanent labour force or the machinery inventory. Changes in labour requirements have been valued at the standard hourly wage rate, and average machinery running costs have been estimated at £60/ha and £25/ha for arable and grassland enterprises respectively.

Shelter and shade

The main reason for retaining non-stock proof hedges on farms, apart from tradition, is the shelter benefit that the hedge provides to both crops and livestock. The semi-permeable structure of a hedge slows the air flow without adverse turbulence. This reduction in air speed has been shown to have an effect on crop yield (Caborn, 1965). The benefit of additional shelter for stock has also been demonstrated, (Blaxter *et al.*, 1964; Caborn, 1965). However, as the in-wintering of beef and dairy stock is now common agricultural practice in the United Kingdom, the economic benefit of additional shelter is slight.

The shading effect of a hedge depends on its height, orientation and the incidence angle of the sunlight. The shading of field plants adversely affects the growth rate and yields (Caborn, 1965).

It has been assumed that the net effect of shelter and shade on the crop is a 50% reduction in yield or stocking density in the area affected (expressed as the field area equivalent to the hedge height multiplied by the hedge length).

Fencing

On stock and dairy farms, back fencing of hedges is commonplace. If a conservation verge is introduced, a fencing barrier is required to exclude stock from the conservation verge. For cattle a single/double barbed wire or electric fence is sufficient but in the case of sheep a netting fence is necessary. For some field boundaries, a conservation verge involves the relocation of the back-fence rather than an additional fence.

Fencing may be used as an alternative field boundary to a hedge. If so, there are beneficial effects in terms of increasing the area available for cultivation and removal of all shade effects. Fencing costs vary with the type of fence. Costs to supply and erect a wire fence eligible for grant are approximately £3.50/metre. If a fence is erected to protect the verge from livestock, the costs are £3.50/metre for sheep netting and £1.20/metre for a single barbed wire for cattle. In the analysis the costs have been amortised over the life of a fence (which is assumed to be 20 years).

HEDGEROW MANAGEMENT PRACTICES

The most common type of managed hedge on farms is 1 m high by 0.75 m wide with no verge. On livestock farms, back-fencing is commonplace.

The majority of hedges are cut annually, using tractor mounted flail cutters. A small number of farmers use reciprocating cutters or circular saw blades. If a verge is present it is normally cut at the same time as the hedge. Regular management is usually undertaken after harvest on arable farms when labour is available. Where labour and machinery are insufficient, contractors are normally engaged.

Hedgerow management costs vary, depending on whether a farmer uses his own machinery and labour or hires a contractor. The costs increase substantially with the size of the hedge and the number of passes necessary, but these are partly offset by less frequent cutting. The cost of cutting the four hedge types, and verge where present, are shown in Table 1. For unmanaged hedges it is assumed verges are cut once every two years.

TABLE 1. Hedgerow and verge cutting costs (£ per 100 m of hedgerow)

	Hedge type (with verge)			
	Type 1	Type 2	Type 3	Type 4
Number of passes	0 (2)	8 (10)	5 (7)	3 (5)
% cut per annum	0 (50)	50 (50)	50 (50)	100 (100)
Own machinery and labour (£3.01)	0 (3)	2 (15)	8 (11)	9 (15)
Contractor (£2.67)	0 (3)	11 (13)	7 (9)	8 (13)

COST BENEFIT ANALYSIS

Farm models

Three models were used to demonstrate the effect of different hedgerow management practices on different farming systems. A 200 ha arable farm; a 75 ha mixed farm; and a 75 ha dairy farm. The average field size was 10 ha, 7.5 ha and 5.7 ha respectively, with an assumed 66% of field boundary hedged. This resulted in hedgerow lengths of 16138 m, 7207 m, and 8151 m for each farm.

Analysis

For each model, the following points were examined with respect to different types of hedgerow practices, with and without verges, and hedge replacement by fencing:

- the land and field effects on gross margins;
- the impact on labour requirements for existing farm operations, and for hedge and verge maintenance;
- the impact on machinery costs for existing farm operations, and for hedge and verge maintenance; and
- the annual capital costs for establishing fencing.

The total annual costs are shown in Table 2.

TABLE 2. Total annual costs (£ per 100m) of different management practices by farm type with respect to hedge type 4.

Farm type	Hedge type (with verge)				
	Type 1	Type 2	Type 3	Type 4	Fence
Arable	17 (21)	21 (25)	8 (12)	0 (11)	17
Mixed	7 (19)	14 (26)	4 (16)	0 (17)	18
Dairy	22 (40)	23 (41)	9 (27)	0 (25)	16

Summary

(i) With respect to the arable and mixed farms, type 2 is more expensive to farmers in cost per 100 metres and has fewer environmental benefits than type 1.

(ii) Type 3 hedge without a verge offers a relatively cheap way of securing some environmental benefits.

(iii) The inclusion of a verge substantially increases the costs since it involves land loss, verge cutting, and, on livestock farms, conservation fencing.

HEDGEROW GRANTS AND INCENTIVES

Grants are available from a number of sources for hedgerows and hedgebanks. However, annual maintenance costs are not usually eligible.

Conclusions

In the context of achieving a transition from existing Type 4 hedgerows without verges to those which enhance environmental qualities, there are two main choices:

- hedge types: Type 3 without a verge is the cheapest option for all farm types but has limited environmental benefits;

- verges: the inclusion of verges adds to environmental quality but can significantly increase costs to farmers, particularly on livestock farms due to the need for conservation fencing.

The preceding analysis has examined the costs to farmers of providing alternative hedgerow characteristics. These costs need to be compared with the environmental benefits provided by the different hedgerow types in order to determine the most cost effective hedgerow management practices.

ACKNOWLEDGEMENTS

The information in the above report was collected under contract to the Directorate of Rural Affairs, Department of the Environment. The authors thank them for allowing publication of the data.

REFERENCES

Blaxter, K.L.; Joyce, J.P.; Webster, A.P.F. (1964) Investigations of environmental stress in sheep and cattle. *Proceedings of MAFF Second Symposium on Shelter Research*, Edinburgh: MAFF, 27 - 35.

Caborn, J.M. (1965) *Shelter Belts and Wind Breaks*, London: Faber and Faber Ltd.

Institute of Terrestrial Ecology (Ed. Hooper, M.D.) (1992). *Hedge Management*, Final Report to the Department of the Environment.

Nix, J. (1990) *Farm Management Pocket-book*, 21st Edition, Wye College, Department of Economics.

CONSERVATION AND RESTORATION OF SPECIES RICH DITCH BANK VEGETATION ON MODERN DAIRY FARMS

W. TWISK, A. J. VAN STRIEN, M. KRUK, H.J. DE GRAAF & W.J. TER KEURS

Environmental Biology, Institute for Evolutionary and Ecological Sciences, University of Leiden, P.O. Box 9516, 2300 RA Leiden, The Netherlands

ABSTRACT

Many ditch banks on modern dairy farms in the western peat district of the Netherlands still contain a species-rich vegetation of international importance. Research was carried out in order to investigate if and how conservation and restoration of this vegetation is compatible with modern dairy farming. The research was started by studying the existing diversity of management in relation to the floristic richness of the ditch bank vegetation. In this way the most important factors determining the species composition were established. Subsequently an experiment was carried out with one of the most important factors: the cleaning frequency of the ditches. The purpose of the experiment was to investigate whether a lower cleaning frequency can be compatible with modern farming and proper water management. That proved to be the case in many situations. In some situations both nature and farmer could even benefit from a lower ditch cleaning frequency.

INTRODUCTION

The plant species diversity of grasslands and ditch banks in the western peat district of the Netherlands is declining because of intensive dairy farming (Provincie Zuid-Holland, 1993). Protecting these nature values by establishing nature reserves is only possible on a limited area because of the high costs. Therefore studies have been carried out to study the possibilities for nature protection within the limits of modern dairy farming (Melman, 1991; Van Strien, 1991; Twisk *et al.*, 1991). The starting-point was that farming and nature conservation don't have to be incompatible, as is commonly assumed (De Boer & Reyrink, 1988). The ditch bank vegetation studies of Van Strien and Twisk will be discussed here.

Study area

The study sites were located in the typical Dutch polder-landscape. The surface soil of these polders consists of peat, intersected by zones of clay and clay-on-peat along the rivers. This landscape was formed by man about a thousand years ago by digging parallel drainage ditches. The result is a landscape with long narrow fields (40-60 m wide and up to 1-2 km deep) separated by ditches (1-7 m wide and 30-60 cm deep). The fields are almost always grasslands with dairy farming as the main form of agricultural land use. The farming intensity is very high with an average N-supply of C. 300 kg N per ha and 1.7 cows per ha.

The polders all lay below the present sea and river level due to drainage and subsequently shrinking of the peat over the last centuries. The water table therefore is man-controlled, in winter 10 to 20 cm lower than in summer. Regional waterboards require the farmers to clean their ditches in order to maintain a proper discharge of water from the polders. Almost all ditches are cleaned mechanically once a year nowadays, often in the autumn. Plants in and near the ditches are removed and dumped on the banks, together with mud from the ditch bottom.

Grassland research

At first the possibilities for nature conservation on the grasslands were investigated (Van Strien, 1991). These possibilities proved to be very limited: grassland vegetation only has some floristic richness when the N-supply is below 100-200 kg N per ha (Figure 1). Such low levels are scarcely attainable in current dairy farming.

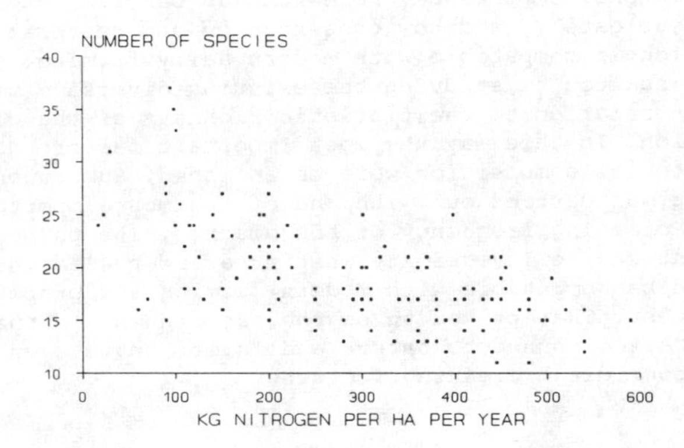

FIGURE 1. The relationship between the number of plant species and the N-supply on grassland (from Van Strien, 1991)

Focus on ditch banks

On the ditch banks along the grasslands however there appeared to be more possibilities to keep and restore species rich vegetation. Firstly, many ditch banks still contain species-rich vegetation, including international quite "rare" species such as *Lychnis flos-cuculi* and *Iris pseudacorus*. Secondly, ditch banks form a marginal part of the farm from a economic point of view. So, management more aimed at nature conservation does not have to result in loss of income. Thirdly, species-rich ditch banks can also be found in situations where the adjacent fields are used intensively, indicating that modern dairy farming can be combined with maintenance of floristic values in the ditch banks. Last but not least, the ditch banks form quite a large potential "nature area", because their lack in width (they are not more than 1 to 2 m wide) is more than compensated by their length (their total length has been estimated at C. 100.000 km for the Dutch peat areas). Therefore, the aim of this research was to consider the possibilities for maintenance or restoration of species-rich ditch banks and its consequences for the farmer.

STUDY DESIGN

A multifactor, transverse study design was used to assess the individual effects of the agricultural factors (Van Strien, 1991). This approach implied the spatial comparison of a great number of plots on ditch banks (>300 spread over >100 dairy farms) differing in management regime. Data on grassland exploitation, ditch management and properties of the banks were obtained from the farmers and from field observations. Only steady-state situations were selected, i.e. situations with a more or less constant management for at least the previous 5 years. The study plots were carefully selected, to yield a data set with an almost independent variation of the factors studied. All factors involved are mentioned in Table 1.

TABLE 1. The influence of some factors on number of plant species in the ditch bank vegetation as well as the expected compatibility of measures with farming practice (based on Van Strien, 1991).

Factor *(+ range)*	Number of species	Expected compatibility
Nitrogen supply on grassland *(0 - 550 kg N ha^{-1} $year^{-1}$)*	--	++[3)]
Type of use *(meadow - pasture)*	ns	-
Level of ditch water *(15 - 80 cm below surface)*	--	--
Slope aspect *(South - West - East - North)*	-[1)]	irrelevant
Slope angle *(0 - 35 degrees)*	++[1)]	-
Soil type *(mesotrophic peat - eutrophic peat - mes. clay-on-peat - eutr. clay-on-peat)*	ns	irrelevant
pH of topsoil *(pH-H_2O 4.0 - 7.2)*	++	+
P and K of topsoil	ns	+
Ditch cleaning frequency *(once every year - less than once every 3 years)*	++[2)]	+?
Ditch cleaning method *(hand - mowing-basket - ditch-scoop - auger)*	ns	++
Peat mud dressing *(less - more than 5 years ago)*	ns	++

1) Effects depend on slope angle and slope aspect respectively.
2) Optimum relationship.
3) Meant is keeping only the banks free from nitrogen supply.

EFFECTS ON VEGETATION

In this paper we will focus on the (distinct) effects on species richness (for effects on other characteristics of floristic richness see Van Strien, 1991). The number of species decreases with increasing nitrogen supply on the parcel (Table 1). The number of species decreases with higher ditch water level (i.e. lower water table). The floristic richness was greater for south-facing banks than north-facing banks, and steep south-facing banks than gentle south-facing banks. The number of species was also larger at a high pH of the bank soil. There appeared to be an optimum relationship between cleaning frequency and number of species (Table 1 and Figure 2). The remaining factors could not be proven to be of importance (but see Melman, 1991 and Melman & Van Strien, 1993 for more detailed information on the effects of type of land use, ditch cleaning method and peat mud dressing).

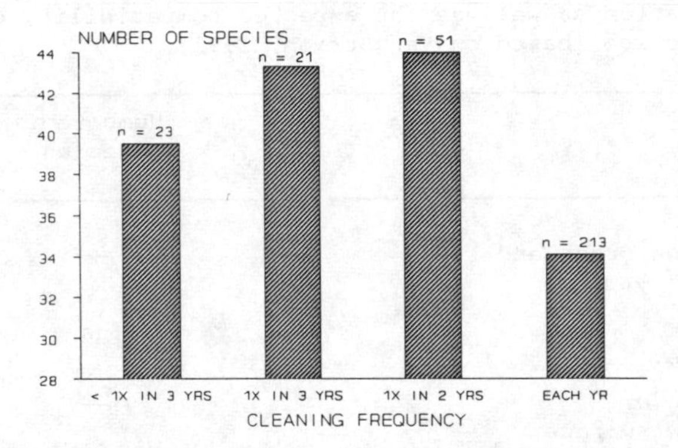

FIGURE 2. Mean number of species on ditch banks in relation to cleaning frequency of ditches (based on Van Strien, 1991)

The relationships above found (with special emphasis on ditch cleaning frequency) will be discussed in relation to aspects that determine the application of measures, i.e. compatibility with farming practice, risk of spreading of noxious weeds and social and political aspects.

COMPATIBILITY WITH FARMING PRACTICE

On the basis of the observed relations, measures can be formulated that may maintain or restore floristic values on the ditch banks. The technical and financial compatibility with farming practice will determine to what extent farmers are able and willing to adopt these measures. This compatibility has been estimated on the basis of knowledge of the farming practice (Table 1). For one of the most promising measures, reducing the cleaning frequency of the ditches, an experiment was performed to study the compatibility more closely (see Melman & van der Linden, 1988 for a similar experiment keeping the banks free from fertilizer).

The most beneficial effect on species richness when lowering the cleaning frequency is the change from once a year to once every two years (Figure 2). The effects of this change on the dairy farming practice were studied in an experiment (Twisk *et al.*, 1991). The effects on the water board tasks (e.g. controlling the water table) were also studied, but will not be discussed here. Eighteen farmers volunteered to skip one regular ditch cleaning in one ditch. The changes in ditch width, depth and filling grade with plants were measured twice a year. In a few ditches the changes in water table (due to rainfall and water discharge) was measured constantly. In addition the time needed to clean the ditch was determined in order to investigate the effects of restricting the ditch cleaning frequency on the amount of labour or costs.

Measurements in combination with model computations showed that water discharge was always sufficient if the ditch was wide enough (>3 m) and deep enough (>30 cm). Most ditches in the peat district exceed this size and are in fact oversized for their functon as drainage channels. One of the reasons for this is that the ditches in former days were used for transportation of e.g. cattle by boat. In some cases the next ditch cleaning took more time, undoing the advantage of only having to clean the ditches once every two years instead of every year. In most cases however, the next ditch cleaning took no more time than usual, giving the farmer the benefit of only having to clean the ditches at half the previous frequency. Considering that an average farm has at least 5 km of ditches, this can mean quite a reduction of labour. So, both farmer and nature could benefit from a lower ditch cleaning frequency.

NOXIOUS WEEDS

The most noxious weeds such as *Cirsium arvense* and *Rumex obtusifolius* rarely occur on the banks, but instead prefer the high-lying parts. Nevertheless the risks of weeds spreading from the banks to the adjacent fields was studied because farmers fear such an effect (Van Strien, 1991). Comparing the amount of weeds on the ditch banks with the floristic richness of the ditch banks (Figure 3) showed that management of the vegetation on the low-lying parts of the banks aimed at species diversity should not increase weed problems, but instead should reduce them.

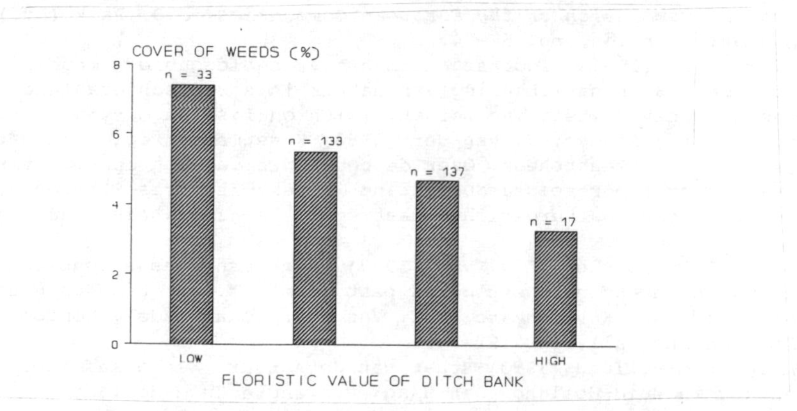

FIGURE 3. The relationship between the summed cover of 12 weed species on the ditch bank and the floristic richness of the bank vegetation (from Van Strien, 1991)

SOCIAL & POLITICAL ASPECTS

The prospects of the ditch bank vegetation do not depend solely on the compatibility of the measures with the technical and economical aspects of farming. There are also psychological and socio-cultural obstacles to vegetation management.

Many Dutch farmers associate vegetation management with poorly productive grasslands, with neglect and with an increase of noxious weeds (Van Strien & Ter Keurs, 1988). However, these opinions mainly arise because farmers are unfamiliar with ditch management. As shown above, for ditch banks none of these associations are correct. Increased information on vegetation management in recent years and practical demonstration of results appear to have had impact, because more and more farmers incorporate this management into their farming practice.

Ditch bank management for conservation purposes could be further promoted if grants are given for the purchase of machines that are necessary for that kind of management and if farmers are rewarded financially for the "production" of species-rich banks on their farms ("nature production payment"). Experiments with this kind of financial stimulation are currently under investigation (see papers from Melman and Kruk *et al.* elsewhere in this issue).

CONCLUSIONS

The results of the above discussed studies prove that farming and nature conservation (in the form of field margin management) can indeed be compatible, as was our assumed starting point. The studies also show that in some cases farming and nature conservation can even benefit from each other.

REFERENCES

Boer, T.F. de; Reyrink, L.A.F. (1988) National Reports. The Netherlands II. In: *Environmental Management in Agriculture, Proceedings Workshop of Commission of the European Communties*, J.R. Park (Ed.), London: Belhaven, pp. 67-74.

Melman, Th.C.P. (1991) Slootkanten in het veenweidegebied; mogelijkheden voor behoud en ontwikkeling van natuur in agrarisch grasland. PhD Thesis, Universiteit van Leiden. (with english summary)

Melman, Th.C.P.; Linden, J. van der (1988) Kunstmeststrooien en natuurgericht slootkantbeheer. Over de betekenis van het opnemen van voorwaarden over perceelsrandbemesting in beheersovereenkomsten als praktische en natuurgerichte maatregel. *Landinrichting*, **28** (1), 37-43.

Melman, Th.C.P.; Strien, A.J. van (1993) Ditch banks as a conservation focus in intensively exploited peat farmland. In: *Landscape Ecology of a Stressed Environment*, C.C. Vos & P. Opdam (Eds), London: Chapman and Hall, pp. 122-141.

Provincie Zuid-Holland (1993) Staat van de natuur. Een graadmeter voor de natuur in Zuid-Holland. Den Haag: Provincie Zuid-Holland.

Strien, A.J. van (1991) Maintenance of plant species diversity on dairy farms. PhD Thesis, Universiteit van Leiden.

Strien, A.J. van; Keurs, W.J. ter (1988) Kansen voor soortenrijke slootkantvegetaties in veenweidegebieden. *Waterschapsbelangen*, **73** (14), 470-478.

Twisk, W; Brussel, N.A. van; Strien, A.J. van; Keurs, W.J. ter (1991) Mogelijkheden voor een lagere slootschoningsfrequentie in veenweidegebieden. Betekenis van een natuurvriendelijke maatregel voor boer en waterschap. *Landinrichting*, **31** (6), 11-17.

VEGETATION PATTERNS AND CHANGES IN FIELD BOUNDARIES AND CONSERVATION HEADLANDS IN SCOTTISH ARABLE FIELDS

D.H.K. DAVIES

Crop Systems, SAC, Bush Estate, Penicuik, Midlothian. EH26 OPH, U.K.

H.M. CARNEGIE

Land Resources, SAC, Craibstone, Aberdeen, AB2 9TQ, U.K.

ABSTRACT

A survey of 50 field margins in Game Conservancy guidelines conservation headland managed fields in eastern Scotland is described. A total of 101 plant species were recorded; of which 87 were in the boundary, 58 within 1m, 36 at 2.5m and 34 at 5m from the boundary. There were fewer species noted in the boundary than in earlier southern England surveys, but the vegetation was generally regarded as being diverse, with tussock and creeping grasses and a range of broad-leaf species. The boundary strip/crop margin had a wide range of annual species. There was some indication of ingress of creeping species from the boundary, in fields that had been managed in this way for more than 1 year; most noticeably *Cirsium arvense*. Spread of another serious weed of arable crops *Elymus repens* was not evident, but it was present, together with the invasive species *Bromus sterilis* and *Galium aparine*, and care should be taken in managing fields to prevent their spread.

INTRODUCTION

Field margins can be an important wildlife habitat on arable farmland. Regular applications of herbicides to arable crops have reduced the population of many arable weeds, and have adversely affected the numbers of associated insect species, many of which are important food sources for game birds and other birds (Sotherton, *et al.* 1991). There is evidence that herbicides could affect the composition of boundary flora (Marshall & Birnie, 1985); with similar consequences. Observations from SAC farm advisory work indicate that many farmers have used non-selective herbicides to eradicate perennial creeping weeds such as *Elymus repens*, so changing the boundary flora leaving a species poor habitat. The Game Conservancy has developed management guidelines for arable field margins, known as conservation headlands, to reduce the impact of crop management on birds by reducing or preventing herbicide and insecticide use in the field margin (Sotherton, 1990). A boundary strip, free of crop is recommended, and the crop margin up to the headland tramline receives no broad-leaf weed herbicide or insecticide during spring or summer.. The work upon which these recommendations were based was undertaken in southern England. The aim of the investigation described here is to examine the development of the conservation headland field margin flora under northern arable conditions, and to advise farmers of weed problems that may arise from such an approach. The data is also perceived to be of assistance in developing potential set-aside policies.

MATERIALS AND METHODS

Fifty arable fields (29 spring barley, 19 winter wheat, 1 winter barley, 1 triticale) from Eastern Scotland, managed to some extent as conservation headlands, were selected for the survey in summer 1993. A representative headland was selected in each field from which a 100 m strip was chosen avoiding field entrances. At 10m intervals, a total of 11 short transects were run from the field boundary into the crop edge. A 0.25m² quadrat was placed in the boundary vegetation at the start of the transect and at 1m, 2.5m and 5m into the crop. A total of 44 quadrats were therefore assessed in each strip. The percent ground cover of each species of higher plant occurring in each quadrat was recorded.

RESULTS

Table 1 lists the species occurring in the survey, comparing occurrence at each sampling point on the transect. The table also includes lists of species seen in the 1992 survey (Carnegie and Davies, 1993), but not found in 1993 to give a complete flora from the series. There were 101 species in 1993, of which 87 were found in the boundary (38 exclusive), 58 at 1m (3 exclusive), 36 at 2.5 m and 34 at 5 m (crop areas), none of which were exclusive to these sampling points. The species with the greatest frequency and overall ground cover are listed in Table 2. The grasses, *E. repens, Dactylis glomerata* and *Holcus mollis* tended to dominate the boundary of the fields, with *Cirsium arvense, Galium aparine, Heracleum sphondylium* and *Urtica dioica* the most frequent broad-leaf species. Other common species were *Poa pratensis, Arrhenatherum elatius, Agrostis stolonifera, Festuca rubra* and *Poa trivialis* plus the annuals *Poa annua* and *Stellaria media.* The field boundary strip tended to have *S. media, Tripleurospermum/Matricaria* spp., *Polygonum aviculare* and *P. annua* present plus a range of both boundary and crop margin species. The boundary species with a spreading habit were present in many boundary strips at moderate levels, notably *C. arvense* and *G. aparine. E. repens* was observed at low density in the crop in this survey. Occasional plants of *H. sphondylium, H. mollis, U. dioica, P. pratensis, P. trivialis* and *R. obtusifolius* were recorded in the crop margin. Comparison was made between fields which had had greater than one year conservation headland management out of three seasons, and those which had only been under such management in 1993 (Table 3). There was little difference between the frequency and ground cover of the listed species, except for a general increase in *R. obtusifolius,* an increase in ground cover, although not incidence, of *C. arvense,* and possibly of *T./Matricaria* spp.

DISCUSSION

A similar number of species were seen in this survey to the 1992 survey of fewer fields (Carnegie and Davies, 1993). There were a few differences noted in the minor species

TABLE 1. Species lists for 1993 field margin survey, and additional species seen in 1992 survey [ ]

Species only in field boundary		*Agrostis stolonifera*	***Species in crop + boundary strip only***	*Holcus mollis*
*Acer pseudoplatanus**	*Stellaria holostea*	*Arrhenatherum elatius*	*Capsella bursa-pastoris*	*Lamium amplexicaule*
*Achillea millefolium**	*Trifolium dubium**	*Cerastium vulgatum*	*Chenopodium album*	*Lamium purpureum*
Alopecurus pratensis	*Trifolium pratense*	*Chamerion angustifolium*	*Euphorbia helioscopia*	*Lapsana communis*
Amsinckia micrantha	*Ulex europaeus*	*Dactylis glomerata*	*Fallopia convolvulus*	*Myosotis arvensis*
Anthriscus sylvestris	*Veronica chamaedrys*	*Equisetum arvense*	*Filaginella uliginosa**	*Papaver rhoeas*
Atriplex patula	*Vicia cracca*	*Festuca pratensis**	*Lycopsis arvensis**	*Phleum pratense*
*Bromus hordeaceus**	*Vicia sativa**	*Holcus lanatus*	*Raphanus raphanistia**	*Poa annua*
*Centaurea nigra**	*Vicia sepium*	*Lolium perenne*	*Solanum tuberosum*	*Poa pratensis*
*Cerastium tomentosum**	*[Agrostis gigantea*	*Mentha arvensis*	*Spergula arvensis*	*Poa trivilis*
Crepis spp.	*Asperula arvensis*	*Senecio jacobaea**	*Viola arvensis*	*Polygonum aviculare*
*Deschampsia cespitosa**	*Carduus personata*	*Stellaria graminae*	*[Volunteer cereals*	*Polygonum persicaria*
Digitalis purpurea	*Festuca ovina*	*Torilis japonica**	*Aphanes arvensis*	*Ranunculus repens*
Festuca rubra	*Filipendula ulmaria*	*Urtica dioica*	*Avena fatua*	*Rumex obtusifolius*
Galium cruciata	*Impatiens sp.*	*Veronica hederifolia*	*Borago officinalis*	*Sinapis arvensis*
Galium verum	*Knautia arvensis*	*[Cardamine hirsuta*	*Polygonum persicaria*	*Stellaria media*
*Geranium dissectum**	*Moenchia erecta*	*Galium saxatile]*	*Sonchus arvensis*	*Taraxacum officinalis*
*Geranium sp.**	*Sarothamnus scoparius*		*Veronica serpyllifolium]*	*Trifolium repens*
*Hydrocotyle vulgaris**	*Senecio vulgaris*	***Species only in boundary strip***	***Species found at all transect positions***	*Tripleurospermum/Matricaria spp.*
Lathyrus pratensis	*Silene latifolia*	*Sonchus oleraceus*	*Aira praecox**	*Veronica persica*
*Lotus corniculatus**	*Symphytum sp.*	*Veronica agrestis**	*Brassica napus olifera*	
Plantago lanceolata	*Taraxacum officinale*	*Urtica urens*	*Bromus sterilis*	
Rosa canina	*Trifolium campestre]*	*[Anchusa arvensis*	*Chamomilla suaveolens**	
Rubus fruticosus		*Glechoma hederacea*	*Cirsium arvense*	
Rubus idaeus	***Species in field boundary + boundary strip (1m)***	*Sisymbrium officinalis]*	*Cirsium vulgare*	
Rumex acetosa	*Aegopodium podagraria*		*Elymus repens*	
Rumex crispus	*Aethusa cynapium*		*Fumaria officinalis*	
*Sambucus nigra**	*Agrostis capillaris*		*Galeopsis speciosa*	
*Silene dioica**			*Galium aparine*	
Sonchus arvensis			*Heracleum sphondylium*	

*** Seen only in 1993 survey***

TABLE 2. Ranking of major species based on % frequency (F) by sites, and % overall ground cover (C), where F= >19% of sites, and/or C = >0.99%, at one or more survey points from the field margin (0, 1, 2.5 or 5m)

	F				C			
Species	0m	1m	2.5m	5m	0m	1m	2.5m	5m
E. repens	90	48	4	2	13.2	2.9	0.1	T
D. glomerata	84	6	-	-	15.0	0.7	-	-
H. mollis	80	28	4	-	19.9	2.3	0.6	-
C. arvense	74	62	14	2	3.7	2.7	0.6	T
G. aparine	72	28	20	10	4.4	1.7	0.6	0.4
H. sphondylium	70	12	-	2	5.9	0.2	-	T
U. dioica	66	4	-	2	3.7	0.2	-	T
P. pratensis	60	8	-	2	3.5	0.4	-	T
A. elatius	56	-	-	-	7.5	-	-	-
P. annua	54	50	90	90	5.5	6.8	6.1	8.5
A. stolonifera	46	10	-	-	4.0	0.3	-	-
F. rubra	42	-	-	-	3.7	-	-	-
S. media	40	88	80	62	1.7	14.2	11.4	4.3
P. trivialis	38	22	6	4	2.4	1.3	0.2	0.1
A. sylvestris	28	-	-	-	1.5	-	-	-
R. obtusifolius	26	18	2	-	1.5	1.1	0.1	-
P. pratense	22	2	-	2	1.1	0.2	-	T
C. angustifolium	16	8	-	-	1.8	0.4	-	-
B. sterilis	8	6	2	2	1.1	0.2	0.1	0.2
V. cracca	26	-	-	-	0.4	-	-	-
C. vulgatum	22	8	-	-	0.4	0.2	-	-
H. lanatus	20	6	-	-	0.9	0.2	-	-
R. repens	20	18	4	2	0.6	0.5	0.1	0.1
Tripl./Matricaria	12	62	58	40	0.3	4.7	3.8	1.8
P. aviculare	2	48	38	38	0.2	2.3	1.7	1.3
G. speciosa	4	34	24	14	0.1	0.8	0.3	0.2
V. arvensis	-	34	28	20	-	0.6	0.6	0.5
M. arvensis	12	28	24	2	0.7	0.6	0.3	T
S. arvensis	-	26	20	18	-	0.5	0.5	0.3
C. bursa-pastoris	-	24	26	16	-	0.5	0.9	0.6
L. purpureum	6	22	16	8	0.3	0.5	0.9	0.1

T: <0.05% ground cover

lists between the surveys, but the major species were common, and tended to show a similar pattern of distribution. The species lists contain many similarities with those of Chancellor & Froud-Williams (1984) and Marshall (1985a; 1985b), who carried our surveys of field margins in southern England, the major difference being the lack of *Alopecurus myosuroides* and *Fallopia convolvulus* in this survey. Marshall (1985a/b) indicates that few species are capable of spreading from the boundary to the crop, but the list of species from this boundary survey includes a number which are known to spread, including *E. repens, C. arvense* and *G. aparine.* These were found in the boundary and crop of a number of fields, and there was some indication from fields that had had more than one year of field margin management with no herbicide use of an increase in ground cover of *C. arvense.* There was less evidence of a general increase in the other species.

TABLE 3. Frequency (F) (% all quadrats) and mean % ground cover (C) of major plant species in 9 field margins having been managed >1 y of 3 y as conservation headlands, compared with 1 y only managed headlands

	>1 y sites		**1 y only**	
	%F	**%C**	**%F**	**%C**
C. arvense	38.9	14.4	37.8	4.4
E. repens	41.7	9.9	34.8	11.5
G. aparine	27.8	4.9	33.5	5.6
H. mollis	36.1	16.6	26.2	21.1
P. annua	86.1	6.9	82.9	8.3
P. aviculare	36.1	2.3	30.5	4.8
R. obtusifolius	30.6	9.5	12.8	1.6
S. media	72.2	13.7	66.5	11.2
Tripl./Matricaria	41.7	11.1	43.3	5.5

In part the lack of major problem of weed spread from boundaries in this survey must be related to a selective choice of boundaries by farmers, and it was evident on some fields that *C. arvense* in headlands had been treated with a clopyralid based herbicide treatment during the early summer to reduce spread. The boundary vegetation in the survey was dominated by tussock grasses such as *D. glomerata, A. elatius* and *Phleum pratense,* mixed with creeping grasses such as *E. repens* and *H. mollis.* There was also a range of perennial broad-leaf species, in particular *C. arvense, H. sphondylium* and *U. dioica,* with *G. aparine*. This canopy limited the development of a wide range of other plant species, and the average number recorded per field boundary was 15.2. This compares with an average of 24.6 from 17 fields surveyed at two farms in southern England by Marshall (1985a). However, the dense vegetation was considered to provide good quality wildlife habitat.

A range of weed species have been identified by Sotherton et. al. (1985) as to be important hosts to insect fauna which are a food source, along with the weeds themselves, for gamebirds and songbirds. The data is mostly in association form, with the removal of weeds leading to reductions in fauna rather than linking specific species with specific fauna. Nevertheless, there are indications that *P. aviculare* and related species, and *Tripleurospermum/Matricaria* spp. are important examples. These are common species in this survey, present throughout the transects. The boundary strip points contained 58 species in total, with no one species dominating, although *S. media* and *P. annua* were very common, indicating the availability of a large food resource.

In the absence of weeds such as *B. sterilis* and *G. aparine*, the effect of weed cover on crop yield from the headland has been shown to be small (Fisher, et al., 1988), Nevertheless, Davies and Whiting (1990) have shown that *P. aviculare* and *S. media* can affect harvesting and *Lolium perenne* can contaminate grain. Carnegie and Davies (1993) listed some of these species that have shown similar effects in the literature, and others that they considered have an equal effect from advisory experience. A number of these species are also difficult to control in other parts of the rotation; notably *C. arvense*.

It is concluded that these conservation headlands show a wide diversity of plant species, which although not as diverse as some southern field margins, could still be a major wildlife habitat resource. The sites were selected to avoid severe *B. sterilis, E. repens* and

G. aparine problems, but where these species are present, care should be taken to avoid spread into crops. The fields where conservation headlands had been repeated did not show large increases in these species, but rotation of the headlands should be encouraged to reduce the risk. There is some evidence, however, that *C. arvense* may increase, and herbicide options may eventually be required to maintain control.

ACKNOWLEDGEMENTS

N.M. Fisher and G.W. Wilson are thanked for help in data handling, the Game Conservancy for assisting in finding sites, and the farmers, estate owners and factors for allowing us access.

REFERENCES

Carnegie, H.M. and Davies, D.H.K. (1993) A survey of the vegetation of field boundaries and conservation headlands in three arable areas of Scotland. *Proceedings Crop Protection in Northern Britain 1993,* 217-222.

Chancellor, R.J. and Froud-Williams, R.J. (1984) A second survey of cereal weeds in central southern England. *Weed Research* 24, 29-36.

Davies, D.H.K.; Whiting, A.J. (1990). Effect of reducing herbicide dose on weed growth and crop safety in cereals and consequences for grain quality and harvesting. *Proc. European Weed Research Society Symposium 1990 - Integrated Weed Management in Cereals,* 331-338.

Fisher, N.M.; Davies, D.H.K.; Richards, M.C. (1988) Weed severity and crop losses in conservation headlands in S.E. Scotland. *Aspects of Applied Biology 18,* 37-46.

Marshall, E.J.P. (1985a) Weed distributions association with cereal field edges - some preliminary observations. *Aspects of Applied biology 9, 1985, The Biology and Control of Weeds in Cereals,* 49-58.

Marshall, E.J.P. (1985b) Field and field edge floras under different herbicide regimes at the Boxworth E.H.F. - initial studies. *Proceedings 1985 British Crop Protection Conference - Weeds,* 999-1006.

Marshall, E.J.P.; Birnie, J.E. (1985) Herbicide effects on field margin flora. *Proceedings 1985 British Crop Protection Conference - Weeds,* 1021-1028.

Sotherton, N.W. (1990) Conservation headlands: a practical combination of intensive cereal farming and conservation. In: *The Ecology of Temperate Cereal Fields,* 32 Symposium of the British Ecological Society with the Association of Applied Biologists, 373-397.

Sotherton, N.W.; Rands, M.R.W.; Moreby, S.J. (1985) Comparison of herbicide treated and untreated headlands on the survival of game and wildlife. *Proceedings 1985 British Crop Protection Conference - Weeds,* 991-998.

BOUNDARY STRIPS IN CEREAL FIELDS: DYNAMICS OF FLORA, WEED INGRESS AND IMPLICATIONS FOR CROP YIELD UNDER DIFFERENT STRIP MANAGEMENT REGIMES

T.P. MILSOM

Central Science Laboratory, Ministry of Agriculture, Fisheries & Food, Tangley Place, Worplesdon, Surrey, GU3 3LQ.

D. TURLEY, P. LANE

ADAS High Mowthorpe, Duggleby, Malton, North Yorkshire YO17 8BP

B. WRIGHT, S.J. DONAGHY, P. MOODIE

ADAS Leeds, Block 2, Government Buildings, Otley Road, Lawnswood, Leeds, LS16 5PY

ABSTRACT

An experiment was carried out to evaluate the effects, over five years, of three types of uncropped boundary strip on weed ingress and cereal yield. The treatments (sown perennial ryegrass sward, rotovated strip and sterile strip receiving regular herbicide treatments) and a control (winter wheat) were arranged along the margin of a cereal field using a randomised block design with four replicates.

Four weed species, characteristic of field margins (*Galium aparine*, *Bromus terilis*, *Elymus repens* & *Poa trivialis*), were used as indicators of weed ingress. All were scarce initially, but became more abundant as the experiment progressed. The boundary strips influenced the rate of weed spread but did not halt it. Yields of winter wheat in the crop margin decreased considerably as weed populations increased. The association between yield loss and weed ingress is discussed.

INTRODUCTION

Field margins are important refuges or corridors for wildlife on intensively managed arable farms but are often perceived by farmers as reservoirs of pernicious weeds and are sprayed accordingly with broad-spectrum herbicides. However, repeated applications of non-selective herbicides, combined with misplaced fertiliser applications, may exacerbate the weed problem over the long-term by favouring competitive annuals, such as *Galium aparine* (Cleavers) and *Bromus sterilis* (Barren Brome), over less invasive perennials (Boatman 1992). Uncropped boundary strips, which receive no fertilisers and have modified herbicide programmes, show promise as weed barriers (Lainsbury *et al* 1992, Rew *et al* 1992) and may have additional environmental benefits by encouraging perennial herbaceous species which comprise suitable habitat for valuable insects and gamebirds (Boatman 1992). However, the efficacy of this alternative strategy for weed management has yet to be demonstrated fully (Marshall 1988).

This paper describes the effects, over the medium-term, of boundary strip management on the dynamics of four weed species: *Galium aparine*, *Bromus sterilis*, *Poa trivialis* (Rough Meadow-grass) and *Elymus repens* (Couch-grass). All show distribution patterns which imply that they originate from field margins (Marshall 1989). They are, therefore, good indicators for evaluating the effectiveness of boundary strips as weed barriers.

METHODS

A boundary strip, two metres wide, was maintained over five growing seasons (1988-1992) along the margin of a cereal field at ADAS High Mowthorpe. Three treatments plus a control were applied to 20 metre sections of the strip in a randomised block design with four replicates. The treatments were as follows: (i) a *Lolium perenne* (Perennial Rye-grass) sward, sown in 1988 and mown twice a year to prevent seeding, (ii) a bare strip rotavated twice a year, (iii) a sterile strip maintained by annual applications of a residual herbicide (propyzamide) in winter and a foliar systemic herbicide (glyphosate) during summer. The control plots contained winter wheat which received the same pesticide and fertiliser treatments as the main cropped area, where winter wheat was grown throughout.

Yield assessments were made annually within the crop margin and were matched to the plots in the boundary strip. From 1990, a 6m band of the crop margin adjacent to the boundary strip was not sprayed with herbicides to allow weed ingress into the crop.

Weed assessments were made annually in July from fixed transects, three per plot. Each transect line lay at 90 degrees to the field boundary and comprised 50 contiguous 10*10cm quadrats, numbered sequentially from 1 at the mid-line of the boundary to 50 in the crop margin. The presence or absence of each weed species was noted in each quadrat.

Weed ingress was assessed by comparing the distribution and abundance of each species between years. The farthest occupied quadrat from the mid-line of the hedge was defined as the weed front. Its position in each plot was taken as the median from the three transect lines. The number of occupied quadrats from the three transects, expressed as a proportion of the total number available, was used as an index of relative abundance in the strip and crop margins of each plot. Proportions were used because the number of quadrats in the strip and crop margin varied slightly between years due to shifts in the plough line that separated the two zones.

The analyses focussed upon trends in the position of the weed front and relative abundance, and whether they differed between treatments. Trends were quantified by subtracting values for a given year of the experiment from those for 1988. These differences were used as dependent variables when assessing treatment effects by analysis of variance.

RESULTS

The weed fronts of all four species moved out from the boundary during the course of the experiment but at different rates (Fig. 1). *Galium*, *Bromus*, and *Elymus* were restricted to the boundary in 1988. *Galium* had reached the crop margin by 1990, whereas the other two did not do so until 1991. In contrast, the weed front of *Poa* lay near the outer margin of the boundary strip in 1988 and had moved well beyond it by 1989.

The rate of movement of the weed fronts of *Bromus* and *Elymus*, differed between treatments in 1990 (ANOVA, $F_{3,12}$=5.84; P=0.011) and 1992 (ANOVA, $F_{3,12}$=6.65; P=0.007) respectively but not in other years (Fig. 1). There were no significant differences between treatments in the rate of spread of either *Galium* or *Poa*.

The relative abundance of each species in the boundary strip increased between 1988 and 1991, as their weed fronts advanced, but declined between 1991 and 1992 (Fig. 2). There were few significant treatment effects on trends in weed abundance in the boundary strips and effects were not consistent between species. In the case of *Galium*, the rate of increase in abundance between 1988 and 1990 differed between treatments (ANOVA, $F_{3,12}$=7.53; P=0.004) but peak abundance and the rate of decrease did not (Fig. 3). In contrast, *Poa* abundance differed between treatments in 1991, the peak year (ANOVA, $F_{3,12}$=5.45; P=0.014), but not when the species was increasing or decreasing. The rate of increase of *Elymus* between 1988 and 1992 differed between treatments (ANOVA, $F_{3,12}$=8.507; P=0.003) but no treatment effects were detected in other years. Trends in *Bromus* abundance were not related to boundary strip treatments in any year.

All species became more abundant in the crop margin, in step with the increases in the boundary strip (Fig. 2). In general, however, the abundance of each species in the boundary strip plots did not correspond with that in adjacent sections of the crop margin. The exception was *Elymus*, in 1992, when the proportion of occupied quadrats in strip and crop respectively were positively correlated (r = +0.686, $F_{1,14}$=12.413, P=0.003).

The mean annual wheat yield decreased markedly as the combined abundance of the four weed species in the crop margin increased but it recovered slightly in the last year of the experiment when weed abundance declined (Fig. 4). This inverse relationship was explored further by using multiple regression analysis to fit a linear model of weed abundance to the yield data from all crop margin sections in all years. The proportion of quadrats in each crop margin section occupied by the four weed species accounted for about 55% of the variance in yield. When year factors were entered as dummy variables, the percentage of variance in yield explained by the model increased to just over 75% (R^2=0.753, $F_{4,71}$=58.30, P<0.001).

The relationship between yield depression from 1988 levels and increase in weed abundance was of particular interest but the two variables were not correlated in either 1990, 1991 or 1992. However, in 1991, yield depression did differ between sections of the crop margin according to the treatments applied to adjacent plots in the boundary strip (ANOVA $F_{3,8}$=6.003, P=0.019; Table 1).

TABLE 1: Depression in winter wheat yield (t/ha) in crop margin between year shown and the 1988 reference levels, in relation to boundary strip treatment.

Year	Adjacent treatments							
	Mown strip		Rotivated strip		Sterile Strip		Control	
	mean SE [1]	n	mean SE	n	mean SE	n	mean SE	n
	Reference yields							
1988	7.02±0.39	4	7.19±0.31	4	6.69±0.23	4	6.82±0.31	4
	Yield depression							
1990	3.97±0.46	4	4.59±0.22	4	2.95±0.39	4	2.98±0.89	4
1991[2]	3.98±0.43	3	4.87±0.39	3	3.40±0.38	3	2.45±0.46	3
1992	2.92±0.86	4	2.46±0.82	4	2.09±0.78	4	2.42±0.91	4

Notes: 1. SE = standard error of the mean. 2. One block of replicates was deleted from analysis because of extensive damage due to wheel rutting.

DISCUSSION

In all cases, where trends in weed abundance differed between treatments, *a posteriori*, comparisons of the treatment means suggested that increases or peaks in abundance were greater in the winter wheat control plots than in the boundary strips. Analagous comparisons of the weed front data showed that the rate of spread was faster in the cereal control plots than in the strip treatments. These findings imply that the boundary strips partially regulated the spread of the four weed species though they were not able to halt it. There was no evidence that a particular treatment performed better than the others.

The lack of correspondence between the weed distributions in the boundary strip and crop margin respectively suggests that the species did not simply advance across the strip and into the crop margin. Indeed, there was some evidence to suggest that, in some plots in the boundary strip, the weeds had reached the crop margin at an early stage of the experiment and then spread laterally along the crop margin which had not been sprayed with herbicides. This lateral spread may well have been facilitated by cultivation, particularly in the case of *Elymus repens*.

The depression of wheat yield in the middle years of the experiment was considerable and coincided with the removal of herbicide treatments on the crop margin and the build up of weed populations from the hedge and boundary strip. As the depression of yield ran counter to trends elsewhere on the Research Centre, it seems likely that it was due to conditions specific to the experimental site. Though the weed species considered in this paper may have been partially responsible, the poor fit of the regression model and the lack of a correlation between yield depression and weed abundance suggest that other factors, perhaps other weed species, played a significant role. The association between yield depression in 1991 and the boundary strip treatments is puzzling, particularly as the yield depression was smallest adjacent to the winter wheat control plots which were least effective as weed barriers.

ACKNOWLEDGEMENTS

The work was funded by the Land Use Conservation and Countryside policy group of MAFF. Tim Marczylo and Rachel Chivers (CSL) processed the field data. Steve Langton (CSL) advised on statistical matters. John Marshall (IACR) gave valuable advice at the planning stage. Chris Feare, Roger Trout (CSL) and Carey Coombs (ADAS) co-ordinated the project and commented on the paper. We are very grateful to them all.

REFERENCES

Boatman, N.D. (1992): Improvement of field margin habitat by selective control of annual weeds. Aspects of Applied Biology, **29**, 431-436.

Lainsbury, M.A., Cornford, P.A. and Boatman, N.D. (1992). The use of quinmerac to control *Galium aparine* in field boundaries. Aspects of Applied Biology, **29**, 437-442.

Marshall, E.J.P. (1988). The ecology and management of field margin floras in England. Outlook on Agriculture, **17**, 178-182.

Marshall, E.J.P. (1989). Distribution patterns of plants associated with arable field edges. Journal of Applied Ecology, **26**, 247-257.

Rew, L.J., Froud-Williams, R.J. and Boatman, N.D. (1992). Implications of field margin management on the ecology of *Bromus sterilis*. Aspects of Applied Biology, **29**, 257-263.

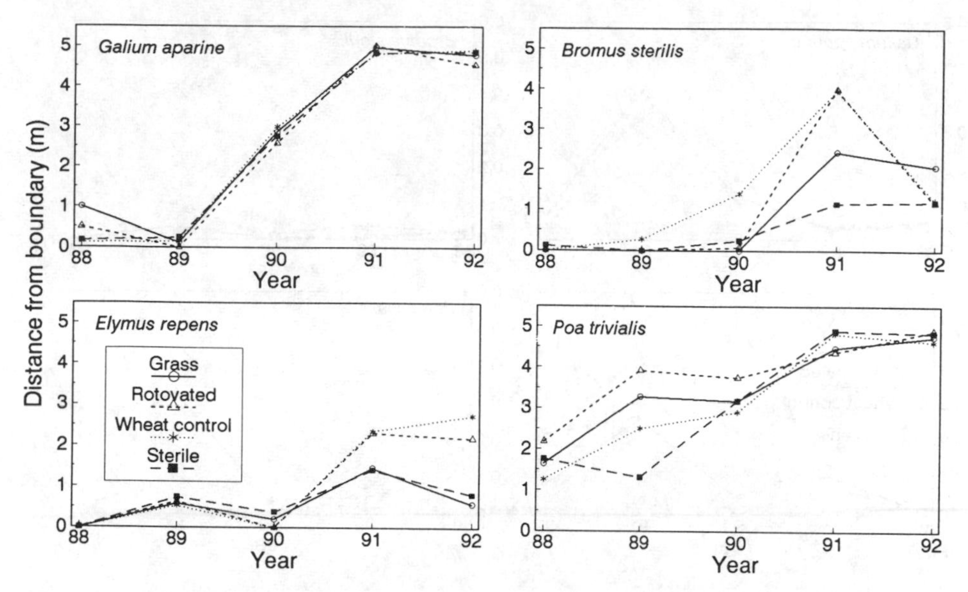

Fig. 1. Trends in the distance of the weed front from the mid-line of the hedge in relation to boundary strip treatment.

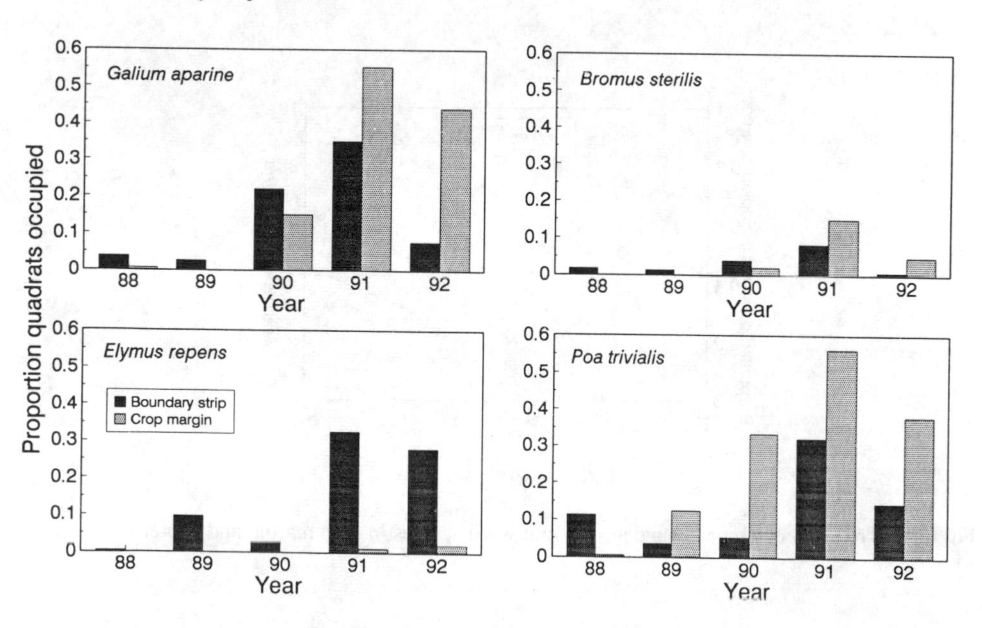

Fig. 2. Trends in overall weed abundance in boundary strip and crop margin.

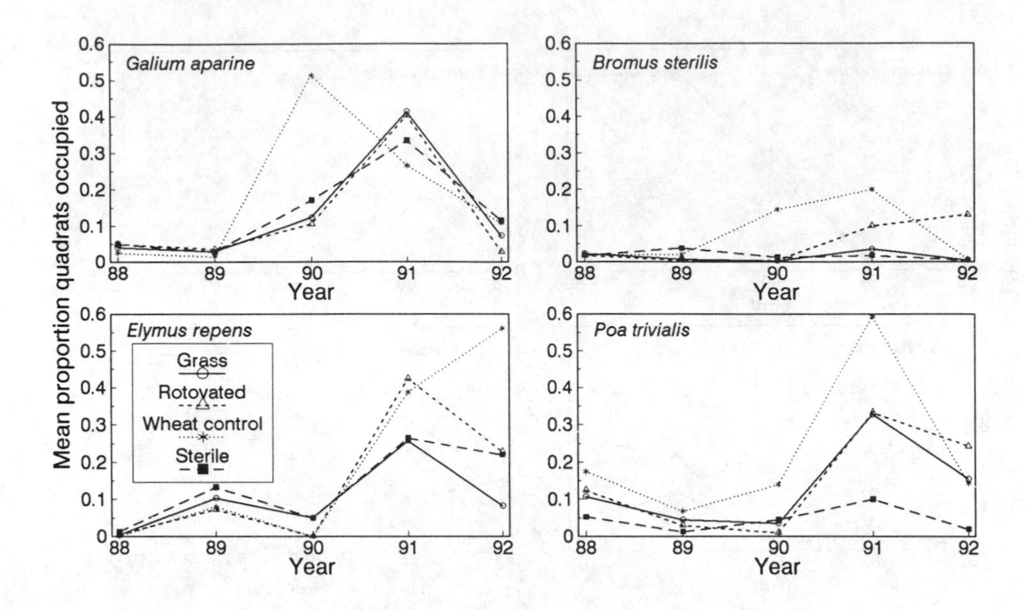

Fig. 3. Trends in abundance of weeds in boundary strip in relation to treatment.

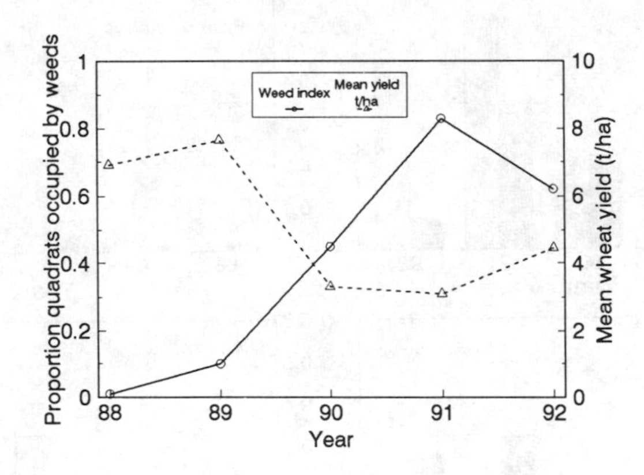

Fig. 4. Trends in combined abundance of four weed species in crop margin and winter wheat yields.

LARGE-SCALE FIELD TRIALS WITH CONSERVATION HEADLANDS IN SWEDEN

P.A. CHIVERTON

Department of Plant and Forest Protection, Swedish University of Agricultural Sciences, Box 7044, S - 750 07, Uppsala, Sweden.

ABSTRACT

The Conservation Headland technique, pioneered by The Game Conservancy U.K., has been tested in Sweden in large-scale field trials during 1991 - 1993. Ten pairs of farms in central and southern Sweden were chosen. One farm in each pair was sprayed normally, whilst on the other farm cereal field headlands received no pesticides. Unsprayed headlands had significantly more weed species, higher weed densities and a higher percentage weed cover. Unsprayed headlands supported higher densities of non-target arthropods, particularly the non-pest species which are important in the diet of insect-eating game-bird chicks. Mean brood sizes and chick survival rates of both partridges and pheasants were higher on farms with unsprayed headlands. The benefits of unsprayed headlands to the fauna associated with arable land are discussed.

INTRODUCTION

In 1986, the Swedish government launched a plan of action aimed at reducing the amounts of pesticides used in agriculture by 50% over a period of five years from 1986 - 1991 (Bernson, 1988). Following this a new target was set to reduce the 1991 level by a further 50% during the period 1991 - 1995 (Bernson & Ekström, 1991). As part of this latest programme, the Ministry of Agriculture commissioned research aimed at protecting the flora and fauna in agricultural land, including margins between cultivated and uncultivated areas (Jonsson, 1991).

Work was already in progress at our department testing the technique known as Conservation Headlands pioneered by The Game Conservancy in the U.K. (Chiverton, 1991). This involves the modification of pesticide use on cereal crop margins to encourage the growth of certain broadleaved weeds and their associated insect faunas (Sotherton et al., 1985).

The method was originally designed to improve survival of game-bird chicks, particularly those of the Grey Partridge (*Perdix perdix*). This species is associated with arable field margins and has declined dramatically since the 1950's both in England (Potts, 1986) and Sweden (Dahlgren, 1987). Partridge chicks are dependent for food on insects, many of which live on arable weeds (Potts, 1986). The use of broad-spectrum herbicides can indirectly reduce food-item insects by removing the host plants on which they depend. Large-scale field trials were started in 1991 to examine the benefits of Conservation Headlands to the flora and fauna in areas under intensive agriculture in central and southern Sweden. This paper describes these trials and presents some of the results obtained to date. The benefits of this method to other wildlife are discussed.

METHODS AND MATERIALS

Study sites.

In 1991, ten pairs of farms in central and southern Sweden were chosen for their similarity in size (mean = 107.1 ha, S.E. 8.38), cropping and agricultural practice. Paired farms were a minimum of 5 km apart and all had previously reported resident populations of partridges. All cereal fields on one farm in each pair ("control" farms) were sprayed normally (i.e. c. 100% of fields treated with herbicides; insecticides and fungicides usually after threshold levels are exceeded), whilst on the other farm ("experimental" farms) Conservation Headlands were employed. In contrast to the Game Conservancy's guide-lines for Conservation Headlands which allow the use of certain preparations at certain times, no pesticides are used at any time in our Conservation (unsprayed) Headlands. During routine applications of herbicides and other pesticides in cereal fields, spray nozzles on the outer boom of tractor mounted sprayers on the experimental farms were turned off so that the outer 6 m of crop on one half of the headlands in each field received no pesticides. In 1992, where crop rotation allowed, the headlands in the opposite half received no pesticide. In other cases, headlands on adjacent fields with cereals were used. In 1993 a similar 'rotation' occurred. This within-field, between-year rotation allowed farmers on experimental farms to treat the excess of weeds in the year following an unsprayed headland.

On individual experimental farms the actual positioning of the unsprayed headlands was determined by the location of pairs of partridges found during the spring survey (see below).

Partridge surveys.

In spring and after harvest each year, partridge counts were done using highly trained gundogs to flush the birds. The spring count established the number of pairs per farm and their locations and, in the case of the experimental farms, where to position the unsprayed headlands. On each farm, the number of cock and hen pheasants flushed in spring, and the size of pheasant broods flushed in autumn was also recorded. Autumn brood counts were conducted to estimate the productivity of game-birds on both experimental and control farms.

Weed assessments.

Weeds were assessed in the sprayed and unsprayed headlands on experimental farms and on sprayed headlands in corresponding cereal crops on control farms. Within each headland plot assessments of weed density, number of species and weed cover (on the Domin scale) were made in ten x 0.25 m^2 quadrats during the first weeks of July each year. Analysis of weed cover was conducted on an arcsine transformation of the percentage cover corresponding to the Domin measurements (Currall, 1987). Only data concerning weed cover will be presented here.

Arthropod assessments.

Vacuum-suction (D-vac) samples were taken from all headland plots at the same time as the weed assessments. In each plot ten samples of 0.5 m^2 were taken to extract the small, diurnal epigeal fauna and the crop/weed fauna. Some of the group or guilds of arthropods selected for analysis include a) Polyphagous predators e.g. Carabidae, Staphylinidae and Lycosid and Liniphiid spiders. b) Aphid specific predators e.g. adults and larvae of Coccinellidae, and the larvae of Neuroptera and Syrphidae. c) Chick food insects: Heteroptera and Homoptera (except Aphididae), Curculionidae, Chrysomelidae, and larvae of both Lepidoptera and Tenthredinidae. Only the results concerning group c) for 1991 - 1993 will be presented here.

RESULTS

Weed assessments.

Percentage weed cover was significantly greater on unsprayed headland plots in each cereal crop (winter- or spring-sown) and in each year (Fig. 1). For data regarding weed densities, and number of weed species see Chiverton (1993).

Arthropod assessments.

Two to four-fold increases in mean densities of chick food insect groups were observed in unsprayed headlands of winter and spring sown cereals on experimental farms compared to densities on sprayed headlands on experimental and control farms in each year (Fig. 2). The majority of these differences were statistically significant.

Partridge surveys.

A total of 21 pairs of partridges were found on experimental farms, and 15 pairs on control farms in 1991. Mean brood size on experimental farms in autumn in 1991 was found to be twice as large as the mean brood size on the control farms (Table 1). This difference was however not statistically significant (t_{12} = 1.17, n.s.). Chick survival rates (CSR = 3.665 $x^{1.293}$, where x is the geometric mean brood size (Potts, 1986)) were doubled where chicks had access to unsprayed headlands (Table 1). During the spring survey in 1992 a total of 20 pairs were found on the experimental farms and 19 pairs on the control farms respectively. Mean brood sizes were again found to be larger on experimental farms (Table 1) although these differences were not statistically significant (t_{19} = 1.33, n.s.). Chick survival rates, generally higher than the previous year, were larger on the experimental farms (Table 1). A total of 30 pairs were found on experimental farms and 23.5 pairs on control farms during the spring surveys in 1993. Autumn brood sizes were again larger on experimental farms but these differences were not statistically significant (t_{33} = 0.65, n.s.). Chick survival rates were high generally, but highest on experimental farms.

Pheasant brood sizes on experimental farms were almost double those on control farms in 1991, but these differences were not significant (t_{11} = 1.23, n.s.). Pheasant chick survival rates (CSR = 3.665 $(1.5x)^{1.293}$, (Sotherton, pers. comm.)) were correspondingly higher (Table 2). In 1992, there was little difference in pheasant brood size (t_{14} = 0.03, n.s.) and corresponding survival rates (Table 2). Differences in brood size and chick survival in 1993 were similar to those in 1991, though not significant (t_{22} = 0.76, n.s.) (Table 2).

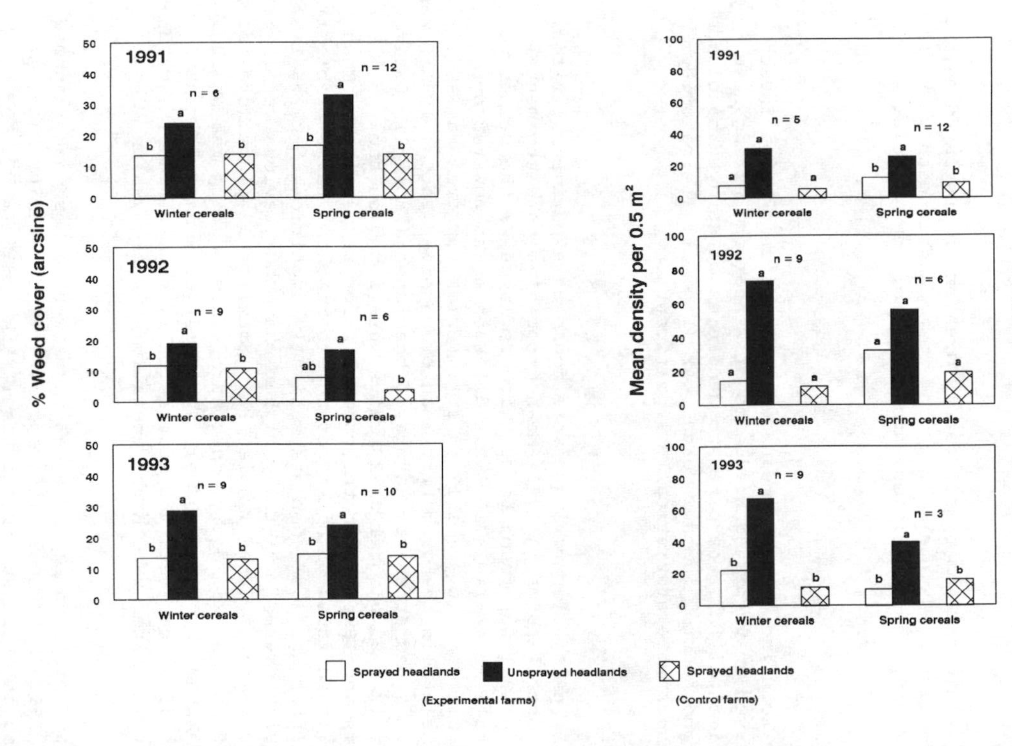

Fig. 1. Estimate of weed cover per 0.25 m^2 in headland plots of winter- and spring-sown cereals either treated with herbicide or remaining untreated, 1991 - 1993.

Fig. 2. Mean densities of insects which are important chickfood items in headland plots of winter- and spring-sown cereals either treated with herbicide or remaining untreated, 1991 - 1993.

(Different letters = $p < 0.05$, SNK; n = no. of farm pairs)

TABLE 1. Mean grey partridge brood sizes and chick survival rates on Swedish farms with unsprayed cereal field headlands and on farms where headlands were fully sprayed, August 1991, 1992 and 1993.

	Unsprayed headlands			Fully sprayed headlands		
	1991	1992	1993	1991	1992	1993
Mean brood size	6.6	8.4	8.9	3.3	6.0	8.0
(S.E.)	(1.5)	(1.1)	(1.1)	(1.4)	(0.9)	(0.8)
Number of broods	10	15	19	4	6	16
% chick survival rate	26.3	47.8	53.6	10.8	34.6	46.7

TABLE 2. Mean pheasant brood sizes and chick survival rates on Swedish farms with unsprayed cereal field headlands and on farms where headlands were fully sprayed, August 1991, 1992 and 1993.

	Unsprayed headlands			Fully sprayed headlands		
	1991	1992	1993	1991	1992	1993
Mean brood size	5.3	4.1	5.0	2.8	4.1	4.1
(S.E.)	(1.2)	(0.6)	(0.9)	(0.6)	(0.9)	(0.8)
Number of broods	7	9	11	6	7	13
% chick survival rate	38.7	34.4	37.1	20.2	31.4	25.7

DISCUSSION

In 1991 the weather during the peak hatch of partridges (last week in June) was very poor with more than double the rainfall, and much lower temperatures compared to 30-year averages (1961 - 1990). Despite this, unsprayed headlands supported significantly greater densities of weeds and, consequently, greater densities of the groups of insects that are vital for game-bird chick survival (Green, 1984). As a result Partridge and Pheasant chick survival rates were higher on experimental farms employing unsprayed headlands.

By contrast, in 1992 many areas in southern Sweden received little or no rainfall throughout the summer months of May, June and the first weeks of July, and temperatures during this period were well above the 30-year average. Several unsprayed headlands in spring sown cereal crops in southern and south eastern Sweden were destroyed by the drought. This probably explains the lack of significant effects of the herbicide treatments (see Chiverton, 1993).

Nevertheless, the trends were similar to those in 1991 with greater densities of both weeds and insects in the unsprayed headlands. Partridge chick survival rates were correspondingly higher on experimental farms employing unsprayed headlands. Chick survival rates for pheasants were however similar on both experimental and control farms. In 1993 favourable weather during, and immediately after, the main period of chick hatch resulted in comparatively high chick survival rates - particularly on the Baltic islands of Gotland and Öland. As in the two previous years however survival rates and brood sizes were higher on farms with unsprayed headlands.

The above results confirm earlier investigations demonstrating the value of unsprayed cereal field headlands for the survival of partridges (see e.g. Rands, 1985; Rands & Sotherton, 1987).

Current work in Sweden is aimed at establishing the agronomic consequences of omitting pesticide treatments on cereal headlands. Preliminary results from 1991 regarding grain yield and quality have shown no significant differences in spring sown cereals, but significant yield reductions in unsprayed headlands of winter wheat. It is anticipated however that Swedish farmers will find that unsprayed headlands offer a practical, realistic and effective method of protecting their game and benefiting other wildlife, at a cost which they find acceptable.

ACKNOWLEDGEMENTS

I thank Lars Johansson for assistance in the laboratory and field; the project is supported by grants from the Swedish National Board of Agriculture.

REFERENCES

Bernson, V. (1988) Regulation of pesticides in Sweden. *Proceedings 1988 British Crop Protection Conference - Pests and Diseases*, 3 , 1059-1064

Bernson, V.; Ekström, G. (1991) Swedish policy to reduce pesticide use. *Pesticide Regulation*, 33 - 36

Chiverton, P.A. (1991) Methods to encourage wildlife and beneficial insects in modern agriculture. (in English) In: *Global resurshushållning - Konsekvenser för svenskt jordbruk*. S.U.A.S., Uppsala. pp 87 - 93

Chiverton, P.A. (1993) Large scale field trials with Conservation Headlands in Sweden. *Proceedings of the Crop Protection in Northern Britain Conference*, Dundee, 207 - 215

Currall, J.E.P. (1987) A transformation of the Domin scale. *Vegetatio*, 72 , 81 - 87

Dahlgren, J. (1987) Partridge activity, growth and survival : Dependence on insect abundance. *Ph. D. Thesis*. University of Lund, Sweden.

Green, R.E. (1984) The feeding ecology and survival of partridge chicks (*Alectoris rufa* and *Perdix perdix*) on arable farmland in East Anglia *Journal of Applied Ecology*, 21, 817 - 830

Jonsson, E. (1991) Behov och prioritering av framtida växtskyddsinformation samt utredning om begränsning av risker vid kemiskbekämpning (in Swedish with English summary) *32 nd Swedish Plant Protection Conference - Pests and Diseases*. Uppsala, 57 - 66

Potts, G.R. (1986) The Partridge : Pesticides, Predation and Conservation. Collins. London

Rands, M.R.W. (1985) Pesticide use on cereals and the survival of grey partridge chicks : a field experiment. *Journal of Applied Ecology*, 22 49 - 54

Rands, M.R.W.; Sotherton, N.W. (1987) The management of field margins for the conservation of game-birds. *1987 BCPC Monograph No. 35 : Field Margins*, pp 95 - 104

Sotherton, N.W.; Rands, M.R.W.; Moreby, S.J. (1985) Comparison of herbicide treated and untreated headlands on the survival of game and wildlife. *Proceedings 1985 BCPC - Weeds*, 3 , 991 - 998

ADOPTION OF CONSERVATION HEADLANDS TO FINNISH FARMING

JUHA HELENIUS

Department of Applied Zoology, P.O.Box 27, FIN-00014 University of Helsinki, Finland

ABSTRACT

Changes in Finnish agricultural practices, including the reduction in pasture area and increase in autumn ploughing, shift to subsurface drainage from open field ditches, and introduction of pesticides have in recent decades altered agricultural biotopes of naturally occuring plants and animals. As a consequence, 25 animal and 14 plant species are listed as endangered. Weed densities decreased by 70% in less than 25 years. Conservation Headlands, i.e. selectively sprayed crop margins, have been studied since 1992 and increased weed and insect densities are reported already.

INTRODUCTION

Finnish agricultural landscapes have undergone similar changes during past decades as other agricultural areas under modern mechanized production. Consequences to wildlife are not well known and management recommendations to farmers for protection of biotopes of naturally occurring plants and animals are either too general or lacking. Conservation Headlands have been shown to offer resources and create habitats for a wider diversity of species than cereal farming would otherwise support (Sotherton, 1991). The aim of the Finnish Conservation Headland study is to assess the suitability of the technique to Finnish spring cereals, and to arable crop production in general.

AGRICULTURAL LAND USE AND FIELD MARGINS

One of the most important changes in Finnish agriculture was the introduction of pesticides some forty years ago. In 1953, agriculture used 235 t AI of pesticides, out of which 72% was herbicides and 16% insecticides (Markkula *et al.*, 1990). In spite of the increase in efficiency, the 1429 t AI sold in 1992 (65% herbicides and 6% insecticides) was six times more than forty years earlier (Hynninen & Blomqvist, 1993). The peak use was 2331 t AI in 1980 (Markkula *et al.*, 1990). The UK use in 1992 was 23,800 t AI.

In 1992 herbicides were used on 0.63 million ha which corresponds to 69% of the area under cereals. During the last thirty years most of the volume (58.5% in 1992) has been phenoxy acids. Insecticides were applied to 0.28 million ha corresponding to 11% of arable land; 48% of the volume was dimethoate. An obvious explanation for the decline in abundance of arable weeds in Finnish cereals is the regular use of herbicides. Total weed density decreased by 70% in less than 25 years, and as an example, the average decline in species that provide prreferred seed food of grey partridge was 29% in terms of frequency and 64% in terms of density (Table 1).

In parallel to introduction of pesticides, two other major forces have reshaped the

TABLE 1. Decline in frequency (% of fields in the survey from which the species was found in the sample) of some common weed species, and decline in density of these and of all species combined in Finnish spring cereal fields in two decades.

Taxon	Frequency %		Density, nos. m^{-2}	
	1962-1964*	1982-1986**	1962-1964	1982-1986
All species	.	.	550	164
Chenopodium album	92	55	60	12
Fallopia convolvulus	59	42	3	2
Galeopsis sp.	94	72	67	11
Polygonum aviculare	42	33	3	1
P. lapathifolium	73	27	16	3
Stellaria media	85	72	83	16

*Mukula *et al.* (1969) (N = 2548 fields)
**Erviö & Salonen (1987) (N = 267 fields) and Mela (1988) (N = 166 fields) combined.

TABLE 2. 'Species of primarily agricultural habitats' listed in the RED BOOK (Rassi *et al.*, 1991) as extinct or endangered, 'primarily because of' changes in agricultural habitats.

BIRDS:	VASCULAR PLANTS:
Crex crex, endangered	*Agrostemma githago*, disappeared
Perdix perdix, declined	*Camelina alyssum*, disappeared
	Cuscuta epilinum, disappeared
LEPIDOPTERA:	*Lolium remotum*, disappeared
Ochsenheimeria taurella, disappeared	*Papaver dubium*, disappeared
Agonopterix laterella, endangered	*Spergula arvensis maxima*, disappeared
	Bromus secalinus, endangered
COLEOPTERA:	*Consolida regalis regalis*, endangered
Aclypea undata, declined	*Fumaria vaillantii*, endangered
Ceutorhynchus pallidactylus, declined	*Odontites verna*, endangered
Longitarsus parvulus, declined	*Vicia villosa*, endangered
Aphthona euphorbiae, poorly known	*Lithospermum arvense*, declined

agricultural landscape. Traditional drainage system is a dense network of open field ditches, which has to a large extent been replaced by subsurface drainage. In the main agricultural area, the proportion of fields that had ditches removed increased from 15.3% in 1960 to 62.2% in 1990 (Anon., 1963, 1991). In thirty years, subsurface drainage resulted in loss of ca. 0.5 million km of field boundary habitat. During the same period, the proportion of arable land that is ploughed in autumn increased from ca. 60% to 80% (Anon., 1963, 1991). This is a consequence of increase in annual crops, especially spring cereals, at the expense of pasture leys. In ploughed fields, boundary habitats with permanent vegetation stand out in sharp ecological contrast to the arable land.

The RED BOOK (Rassi *et al.*, 1991) does not directly attribute any plant or animal species being endangered because of the above mentioned changes. However, out of the four bird, 21 insect and 14 vascular plant species that are included because of 'changes in agricultural habitats' (Table 2), grey partridge (*Perdix perdix*) (see Potts, 1986), possibly corncrake (*Crex crex*), kestrel (*Falco tinnunculus*) and even the now extinct quail (*Coturnix coturnix*) have suffered from indirect effects of herbicides, from disappearance of boundary habitats and from autumn ploughing. Some declines have obvious reasons: e.g. many dung beetles are endangered because cow droppings are no longer available in certain types of meadow pastures (many more endangered or extinct species are associated to disappearance of traditional forest meadows in absence of grazing), and *Cuscuta epilinum* disappeared along with growing of flax.

Decreasing trends in plant and arthropod diversity in arable ecosystems are reported, and among other measures, Conservation Headland techniques to prevent further loss are being developed, elsewhere (e.g. in England: Potts & Vickerman, 1974; Potts, 1986; Rands & Sotherton, 1986; Boatman, 1987; Wilson *et al.*, 1990; in Germany: Schumacher, 1980, 1987; in Sweden: P.A. Chiverton, unpublished). The available data suggest declining trends in diversity in Finland as well.

THE CONSERVATION HEADLAND STUDY

Materials and methods

Altogether 12 experimental Conservation Headlands and control headlands were established for 1992-1994 in four farms in Central and Southern Finland. Pesticides were excluded: not even selective sprayings are allowed in the experimental headlands. The crops are spring cereals barley, oats and wheat. Each year, decision on the crop species and all other management except crop protection is left to the farmer. The control headlands and the main crop always received one spraying of phenoxy acid herbicides, and of these, three received one spraying of dimethoate insecticide in both years and an additional two received one spraying of pyrethroid insecticide in 1992 against bird cherry-oat aphid (*Rhopalosiphum padi*). The pesticides were applied at GS 1 to GS 2. Nine of the headlands are 6m wide and three (in one of the farms) are 4m wide. The length of a Conservation Headland varies from 100m to 200m, the adjacent control headland always being of same length as the experimental headland.

In late July, six 0.25 m^2 quadrats per headland are sampled for above ground phytomass of the crop plant, for AGP of weeds, and for species abundance. The scheme is systematic: the quadrats are evenly spaced into three 'sampling stations' within a crop margin. At each station, two parallel quadrats are sampled, 1.5 m and 3.0 m from the

TABLE 3. Number of weed species or genera (total from all the sites), and mean density and phytomass of weeds in experimental unsprayed Conservation Headlands (CH) and control headlands. (SE in parentheses)

Year	Treatment	Species number	Density nos. m^{-2}	Phytomass g d.m. m^{-2}
1992	CH	31	274.7 (36.7)	13.7 (1.8)
	Control	31	160.0 (20.9)	7.4 (1.5)
1993	CH	38	419.9 (44.6)	36.9 (4.8)
	Control	36	370.7 (63.1)	20.8 (4.3)

crop edge. The stations are the same each year, in order to minimize spatial variation.

In the first year, cereal grain yields were estimated by combining two samples of 4.0 row metres (rows are sown 12.5 cm apart) at each station, for three samples equalling 1.0 m^2 each. Three additional samples were taken within the main crop, 20 m from the crop edge. In order to improve accuracy, in 1993 grain yields were harvested by a plot combine. Three samples representing 10 m^2 to 20 m^2 (varied with the make and model of the harvester) were sampled from each crop margin and from the main crop. The procedure will be repeated in 1994.

In early July, arthropods are sampled by D-Vac® (five x 10 s per sample) around each of the vegetation quadrats described above. A sample consists of five 10s suctions parallel to the crop edge. In addition, a sweep net sample of two x 15 sweeps (method: Heikinheimo & Raatikainen, 1962) is taken from each headland.

First years' results and discussion

Weed densities and phytomass of weeds were significantly higher in the Conservation Headlands than in control headlands (Table 3). Altogether 47 weed species or genera has been identified so far. No difference in total number of species was detectable (Table 3). Twenty species were common to both treatments in both years. Conservation headlands shared 80% of the species with conventional headlands. Further analysis of species composition is required for detecting possible effects on diversity. Comparisons to earlier data (Table 1) suggest greater weed abundance in crop margins than in the main crop.

In most cases, weeds remained as an undergrowth and did not hamper combine harvesting. At one site, an infestation of *Cirsium arvense* and *Sonchus arvensis* was spreading from the Conservation Headland into the main crop. Development of selective spraying schemes (see Sotherton, 1991) will be necessary. Because of weed competition, cereal yields in Conservation Headlands were significantly ($p<0.05$, LSD=0.59t) reduced by 15% from 3.9t ha^{-1} (SE 0.4) to 3.3t ha^{-1} (SE 0.3); in comparison to the main crop's 4.2t ha^{-1} (SE 0.3), conventional headlands produced 7% and Conservation Headlands 21% less (1992 data only). Drought stress during the season may have aggravated the weed problem.

TABLE 4. Mean sweep net catch (nos. per 30 sweeps) of the six most abundant insect orders from experimental unsprayed Conservation Headlands (CH) and control headlands. (SE in parentheses)

Order	1992		1993	
	CH	Control	CH	Control
Thysanoptera	959.6 (669.8)	1846.0 (1670.4)	590.5 (166.1)	745.9 (290.8)
Homoptera	1401.2 (457.5)	706.4 (384.5)	112.4 (30.3)	85.2 (16.8)
Diptera	76.9 (24.9)	69.4 (23.2)	79.8 (16.7)	74.2 (11.8)
Hymenoptera	58.0 (15.1)	46.2 (18.9)	34.3 (9.6)	29.4 (7.6)
Heteroptera	108.5 (91.7)	43.0 (34.0)	8.5 (2.9)	7.3 (3.4)
Coleoptera	13.5 (5.3)	6.5 (2.6)	7.0 (2.6)	5.2 (1.5)

Variation in insect diversity between Conservation Headlands and control headlands was much smaller than between-site variation (sweep net data at Order-level). Overall, thrips (Thysanoptera) were more abundant in conventionally sprayed headlands in both years, whereas other main groups were more abundant in the Conservation Headlands (Table 4). In order to reveal patterns, a closer analysis at lower taxonomic level (preferably at species level) and site by site is needed. Survey data from thirty years ago is available for, e.g. leafhopper (Raatikainen & Vasarainen, 1971) and spider faunas (Raatikainen & Huhta, 1968) in cereals. These can be used for detecting long term changes in abundance and diversity.

CONCLUSIONS

Conservation Headlands are designed to conserve biodiversity in cereal ecosystems. In this respect, the need and importance of Conservation Headlands depend on intensity of management. Earlier survey data and information on endangered species indicate that in Finland, changes in agricultural practices have resulted in loss of diversity. Conservation Headlands would be most appropriate to specialized cereal farms in South and South-West Finland. The results from the ongoing study suggest that the benefits to wildlife from creating Conservation Headlands in Finland would be similar to those obtained in other countries. It is obvious that the effect on biodiversity in not immediate; rather, environmental benefits would accumulate over years. Conservation Headlands should be included as one option of a set of management practices designed to conserve wildlife in arable ecosystems.

ACKNOWLEDGEMENTS

I thank A. Nissinen, P. Nummi, S. Tuomola and M. Westerstråhle for collaboration. The study was funded by the Research Council for Agriculture and Forestry and conducted at South Savo Research Station.

REFERENCES

Anon. (1963) Annual statistics of agriculture and sampling census of agriculture 1960. *Official Statistics of Finland, Agriculture*, **III:56**, 177 pp.

Anon. (1991) Yearbook of farm statistics 1990. *Official Statistics of Finland, Agriculture and Forestry*, **1991:2**, 254 pp.

Boatman, N.D. (1987) Selective grass weed control in cereal headlands to encourage game and wildlife. *1987 British Crop Protection Conference - Weeds*, **4A-2**, 277-284.

Erviö, L-R.; Salonen, J. (1987) Changes in the weed population of spring cereals in Finland. *Annales Agriculturae Fenniae*, **26**, 201-226.

Heikinheimo, O.; Raatikainen, M. (1962) Comparison of suction and netting methods in population investigations concerning the fauna of grass leys and cereal fields, particularly in those concerning the leafhopper, *Calligypona pellucida* (F.). *Publications of the Finnish State Agricultural Research Board*, **191**, 31 pp.

Hynninen, E-L.; Blomqvist, H. (1993) Pesticide sales in Finland 1992. *Kemia-Kemi*, **20**, 535-537.

Markkula, M.; Tiittanen, K.; Vasarainen, A. (1990) Torjunta-aineet maa- ja metsätaloudessa 1953-1987. *Maatalouden tutkimuskeskus, Tiedote*, **2/90**, 58 pp.

Mela, T. (1988) Organic farming in Finland. *University of Helsinki, Department of Plant Husbandry, Publication*, **16**, 220 pp. (in Finnish)

Mukula, J.; Raatikainen, M.; Lallukka, R.; Raatikainen, T. (1969) Composition of weed flora in spring cereals in Finland. *Annales Agriculturae Fenniae*, **8**, 59-110.

Potts, G.R. (1986) *The Partridge: Pesticides, Predation and Conservation*. London, Collins. 274 pp.

Potts, G.R.; Vickerman, G.P. (1974) Studies on the cereal ecosystem. *Advances in Ecological Research*, **8**, 107-197.

Raatikainen, M.; Huhta, V. (1968) On the spider fauna of Finnish oat fields. *Annales Zoologici Fennici*, **5**, 254-261.

Raatikainen, M.; Vasarainen, A. (1971) Comparison of leafhopper faunae in cereals. *Annales Agriculturea Fenniae*, **10**, 119-124.

Rands, M.R.W.; Sotherton, N.W. (1986) Pesticide use on cereal crops and changes in the abundance of butterflies on arable farmland in England. *Biological Conservation*, **36**, 71-82.

Rassi, P., Kaipiainen, H., Mannerkoski, I. & Ståhls, G. (1991) Report on the monitoring of threatened animals and plants in Finland. *Ministry of the Environment, Committee report*, **1991:30**, 328 pp. (in Finnish)

Schumacher, W. (1980) Schutz und Erhaltung gefährdeter Ackerwildkräuter durch Integration von landwirtschaftlicher Nutzung und Naturschutz. *Natur und Landschaft*, **55**, 447-453.

Schumacher, W. (1987) Measures taken to preserve arable weeds and their associated communities in Central Europe. *1987 British Crop Protection Monograph*, **35**, 109-112.

Sotherton, N.W. (1991) Conservation headlands: a practical combination of intensive cereal farming and conservation. In: L.G. Firbank, N. Carter, J.F. Darbyshire & G.R. Potts (ed), *The Ecology of Temperate Cereal Fields*, Oxford, Blackwell, pp. 373-397.

Wilson, P.J.; Boatman, N.D.; Edwards, P.J. (1990) Strategies for the conservation of endangered arable weeds in Great Britain. *Proceedings of the EWRS Symposium 1990, Integrated Management in Cereals*, 93-101.

COST-BENEFITS OF UNSPRAYED CROP EDGES IN WINTER WHEAT, SUGAR BEET AND POTATOES.

G.R. DE SNOO

Centre of Environmental Science, Leiden University,
P.O. Box 9518, 2300 RA Leiden, The Netherlands

ABSTRACT

In winter wheat, potatoes and sugar beet unsprayed crop edges were established on which no herbicides or insecticides were used. The yield (quantity and quality) from the unsprayed crop edges was compared with that from the sprayed edges and the centre of the field. In every crop the yield at the field centre was much greater (by 14%) than at the sprayed field edge. Relative to the sprayed crop edges the reduction in yield on the unsprayed edges (3 metres wide) was about 30% in sugar beet and 2% in potatoes. In winter wheat the reduction in yield was 13% average for the outer 3 metres and 11% for the outer 6 metres of the unsprayed crop edge. The quality of the harvests was unchanged, with the exception of potato size in one year. Balancing the yield losses against the benefits, *viz.* reduced pesticide costs, it is concluded that with winter wheat and potatoes, unsprayed field margins can be well adopted in agricultural practice. In sugar beet, however, the cost is too high.

INTRODUCTION

From an agricultural point of view field margins are economically less valuable than field centres. Management of crop edges often requires additional effort - in the case of wedge-shaped fields, for example - which does not benefit operational efficiency. The yield from crop edges is also often lower, for a variety of reasons (cf. Boatman & Sotherton, 1988): less favourable crop site factors (e.g. fertilizer and water regime, soil structure), competition from off-plot vegetation (tree shade, infestation by perennial weeds), increased feeding damage by pheasants, wild boars, mice and other animals and finally, direct crop damage due to more intensive machinery use on field margins (turning tracks on headlands, and tracks resulting from ditch bank management).

Over the past ten years, in many West European countries there has been a growing interest in establishing unsprayed crop edges for the purpose of nature conservation, biological pest control or reducing pesticide drift to surrounding areas. It has been demonstrated that the creation of unsprayed crop edges has a positive impact on the abundance of rare weed species, invertebrates, birds and mammals (Schumacher, 1984; Rands, 1985; Hald *et al.*, 1988; Tew *et al.*, 1992; De Snoo *et al.*, these proc.). This practice has also been shown to help reduce pesticide drift to surrounding areas (Cuthbertson & Jepson, 1988). There is still little research data available on the costs associated with establishment of unsprayed field margins, however. Boatman & Sotherton (1988) have calculated the cost of maintaining unsprayed cereal crop edges. In the present study the cost has also been calculated for potato and sugar beet crops. The study forms part of the Dutch Field Margin Project being undertaken in the Haarlemmermeerpolder. The aim of this project is to develop a management strategy for promoting nature conservation in arable fields and reducing pesticide drift to non-target areas. Effects on weeds, invertebrates, vertebrates, pesticide drift and costs to the farmer are being studied.

METHODS

Research area

The study was carried out in the Haarlemmermeerpolder (clay soil) during the period 1990-1993 in winter wheat, sugar beet and potato fields (Table 1). Yields were determined on crop edges 100 m long and 3 m wide left unsprayed with herbicides or insecticides and these were compared with yields from sprayed crop edges in the same fields as well as from the field centres. In 1992 and 1993 yields were also measured on 6-m unsprayed edges in winter wheat fields. Because of proliferous weed growth in some plots, in 1990 weeds were cleared by hand on 4 and in 1991 on all sugar beet field margins over a length of 85 m. In the winter wheat crop, too, some of the weeds (*Matricaria recutita* only) were cleared on several farms to prevent seeding; in 1991 on 3 edges, and in 1992 and 1993 on 2 edges. Because of rampant weed growth in the sugar beet crop, in 1991 the scope for differentiated spraying of this crop was considered on a limited scale (strips 25 m long), varying from zero to three herbicide treatments.

TABLE 1. Number of fields investigated. In every field an unsprayed crop edge, a sprayed crop edge and the field centre were investigated.

	winter wheat 3 m wide	winter wheat 6 m wide	potatoes	sugar beet	sugar beet 0-3 sprayings
1990	-	-	5	6	-
1991	6	-	5	8	4
1992	7	5*	7	-	-
1993	5	4**	-	-	-

* = no field centre investigated; ** = incl. 2 spring wheat fields

Yield measurements

In the potato plots, in 1990 the crop was dug up by hand along the crop edge and in the field centre over an area measuring 12 m² (4 x 3 m²), and in 1991 and 1992 over 18 m² (6 x 3 m²). On the edges the samples were taken alternately on the 2nd and 3rd potato ridges and in the centre on the 21st and 22nd ridges. The potatoes were sorted into three size classes: <40 mm, 40-50 mm and >50 mm. In processing the data for the class <40 mm, half of the aggregate weight was taken, as machine harvesting leaves a large proportion of these potatoes behind on the land. The dry-matter content was calculated by determining, for each sample, the underwater weight of the potatoes in the >40 mm fraction (Ludwig, 1972).

In the sugar beet plots, in both years an area measuring 12 m² was manually harvested on each strip (in 1990 4 x 3 m², in 1991 6 x 2 m², including the trial with differentiated spraying). Along the field edges the samples were taken on the 2nd, 3rd and 4th beet rows, and in the centre on the 21st and 22nd rows. In 1990, in the unsprayed, uncleared section (15 m long) an area of only 1 x 3 m² was harvested. In 1990 and 1991 the sugar content as well as the potassium, sodium and alpha-aminonitrogen content of respectively 2 and 4 samples were determined by the Dutch Institute for Rational Sugar Production and the results employed to calculate the sugar extractability index using Van Geyn's formula.

In the winter wheat plots, in 1991 a reed mower was employed to reap a swath 50 m long and 1.20 m wide along the edges and at the centre of the fields, and the harvest was removed and then threshed using a plot combine (Deutz-Fahr M660, type 2116). In 1992 and 1993 a plot combine (Claas, type Cosmos) was used to combine (reap and thresh) a single swath 100 m long and 2.20 m wide. On the 6 m wide strips the outer 3 m and inner 3 m of the crop edges were combined separately. Data on the strips partly cleared of weeds were also processed, and there was found to be no difference in crop yield. On each strip the dry crop weight was determined by drying a 5 x 1 kg sample at 100 °C and reweighing.

RESULTS

Potatoes

Table 2 shows the yields from the various strips for potatoes. Comparison of the sprayed and unsprayed edges indicates a lower yield on the unsprayed edges in 1990 and 1992 (by 3.5% and 8.6%, respectively). This yield loss was significant in 1992 only. In that year there was a lower yield of large potatoes (>50 mm), especially. In 1991, however, on 4 of the 5 farms the yield from the unsprayed crop edges was higher than from the sprayed edges (by 5.8% on average). In 1990 and 1991 no significant effects on potato size were found. In all three years there were significant differences, of about 10.0-12.6%, between yields from the field centre and from the sprayed crop edge. In the field centre the yield of large potatoes (>50 mm), particularly, was significantly higher than on the edges. At the field centre there were fewer potatoes in the class <50 mm. No difference in dry weight was found between potatoes from sprayed and unsprayed crop edges (Table 3). In 1992 potatoes from the field centre generally contained more water.

TABLE 2. Potato yield (kg/m²)

	sprayed edge		unsprayed edge			field centre		
	av.	s.d.	av.	s.d.	T	av.	s.d.	T
1990 (n=5)								
<40 mm	0.44	± 0.12	0.50	± 0.24	n.s.	0.32	± 0.11	***
40-50 mm	1.56	± 0.18	1.51	± 0.29	n.s.	1.47	± 0.40	n.s.
>50 mm	1.95	± 1.33	1.81	± 1.51	n.s.	2.70	± 1.59	***
total	3.96	± 1.22	3.82	± 1.28	n.s.	4.49	± 1.12	**
1991 (n=5)								
<40 mm	0.57	± 0.15	0.54	± 0.10	n.s.	0.46	± 0.12	***
40-50 mm	1.98	± 0.46	2.14	± 0.38	n.s.	1.59	± 0.24	***
>50 mm	1.79	± 1.03	1.92	± 0.64	n.s.	2.77	± 1.02	***
total	4.34	± 0.93	4.59	± 0.67	n.s.	4.82	± 0.75	**
1992 (n=7)								
<40 mm	0.27	± 0.09	0.31	± 0.12	*	0.24	± 0.10	*
40-50 mm	1.71	± 0.32	1.74	± 0.39	n.s.	1.59	± 0.49	n.s.
>50 mm	2.64	± 1.25	2.18	± 1.04	**	3.45	± 1.31	***
total	4.62	± 0.92	4.22	± 0.68	**	5.28	± 0.85	***

Legend: av. = average; s.d. = standard deviation between fields; n = no. of fields; * = $P < 0.05$; ** = $P < 0.01$; *** = $P < 0.001$; n.s. = not significant; T = two-way, two-tailed ANOVA.

TABLE 3. Percentage dry matter in potatoes (>40 mm)

	sprayed edge		unsprayed edge			field centre		
	av.	s.d.	av.	s.d.	T	av.	s.d.	T
1990 (n=5)	21.8	± 0.6	21.8	± 0.8	n.s.	21.0	± 1.2	n.s.
1991 (n=5)	22.4	± 1.2	22.3	± 0.9	n.s.	21.6	± 2.1	n.s.
1992 (n=7)	19.4	± 2.1	19.3	± 1.9	n.s.	18.2	± 2.3	*

Legend: see Table 2; T = one-way, two-tailed ANOVA

Sugar beet

Because by many of the farmers the unsprayed strip is partly weeded, which meant that in the worst case (1990) only one sample could be taken per plot, it was decided to test the average yields from the individual plots non-parametrically. Compared with the sprayed crop edges, the average sugar beet yield from the unsprayed edges was 20.6% lower in 1990 and 39.1% lower in 1991 (Table 4). This reduction in yield was significant in 1991 only. In both years there was found to be a significant difference between the sprayed edges and the field centre (16.4 and 16.0% average yield reduction on the edges).

No significant differences in sugar content were found between beets from the sprayed and unsprayed crop edges, nor between beets from the sprayed edges and the field centre. Although the sugar content in the field centre was slightly lower than on the sprayed edges (Table 5), the difference was not significant (P = 0.054, Wilcoxon matched pairs, in 1991). There was also no difference in sugar extractability index between beets from the sprayed and unsprayed edges. This index was much lower for beets from the field centre than for those from the sprayed edges (P = <0.05, Wilcoxon matched pairs).

TABLE 4. Sugar beet yield (kg/m^2)

	sprayed edge		unsprayed edge			field centre		
	av.	s.d.	av.	s.d.	T	av.	s.d.	T
1990 (n=6)	7.34	± 1.57	5.83	± 2.26	n.s.	8.78	± 0.52	*
1991 (n=8)	5.97	± 1.32	3.63	± 1.74	**	7.11	± 0.94	*

Legend: see Table 2; T = one-tailed Wilcoxon matched pairs.

TABLE 5. Sugar content (%) and extractability index (%) of sugar beet

	sprayed edge		unsprayed edge			field centre		
	av.	s.d.	av.	s.d.	T	av.	s.d.	T
sugar content								
1990	17.15	± 1.07	17.03	± 0.98	n.s.	16.47	± 1.29	n.s.
1991	15.82	± 0.78	15.47	± 1.42	n.s.	15.16	± 0.77	n.s.
winning index								
1990	91.90	± 2.41	92.22	± 3.35	n.s.	87.96	± 6.61	*
1991	92.54	± 2.19	93.13	± 3.86	n.s.	88.26	± 4.74	*

Legend: see Table 2; T = two-tailed Wilcoxon matched pairs

The results of the experiment with differentiated spraying are shown in Table 6. There was a significant difference in yield between the unsprayed (36.3% loss) or once-sprayed strip (9.6% loss) and the fully-sprayed strip (3 treatments). Although there was little difference the double spraying gave a significant increase in yield (8.5% loss).

TABLE 6. Sugar beet yield (kg/m²) with various spraying regimes (n=4)

	av. sd.	T	
unsprayed	4.05 ± 1.21	***	(0 vs 3 sprayings)
1 spraying	5.74 ± 0.70	**	(1 vs 3 sprayings)
2 sprayings	5.82 ± 0.56	*	(2 vs 3 sprayings)
3 sprayings	6.35 ± 1.43		

Legend: see Table 2: T = two-way, one-tailed ANOVA

Winter wheat

The yield losses on the unsprayed crop edges are presented in Table 7. The average losses are significant each year: 11.0% in 1991, 11.1% in 1992 and 17.2% in 1993 for the outer 3 m of the strips. In the inner 3 m of the unsprayed 6-m strips, yield losses were 2.7% in 1992 and 7.4% in 1993. In neither year was this difference significant. Over the full 6 m of the strip, the average yield loss was 6.2% in 1992 and 15.2% in 1993, giving a combined loss of 10.7% for the two years. Each year the yield from the sprayed edge was significantly less than at the plot centre; in 1991 this loss was 17.5%, in 1992 11.3% and in 1993 12.1%. There was no significant difference in the moisture content of the grain between the sprayed crop edge, the unsprayed edge and the plot centre.

TABLE 7. Winter wheat yield (kg/m²; 16% grain moisture)

	sprayed edge		unsprayed edge			field centre		
	av.	s.d.	av.	s.d.	T	av.	s.d.	T
1991 0-3 m (n= 6)	0.729	± 0.079	0.649	± 0.076	*	0.883	± 0.087	* 1)
1992 0-3 m (n=12)	0.808	± 0.139	0.718	± 0.120	**	0.952	± 0.154	* 1)
4-6 m (n= 5)	0.872	± 0.093	0.848	± 0.073	n.s.	----		
1993 0-3 m (n= 9)	0.805	± 0.102	0.667	± 0.112	**	0.916	± 0.171	**
4-6 m (n= 4)	0.971	± 0.054	0.899	± 0.091	n.s.	1.054	± 0.056	*

Legend: see Table 2; T = one-tailed Wilcoxon matched pairs; 1) n = 5

DISCUSSION

In making a economic cost-benefit analysis of the unsprayed field margins, yield losses must be offset against savings in pesticide use. In the potato crop, yield losses were only minor (average 2%). The cost of this crop is average Dfl. 0.02 per m² (based on potato price over last six years of Dfl. 0.17 per kg, *cf.* IKC, 1993). Savings on pesticide use (*cf.* IKC, 1993) amount to Dfl. 0.02 per m², giving no nett extra expenditure. In the case of sugar beet, the direct loss of yield is rather greater (average 30%), at a cost of about Dfl. 0.24 per m² (based on 'mixture price' sugar beet of Dfl. 0.11 per kg, adjusted for sugar content and extractability index, *cf.* IKC, 1993). Savings on pesticide use total Dfl. 0.03, giving average Dfl. 0.21 nett extra expenditure per m². The experiment with differentiated spraying shows that the first, pre-emergence treatment is most important for limiting these yield losses. In winter wheat, finally, yield losses are 13% average for the outer 3 m, i.e. Dfl. 0.03 per m² (based on EC price of Dfl. 0.26 per kg). Savings on pesticide use are approximately Dfl. 0.02 per m², giving about Dfl. 0.01 nett extra expenditure per m². For a strip 6 m wide this means about Dfl. 0.005 per m². In England yield losses on 6-m wide unsprayed crop edges have been found to be only 3% for winter wheat and 6% for spring barley. The cost of yield losses in that country, including the cost of separate harvesting, threshing, drying, storage and possibly extra spraying of edges, is about Dfl. 0.04 per m^2 (Boatman & Sotherton, 1988; Boatman, 1990). As a general conclusion it can be said that maintaining unsprayed crop edges is economically viable for crops of potatoes and winter wheat. In sugar beet, however, the cost appeared to be too high.

Besides savings on pesticide use, benefits from establishing unsprayed field margins may also be accrued in the form of extra income from huntsmen, for example. In England the cost of maintaining unsprayed cereal edges is found to be compensated by the increase in revenue resulting from larger populations of partridges and pheasants (Boatman, 1990). This might be tied in with payment for 'nature production'.

The differences in yield between 3-m wide crop edges and field centres are in reasonable agreement for the various years and crops: for potatoes 11% difference, for sugar beet 16% and for winter wheat 15%. Boatman & Sotherton (1988) found that the yield from the edges of cereal fields (6-m wide) was 18% less, on average, than that from the field centre. In establishing compensation measures for arable farmers, therefore, it is important not to base payments on the average yield per hectare but on a yield loss of say 10% to 15% from the field edge.

ACKNOWLEDGEMENTS

The author wishes to thank F. Mugge, S. van Doorn, P. Hartsuijker, E. Koning, J. de Leeuw, P.J. de Wit and M. van der Wal and all volunteers for their help with the field work, R.J. van der Poll for his help with field work and data processing and C.J. Kloet (IKC) for his help with the cost-benefits calculations.

REFERENCES

Boatman, N.D. (1990) Conservation headlands and the economics of wild game production. In: *De toekomst van de wilde hoenderachtigen in Nederland*. Lumeij, J.T; Hoogeveen, Y.R. Elinkwijk B.V., Utrecht, 198-206.

Boatman, N.D.; Sotherton N.W. (1988) The agronomic consequences and costs of managing field margins for game and wildlife conservation. *Aspects of Applied Biology* **17**, 47-55.

Cuthbertson, P.S.; Jepson, P.C. (1988) Reducing pesticide drift into the hedgerow by inclusion of an unsprayed field margin. *Brighton Crop Protection Conference - Pests and diseases* 1988, 747-751.

Hald, A.B.; Nielsen, B.O.; Samsøe-Petersen, L.; Hansen K.; Elmegaard, N; Kjoholt, J. (1988) *Sprojtefri randzoner i kornmarker*. Miljoprojekt nr. 103. Miljostyrelsen, Kobenhavn.

IKC, 1993. *Kwantitatieve informatie voor de Akkerbouw en de Groenteteelt in de Vollegrond. Bedrijfssynthese 1993-1994*. Informatie- en Kenniscentrum voor de Akkerbouw en de Groenteteelt in de Vollegrond (IKC) Lelystad. Publikatie **69**, 212 p.

Ludwig, J.W. (1972) Bepaling van het droge-stofgehalte van aardappelen via onderwaterweging Instituut voor Bewaring en Verwerking van Landbouwprodukten Wageningen. IBVL publikatie **247**, 12 pp.

Rands, M.R.W. (1985) Pesticide use on cereals and the survival of grey partridge chicks: a field experiment. *Journal of Applied Ecology* **22**, 49-54.

Schumacher, W. (1984) Gefährdete Ackerwildkräuter können auf ungespritzten Feldrändern erhalten werden. *Mitteilungen der LÖLF* **9** (1), 14-20.

Tew, T.E.; MacDonald, D.W.; Rands, M.R.W. (1992) Herbicide application affects microhabitat use by arable wood mice (*Apodemus sylvaticus*). *Journal of Applied Ecology* **29**, 532-539.

RESOURCE USE OF CROPS AND WEEDS ON EXTENSIVELY MANAGED FIELD MARGINS

R. LÖSCH, D. THOMAS, U. KAIB, F. PETERS

Abt. Geobotanik der Universität, D-40225 Düsseldorf, Germany

ABSTRACT

Many weed species have virtually disappeared from the agricultural landscape in Germany. Encouraged by governmental programmes some farmers try to re-establish a species-rich flora on crop margins omitting fertilizer and pesticides. Biodiversity has increased promisingly on such extensively used boundary strips of fields on sandy soils in the Lower Rhine area. An analysis of weed and crop green area indices, biomass, mineral ion contents and water use patterns gives evidence that the crop nevertheless is superior in the balance of competition. Extensively managed crop margins increase therefore the biotic diversity of a landscape with economic losses remaining at a tolerable level.

INTRODUCTION

In Germany plant species diversity decreased drastically during the past decennia. This is particularly obvious with regard to the arable weed flora where 397 species of a total 581 species are ranked as endangered in their existence (Kaule, 1991). The main reason for this decrease in species diversity is intensive management of fields, viz. increased fertilization and pesticide use. To counteract this undesired trend, programmes for extensive management of crop margins have been initiated designed to re-establish a species-rich field flora from the still existing seed bank. Volunteer farmers are engaged who avoid pesticide and high fertilizer use on 5 m wide crop margins. In the Lower Rhine area of North-Rhine-Westfalia, nutrient-poor cereal fields on light sandy soils are managed in this way.

We investigated the results of these practices under two aspects: The increase in species diversity was assessed and the biomass accumulation, nutrient and water use by some of the most prominent weeds and the crops was quantified in order to obtain information concerning competition for resources between crops and weeds under this field margin management.

MATERIAL AND METHODS

Species diversity and floristic richness were determined for cereal fields in the sandy alluvial plains in the Lower Rhine area near the Dutch/German border. They were managed either according to the regulations of the governmental field margins programme or in the traditional way, with intensive fertilizer and pesticide application. Crop and weed biomass, green area index (GAI) and mineral contents were measured with plants from eight extensively managed fields. Transpiration was measured with *Centaurea cyanus* and two *Papaver* species which all contribute reasonably to the aesthetic value of cereal fields.

The diversity index was calculated according to the Shannon formula, $\text{diversity } D = -\Sigma\,(p_i \ln p_i)$ [p_i = frequency of occurrence of species i].

For the determination of biomass, GAI, and mineral contents above- and below-ground parts of the crop and of the two most abundant weeds were harvested from 20 randomly selected 9 dm^2 plots per field. Green area determinations were done with a LI-3100 area meter (LiCor, USA). For dry matter determinations the harvested material was weighed after complete drying at 80°C. K^+, Mg^{++} and Ca^{++} contents were measured with an AAS (Perkin-Elmer 2280) after six hours wet digestion with 65 % HNO_3 at 180°C.

Transpiration was measured porometrically (Li-1600, LiCor, USA) on plants transferred with the original soil from the field into containers where water supply was controlled.

RESULTS

Species richness

After some years the results of the management program are very promising in terms of increased species diversity. In the study area 100 different species were found, on average 44 per field, which makes up 42% of the local cormophytic flora. Nine species, mentioned by the red data book as highly endangered, recovered in their local occurrence. Most prominent among them are *Arnoseris minima*, *Anthoxanthum aristatum*, *Galeopsis segetum*, *Teesdalia nudicaulis*, *Misopates orontium*,

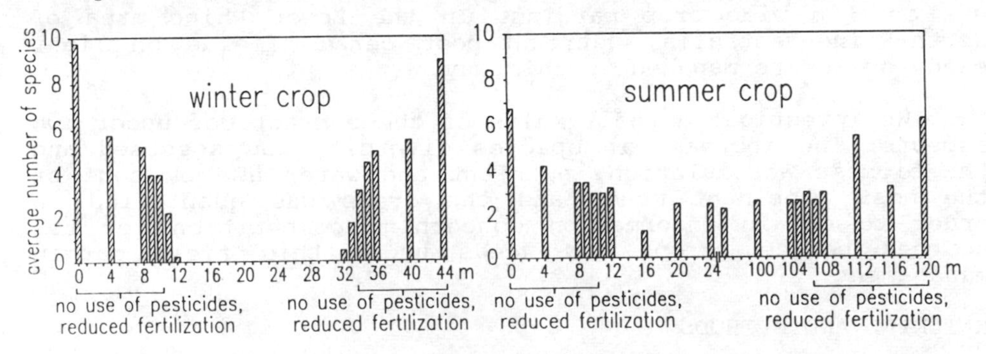

FIG. 1: Species numbers along transects across fields with extensively managed margins and intensively managed centres

and *Gagea arvensis*. Most of them are phytosociological character species of the Teesdalio-Arnoseridetum minimae and the Papaveretum argemonis. Whereas traditionally managed fields had diversity indices around -1 index values of the extensively treated crop margins ranged between -2.1 and -2.4 A steep gradient in species numbers existed between the herbicide- and fertilizer-treated areas and the extensively managed field margins (Fig.1).

GAI and biomass

Maximum GAI values of the crops were measured in June. On extensively managed areas they were between 2 and 4 m^2 m^{-2} for rye, 3 and 7 m^2 m^{-2} for oat and barley, and approx. 3 m^2 m^{-2} for *Triticale*. Maximum GAI values of the weeds ranged between 0.2 and 1.7 m^2 m^{-2} (Fig.2). During spring particularly fall-sown crops showed a much higher GAI than the accompanying weeds. Also biomass productivity of crops was much greater than that of weeds (Fig. 3). Peak values of the above-ground biomass were 1800 g dw m^{-2} soil surface in *Avena*, 1500 g dw m^{-2} in *Triticale*, and 1000 g dw m^{-2} in *Seçale*. The highest values of the weeds came to only 200 g m^{-2}. On average, there was a 22 ± 14 % reduction of crop biomass yield on the margins as compared with the intensively managed central parts of the fields. The highest below-ground biomass was 190 g m^{-2} and 25 g m^{-2} for crops and weeds, respectively. This trend is mirrored also by the relationship between crop and weed root surface areas which was in the extreme case of an oat field 2:1 at the start of the season and 27:1 at harvest in July.

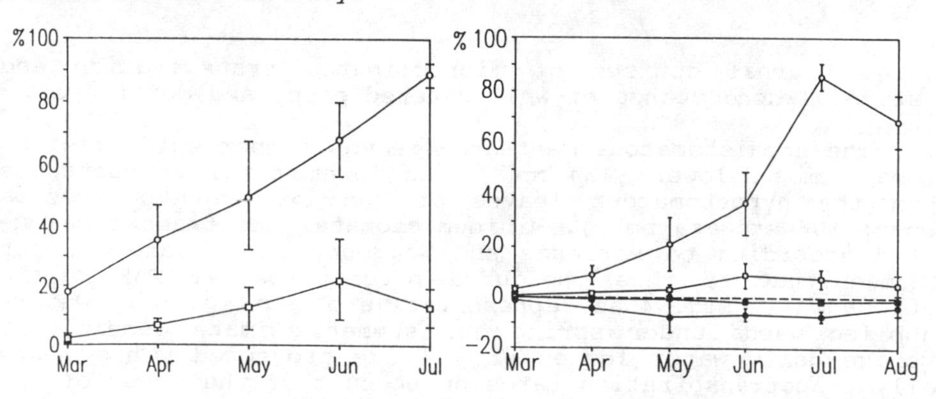

FIG.2: GAI of crops (circles) and weeds (squares) during the season; percentage of average total maximum of the 8 investigated fields.

FIG.3: Above-and below-ground biomass (open and closed symbols, respectively) of crops (circles) and weeds (squares) during the season (Percentage of average total maximum biomass of the eight investigated fields).

Water consumption and mineral uptake

Transpirational water loss of *Centaurea* and *Papaver* was considerably high. Even under high leaf-to-air vapour saturation deficits stomata did not close so that high leaf conductances prevailed throughout the day (Fig. 4).

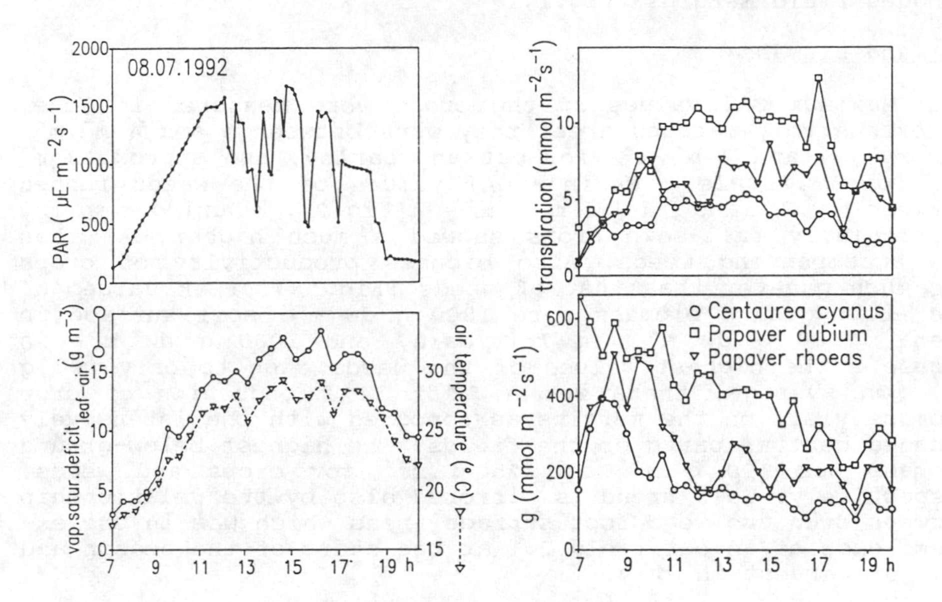

FIG.4: Diurnal courses of microclimate, transpiration and stomatal conductance of well watered poppy and cornflower.

The amphistomatous *Centaurea* leaves (upper epidermis: 85 stomata mm^{-2}, lower: 134 mm^{-2}) had lower total conductances than the hypostomatous leaves of the two *Papaver* species (both: 106 stomata mm^{-2}). Minimum stomatal resistances calculated according to Parlange and Waggoner (1971) are similar in magnitude to those measured porometrically. Taking the data shown in Fig. 4 as representative of sufficiently water-supplied weeds under spring and summer climate conditions, average daily water losses of 4 l are projected. This means daily evapotranspiration rates on account of the weeds of approx. 1 - 6 mm, depending on weed GAI.

Average potassium, calcium and magnesium contents of the crops and the most important weeds as measured in May, June and July are shown in Fig. 5. If related to dry weight, ion accumulation in barley and oat on the one hand and rye and *Triticale* on the other hand are nearly equal. Ion contents of the dicot weeds are, as a rule, higher than those of grasses. However, due to a much higher biomass, the grain crops take up most of the nutrients per soil surface area.

DISCUSSION

Omission of herbicides and reduced fertilizer application increases floristic diversity of cereal fields, and even rare species can be re-established from the soil seed bank. This was evident with the fields on acidic, light sandy soils studied here in the Lower Rhine area, and it has been reported also from calcareous soils (Van Elsen, 1989; Otte, 1990). The weeds compete with the cultivated plants for space, nutrients, and water, so that unavoidably some crop yield reduction occurs. According to Lotz et al. (1990) it will be greater in winter than in summer crops. But generally, the grain crops are rather competitive. As a result, space competition by the weeds, as quantified by the respective GAI and biomass values, is not critical even after 3-4 years extensive management. Nutrient accumulation per dry matter is equal or even higher in weeds as compared with the crops. However, since crop biomass per soil surface area is much bigger than that of the weeds, more than 3/4 of the plant

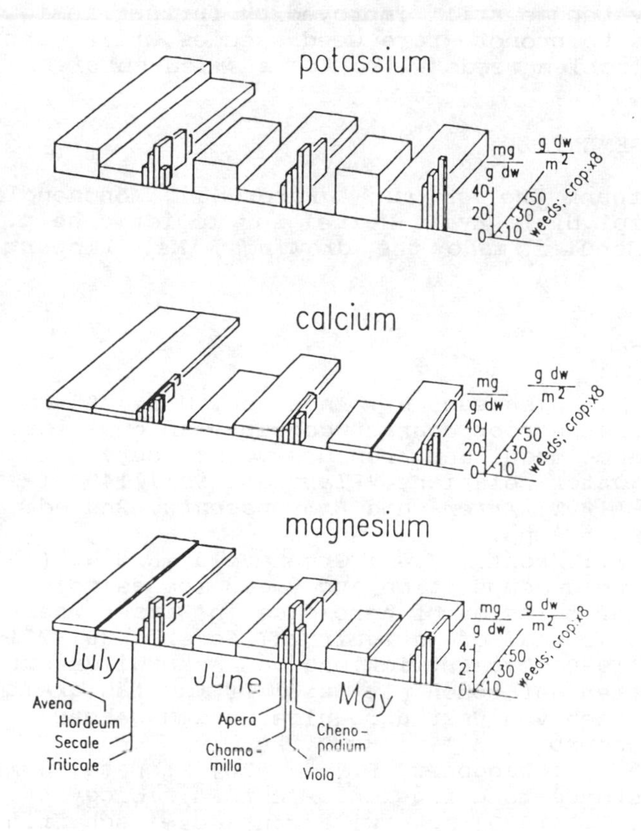

FIG.5: Average potassium, calcium and magnesium contents of grain crops and the most prominent weeds of the fields investigated, *Apera spica-venti*, *Chamomilla recutita*, *Viola arvensis*, and *Chenopodium album*, and average dry matter of these plants per square metre of soil surface area.

nutrient uptake is sequestered in the crop. We did not compare crop and weed transpiration rates directly, but from the measurements with poppies and cornflowers a water loss from weeds under spring and summer fairweather conditions is projected in the order of magnitude of 5 mm day^{-1}. A similar amount has been assessed for densely sown, intensively managed barley under similar edaphic (light sandy soils) and climatic (sub-atlantic) conditions in Jutland (Jensen *et al.*, 1993).

Modern agriculture ceases to be focused simply on yield maximization. Instead, the high costs of agrochemicals and the environmental impact of them, and the benefits for the public of a structured agricultural landscape with high biodiversity are also taken into account. An extensive management of crop margins could be a fair compromise between economic and ecological interests. Yield losses are kept in this case at a tolerable level, particularly if they are compensated to some extent by equalization payment by society. The system may become still improved by further insight into possibilities to promote rare weed species while suppressing undesired "problem weeds" like *Apera spica-venti*.

ACKNOWLEDGEMENTS

We thank the Amt für Agrarordnung Mönchengladbach (ORR Bläser, Dipl.biol. Evelt-Neite) for logistic help. Ms. Kiefer and Ms. Schüler made the drawings, Ms. Limpert typed the text.

REFERENCES

Jensen, C.R.; Svendsen, H.; Andersen, M.N.; Lösch, R. (1993) Use of the root contact concept, an empirical leaf conductance model and pressure-volume curves in simulating crop water relations. *Plant and Soil* **149**, 1-26.

Kaule, G. (1991) *Arten- und Biotopschutz*. 2nd ed., Stuttgart: Ulmer, 519 pp.

Lotz, L.A.P.; Kropff, M.J.; Groeneveld, R.M.W. (1990) Modelling weed competition and yield losses to study the effect of omission of herbicide in winter wheat. *Netherlands Journal of Agricultural Science* **38**, 711-718.

Otte, A. (1990) Die Entwicklung von Ackerwildkrautgesellschaften auf Böden mit guter Ertragsfähigkeit nach dem Aussetzen von Unkrautregulierungsmaßnahmen. *Phytocoenologia* **19**, 43-92.

Parlange, J.Y.; Waggoner, P.E. (1970) Stomatal dimensions and resistance to diffusion. *Plant Physiology* **46**, 337-342.

Van Elsen, T. (1989) Ackerwildkraut-Gesellschaften herbizidfreier Ackerränder und des herbizidbehandelten Bestandsinneren im Vergleich. *Tuexenia* **9**, 75-105.

THE IMPACT OF NITROGEN FERTILISERS ON FIELD MARGIN FLORA

N D BOATMAN

The Game Conservancy Trust, Fordingbridge, Hampshire, SP6 1EF

L J REW, A J THEAKER AND R J FROUD-WILLIAMS

Department of Agricultural Botany, University of Reading, 2 Earley Gate, Whiteknights, PO Box 239, Reading, RG6 2AU

ABSTRACT

This paper examines the potential role of misplaced nitrogen fertiliser as a causal factor in the degradation of field margin floras and encouragement of weed species. Evidence from the literature is reviewed and experimental evidence described. The weed species causing greatest concern in relation to field boundaries, *Bromus sterilis, Galium aparine* and *Elymus repens*, are all highly responsive to applied nitrogen. Experiments showed that *B. sterilis* is an effective competitor for nitrogen, but the response of *G. aparine* is inversely related to the competitive ability of competitor species. Application of nitrogen fertiliser to hedge bank vegetation did not alter botanical composition over a three year period, but increased vegetative and reproductive growth of transplanted *B. sterilis*. It is suggested that misplaced nitrogen fertiliser is more likely to affect the herbaceous composition of field boundaries where disturbance maintains a large number of safe sites for germination of annual plant seeds.

INTRODUCTION

It has been suggested that the vegetation of any given habitat is broadly determined by levels of resource availability, or stress, and levels of disturbance (Grime, 1977). Arable field boundaries generally have species-poor floras compared to those adjacent to other land use types, and in recent years there has been a general reduction in the species diversity of the herbaceous vegetation of hedgerows with a shift towards the "arable" type communities (Bunce *et al*, 1994). Disturbance, caused by close cultivation, spray drift, or deliberate herbicide application, and resource augmentation via fertiliser misapplication have been implicated as contributing to the decline in diversity and concomitant increase in weedy species capable of infesting neighbouring crops (Marshall, 1988; Boatman, 1989; Smith & MacDonald, 1989). However, the actual or relative importance of these different factors in field margins have been little studied until recently.

This paper reviews evidence and presents data from some recent studies on the effect of a single nutrient, nitrogen, on the herbaceous vegetation of field margins. Nitrogen fertiliser use has increased dramatically in the last 50 years, nitrogen being the nutrient generally limiting crop growth. For example, the average amount applied to winter wheat increased tenfold from 19 kg/ha in 1943 to 192 kg in 1985 (Church, 1981; Burrell *et al*, 1990). Most of this fertiliser has been applied with broadcast -type distributors, which rely on overlapping bouts to achieve uniform application, though in a questionnaire survey of 120 farmers, 26% used pneumatic or liquid fertiliser applicators, which spread evenly over the whole width (Rew *et al*, 1992a). Broadcasters, whilst capable of accurate application over most of the field, are inherently inaccurate at field margins, and deposit fertiliser outside the cropped area if no measures are taken to prevent this. Various methods of countering this problem are available, but none are totally satisfactory, and often cause uneven application on the outer few metres of crop (Rew *et al*, 1992a).

Although very few studies have previously been carried out in field boundaries, there is a wealth of evidence that application of nitrogen fertilisers reduces floristic diversity in grassland communities (e.g. Green, 1972; Tilman, 1982; Mountfield *et al*, 1993). This may be considered undesirable in terms of conservation status, but also important in the context of field boundaries in its effect on the relationship between weedy and non-weedy species. Of particular concern are the species *Galium aparine*, *Bromus sterilis* and *Elymus repens*, which are common in field boundary vegetation and are aggressive and damaging weeds in crops (Boatman & Wilson, 1988; Marshall, 1989; Boatman, 1989). Previous studies have shown that *E. repens* (formerly *Agropyron repens*) increased under high nitrogen conditions in competition with other species (Cussans, 1973; Tilman, 1988; Marshall, 1990). There are also numerous studies which indicate that the growth and competitive ability of *G aparine*, in competition with wheat, is greater at higher levels of nitrogen supply (Mahn, 1984; Rooney *et al*, 1990; Baylis & Watkinson, 1991; Wright & Wilson, 1992). However, the effect of nitrogen on the competitive ability of *G. aparine* in hedgerows has not been investigated. The effect of nitrogen on *B. sterilis* has not been widely studied, but Lintell-Smith *et al* (1991; 1992) observed greater reductions in yield of wheat and greater seed production of *B. sterilis* when grown in competition with wheat at a higher level of nitrogen.

To obtain further information about the effects of misplaced nitrogen fertiliser on plants in field boundaries, we have carried out several pot, semi-field and field experiments.

METHODS

i) Pot experiment

Four grass species, *Bromus sterilis, Poa trivialis, Holcus lanatus* and *Festuca rubra*, were grown in pots in monoculture and 1:1 additive mixtures with *B. sterilis* at two densities and four levels of nitrogen supply (equivalent to 0, 40, 80 and 160 kg N/ha). The growing medium was John Innes No 1 compost (1:1:1 sand:loam:peat), and pots were maintained under glass house conditions. Seeds were sown within the pots in two concentric circles, at densities of eight or sixteen plants per pot in monoculture (16 or 32 in mixture). There were three replicates of each treatment and two harvest dates. Aggressivity (A), a measure of competitive ability (McGilchrist & Trenbath, 1971) was calculated for each species as $A = (W_{ij}/W_{ii})-W_{ji}-W_{jj})$ where W_{ii} and W_{jj} are dry weights per pot of species i and j in pure stands; W_{ij} and W_{ji} are weights of species i and j in mixture with each other. Further details are given in Rew (1993).

ii) Semi-field experiment

Seeds of *B. sterilis, Arrhenatherum elatius, Dactylis glomerata, Aethusa cynapium, Scandix pecten-veneris, Galium aparine, Anthriscus sylvestris* and *Heracleum sphondylium* were sown separately in November 1990 into the central $0.002m^2$ of $0.1m^2$ plots sown at the same time, of winter wheat, *P. trivialis*, or *H. lanatus*, or into unvegetated control plots. There were six replicates of each test species. Nitrogen fertiliser was applied at one of three levels, 0, 80 and 160 kg N/ha in the following spring. Further details are given in Rew *et al* (1992b) and Rew (1993). In autumn 1991, the central portions of plots which had contained annual species were re-sown with *B. sterilis* or *A. sylvestris*, or pot grown *B. sterilis* were transplanted into them in spring 1992. Annual plants were harvested between June and August in 1991 (when seeds were ripe but not shed), and the re-sown annuals and longer lived species in July 1992.

iii) Field experiments

Two experiments were carried out along hedgerows on chalk soil in Hampshire. The study farm had used a pneumatic fertiliser applicator for over ten years prior to the experiment, in combination with one metre wide bare "sterile strips" between crop and field boundary vegetation, so hedge banks had not recently been subject to fertiliser input. Nitrogen fertiliser was applied to 25 metre long plots of herbaceous vegetation next to the hedges at 0, 65 or 130 kg N/ha in March of each year for three years, starting in 1990. There were three replicates of each treatment per experiment. In February 1992, four pot-grown seedlings of *B. sterilis* were transplanted into each plot and marked.

Percentage cover was assessed in May 1990 and 1992, and numbers of naturally occurring *B. sterilis* panicles per unit length of hedge bank counted in all three years. Transplanted *B. sterilis* plants were harvested in June, and various vegetative and reproductive attributes measured. Soil samples were taken in November 1990 and March, April and June 1992 for determination of available nitrate and total nitrogen content.

RESULTS

i) Pot experiment

Bromus sterilis was the most competitive of the four grass species, *Holcus lanatus* being the next most competitive and *Festuca rubra* the least. Applied nitrogen significantly increased the competitive ability of *B. sterilis* at the first harvest (when flowering heads started to appear), but by the time of the second harvest, there was no significant effect (Table 1).

TABLE 1. Mean aggressivity of *Bromus sterilis* at four levels of nitrogen application

	Nitrogen applied (Kg N/ha)					
	0	40	80	160	SED	P
First harvest	0.42	0.55	0.51	0.54	0.04	<0.01
Second harvest	0.48	0.43	0.51	0.50	0.04	NS

ii) Semi-field experiment

Only *B. sterilis, Galium aparine, Scandix pecten-veneris* and *Arrhenatherum elatius* established successfully in all treatments. *Dactylis glomerata* and the spring-germinating umbellifers *Heracleum sphondylium, Anthriscus sylvestris* and *Aethusa cynapium* did not establish in competition with *Holcus lanatus*, and made very little growth in the other treatments where competitors were present.

B. sterilis was relatively unaffected by competition with *H. lanatus* and *P. trivialis* sown at the same time and showed a pronounced response to nitrogen application in all treatments (Figure 1(a)). For *G. aparine* there was a significant nitrogen x competitive interaction, with response to nitrogen being greater in the absence of competition and declining as the competitive ability of the competitor species increased, in the order *Poa trivialis* < *Holcus lanatus* < wheat (Figure 1(b)). Seeds sown into one year old swards in autumn 1991 did not establish successfully. Transplanted *B. sterilis* seedlings established but made little growth

FIGURE 1. Response of annual weeds of field margins to nitrogen fertiliser in the presence of different competitors

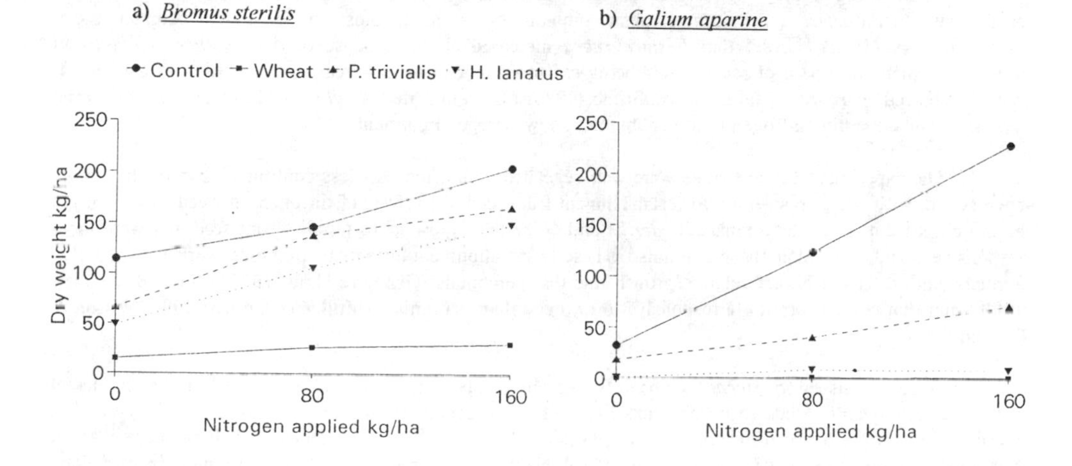

iii) Field experiment

Application of nitrogen fertiliser increased the level of available nitrate in the soil, but did not affect the percentage cover of any of the species present in either hedgebank over the three year period during which measurements were made. The number of *B. sterilis* panicles and spikelets per 0.5m of hedgebank was unaffected by fertiliser, but the number of spikelets per panicle was significantly increased by fertiliser application in the second year. Transplanted *B. sterilis* were larger and had more and larger panicles when nitrogen was applied (Table 2), though in this case there were not significantly more spikelets per panicle.

TABLE 2. Mean plant dry weight, number of tillers, tiller length and reproductive parameters (with one standard error) for *Bromus sterilis* transplanted into hedgebank vegetation (combined data from two hedgerows)

	Nitrogen applied			
	0	65	130	P
Plant dry weight (g)	1.79 (0.45)	3.94 (0.93)	5.64 (1.89)	<0.05
No. tillers/plant	2.3 (0.4)	4.0 (0.7)	5.3 (1.7)	<0.05
Tiller length (mm)	832 (87)	922 (31)	953 (12)	<0.01
Spikelets/panicle	15.4 (2.3)	18.0 (0.5)	18.0 (1.8)	NS
Florets/spikelet	5.64 (0.41)	6.05 (0.28)	6.94 (0.55)	<0.05

DISCUSSION

These experiments have shown that *Bromus sterilis* increases growth and reproductive output in response to nitrogen fertiliser when grown with other species in pots or in field plots, or when transplanted into established vegetation. When seeds were sown into established field plots of *Poa trivialis* and *Holcus lanatus* however, none survived to maturity. *Galium aparine* also responded strongly to nitrogen application, but the response was reduced in proportion to the degree of competition from other species.

Treatment of hedgebank vegetation with nitrogen did not affect the relative species composition over the three-year period of the experiments, in contrast to the results reported by Melman & van Strien (1993), who found significant differences in the botanical composition of nitrogen-treated and untreated ditch banks. However, their nitrogen rates were higher that those in our study. In longer term experiments, fertiliser has been shown to influence composition of plant communities similar to those found in field boundaries e.g. Berendse *et al* (1992) showed that *Festuca rubra* increased at the expense of *Arrhenatherum elatius* in unfertilised plots, the speed of replacement being enhanced by cutting, whereas in the fertilised treatment *A. elatius* replaced *F. rubra*. Mahmond & Grime (1976) also found that *A. elatius* eliminated *F. rubra* and *Agrostis tenuis* in a high nitrogen treatment but not a low nitrogen treatment.

The experimental hedgebanks were well vegetated with a more or less continuous sward. In a more open sward, with more opportunity for establishment from seed, the effects of nitrogen on weed species might be more apparent, since the annuals *B. sterilis* and *G. aparine* grew more rapidly from seed and were more responsive to nitrogen than the perennials. These observations conform with previous work showing that annuals tend to have a higher relative growth rate that perennials (Grime & Hunt, 1975; Muller & Garnier, 1990), and that annuals are more responsive to nitrogen than perennials (Muller & Garner, 1990; Wilson & Tilman, 1991).

Once established, *B. sterilis* seems to be a strong competitor, but it cannot establish in the absence of gaps and consequently tends to become marginalised by perennial grasses in undisturbed field boundaries, towards the interface between the boundary vegetation and the cultivated land (Dunkley & Boatman, 1994). It is commonly observed in a strip along this interface. Nevertheless, annual bromes were found to be capable of establishing in smaller gaps than other grassland herbs (Watt & Gibson, 1988), and this may account for the ability of *B. sterilis* to persist at low levels in readily undisturbed hedgebanks. *G. aparine*, establishes readily in bare ground beneath hedgerow shrubs, where *B. sterilis* does not occur, but it may be more susceptible to competition in a perennial sward.

In conclusion, misplacement of nitrogen fertiliser seems unlikely to encourage annual weeds in a dense closed field boundary sward, but where frequent disturbance creates gaps of sufficient size for establishment, nitrogen fertiliser may stimulate weed growth and increase the likelihood of spread into the cropped area. This paper has only considered nitrogen, but misplacement of fertilisers containing phosphorus and potassium may also be important.

ACKNOWLEDGEMENTS

This work was funded by the Natural Environment Research Council as part of the Joint Agriculture and Environment Programme.

REFERENCES

Baylis, J.M.; Watkinson, A.R. (1991) The effect of reduced nitrogen fertiliser inputs on the competitive effect of cleavers (*Galium aparine)* on wheat (*Triticum aestivum*). *Brighton Crop Protection Conference - Weeds 1991*, **1**, 129-134.

Berendse, F.; Elberse, W.T.; Geerts. R.H.M.E. (1992) Competition and nitrogen loss from plants in grassland ecosystems. *Ecology*, **73**, 46-53.

Boatman, N.D. (1989) Selective weed control in field margins. *Brighton Crop Protection Conference - Weeds 1989*, **2**, 785-794.

Boatman, N.D.; Wilson, P.J. (1988) Field margin management for game and wildlife conservation. *Aspects of Applied Biology 16, The practice of weed control and vegetation management in forestry, amenity and conservation areas*, 53-61.

Bunce, R.G.H.; Cummins, R.P.; French, D.D. (1994) Botanical diversity in British hedgerows. In: *Field Margins - Integrating Agriculture and Conservation*, N.D. Boatman (Ed.), *BCPC Monograph No. 58*, Farnham: BCPC Publications (this volume).

Burrell, A; Hill, B; Medland, J. (1990) *Agrifacts - A handbook of UK and EEC Agricultural and Food Statistics*. Hemel Hempstead: Harvester Wheatsheaf.

Church, B.M. (1981) Use of fertiliser in England and Wales, 1980. *Report of Rothamsted Experimental Station fo 1980*, part 2, pp. 115-122.

Cussans, G.W. (1973) A study of the growth of *Agropyron repens* (L.) Beauv in a grass ley. *Weed Research*, **13**, 283-291.

Dunkley, F.A.; Boatman, N.D. (1994) Preliminary findings from a study of sown and unsown management options for the restoration of perennial hedge-bottom vegetaion. In: *Field Margins - Integrating Agriculture and Conservation*, N.D. Boatman (Ed.), *BCPC Monograph No. 58*, Bracknell: BCPC Publications (this volume).

Green, B.H. (1972) The relevance of seral eutrophication and plant competition to the management of successsional communities. *Biological Conservation*, **4**, 378-384.

Grime, J.P. (1977) Evidence for the existence of three primary startegies in plants and its relevance to ecological and evolutionary theory. *American Naturalist*, **111**, 1169-1194.

Grime, J.P.; Hunt, R.H. (1975) Relative growth rate: its range and adaptive significance in a local flora. *Journal of Ecology*, **63**, 393-422.

Lintell-Smith, G.; Watkinson, A.R.; Firbank, L.G. (1991) The effects of reduced nitrogen and weed-weed competition on the populations of three common cereal weeds. *Brighton Crop Protection Conference - Weeds 1991*, **1**, 135-140.

Lintell-Smith, G.; Baylis, J.M.; Watkinson, A.R.; Firbank, L.G. (1992) The effects of reduced nitrogen and weed competition on the yield of winter wheat. *Aspects of Applied Biology 30, Nitrate and Farming Systems*, 367-372.

Mahmond, A.; Grime, P.J. (1976) An analysis of competitive ability in three perennial grasses. *New Phytologist*, **77**, 431-435.

Mahn, E.G. (1984) The influence of different nitrogen levels on the productivity and structural changes of weed communites in agro-ecosystems. *Seventh International Symposium on Weeds Biology, Ecology and Systematics, Paris*, pp 421-428.

McGilchrist, C.A.; Trenbath, B.R, (1971) a revised analysis of plant competition experiments. *Biometrics*, **27**, 659-671.

Marshall, E.J.P. (1988) The ecology and management of field margin floras in England. *Outlook on Agriculture*, **17**, 178-182.

Marshall, E.J.P. (1989) Distribution patterns of plants associated with arable field edges. *Journal of Applied Biology*, **26**, 247-257.

Marshall, E.J.P. (1990) Interference between sown grasses and the growth of rhizomes of *Elymus repens* (couch grass). *Agriculture, Ecosystems and Environment*, **33**, 11-22.

Melman, Th. C.P.; van Strien, A.J. (1993) Ditch banks as a conservation focus in intensively exploited peat farmland. In: *Ecology of a stressed landscape*, C.C. Vos and P. Opdam (Eds.), London: Chapman and Hall, pp 122-141.

Mountfield, J.O.; Lakhani, K.H.; Kirkham, F.W. (1993) Experimental assessment of the effects of nitrogen addition under haycutting and aftermath grazing on the vegetation of meadows on a Somerset peat moor. *Journal of Applied Ecology*, **30**, 321-331.

Muller, B.; Garnier, E. (1990) Components of relative growth rate and sensitivity to nitrogen availability in annual and perennial species of *Bromus*. *Oecologia*, **84**, 513-518.

Rew, L.J. (1993) Spatial and competitive aspects of arable field margin flora. Unpublished PhD Thesis, University of Reading.

Rew, L.J.; Theaker, A.J.; Froud-Williams, R.J.; Boatman, N.D. (1992a) Nitrogen fertiliser misplacement and field boundaries. *Aspects of Applied Biology 30, Nitrate and Farming Systems,* 203-206.

Rew, L.J.; Froud-Williams, R.J.; Boatman, N.D. (1992b) Implications of field margin management on the ecology of *Bromus sterilis*. *Aspects of Applied Biology 29, Vegetation Management in Forestry, Amenity and Conservation Areas,* 257-263.

Rooney, J.M.; Clarkson, D.T.; Highett, M.; Hoar, J.J.; Purves, J.V. (1990) Growth of *Galium aparine* (cleavers) and competition with *Tritcum aestivum* (wheat) for nitrogen. *Proceedings European Research Society Symposium: Integrated Weed Management in Cereals, Helsinki,* pp 271-280.

Smith, H.; MacDonald, D.W. (1989) Secondary succession on extended arable field margins: its manipulation for wildlife benefit and weed control. *Brighton Crop Protection Conference - Weeds 1989*, 1063-1068.

Tilman, D. (1982) *Resource competition and community structure.* Princeton: Princeton University Press.

Tilman, D. (1988) *Plant strategies and the dynamics and structure of plant communities.* Princeton: Princeton University Press.

Watt, T.A.; Gibson, C.W.D. (1988) The effects of sheep grazing on seedling establishment and survival in grassland. *Vegetatio*, **78**, 91-98.

Wilson, S.D.; Tilman, D. (1991) Interactive effects of fertilisation and disturbance on community structure and resource availability in an old-field plant community. *Oecologia,* **88**, 61-71.

Wright, K.J.; Wilson, B.J. (1992) Effects of nitrogen fertiliser on competition and seed production of *Avena fatua* and *Galium aparine* in winter wheat. *Aspects of Applied Biology 30, Nitrate and Farming Systems,* 381-386.

PHACELIA TANACETIFOLIA FLOWER STRIPS AS A COMPONENT OF INTEGRATED FARMING

J.M. HOLLAND, S.R. THOMAS, S. COURTS

The Game Conservancy Trust, Fordingbridge, Hampshire, SP6 1EF

ABSTRACT

The impact of a *Phacelia tanacetifolia* field margin on aphid-specific predators in winter wheat was investigated at one of the MAFF/LINK Integrated Farming Systems (IFS) sites. Syrphidae were attracted to the *Phacelia* strip and gut dissections confirmed they were feeding on the pollen before distributing up to 100m into the field. No evidence of improved fecundity as a result of feeding on *Phacelia* was found. Ichneumonidea were also more abundant in the *Phacelia* strip compared to the adjacent wheat crop. There were trends towards higher numbers of Braconidae, Proctotrupoidea and *Platypalpus* spp. in the IFS plots whereas Chalcidoidea were more numerous in the conventional farming system plots. Aphid abundance and percentage parasitism were unaffected by differences in aphid-specific predator distribution.

INTRODUCTION

An option for arable crop production currently being investigated throughout Europe is the adoption of integrated farming methods (Holland *et al.*, 1994). One of the main objectives of integrated farming is to utilise natural regulatory mechanisms and thereby reduce the need for inputs of pesticides, fertilizers and fuels (Anon, 1993). The establishment of a flower border strip which is attractive to cereal aphid predators, for example hoverflies (Diptera:Syrphidae), is one technique which may enhance cereal aphid control and so reduce the need for aphicides. Adult hoverflies require proteins from pollen to mature their reproductive systems (Schreiber, 1948). Eggs are then laid in association with aphid colonies on which the larvae subsequently feed (Dean, 1974). Many hymenopteran parasitoids also feed on non-host flowers to obtain nectar or pollen (Jervis *et al.*, 1993) which can increase fecundity and longevity (Jervis and Kidd, 1986). A number of plant species have been evaluated for their attractiveness to hoverflies (MacLeod, 1992) and hymenopteran parasitoids (Jervis *et al.*, 1993). In addition *Phacelia tanacetifolia* pollen is very distinctive and can be readily identified within the guts of dissected hoverfly specimens so aiding experimental evaluations (Hickman, 1994).

A new programme of research was started in 1992 to investigate integrated farming - the MAFF/LINK Integrated Farming Systems project (Holland, 1994). As part of a programme of measures to reduce the inputs of agrochemicals, strips of *Phacelia* were established around the edge of cereal fields at some of the experimental sites. Their impact on hoverfly, parasitoid and aphid populations was evaluated at one of these sites.

MATERIALS AND METHODS

At the MAFF/LINK IFS site in north Hampshire, conventional farm practice (CFP) and integrated farming system (IFS) were compared using adjacent pairs of plots (min. size 5 ha, min. width 100 m). Detailed descriptions of the experimental design are available in Holland

(1994). In this study four pairs of winter wheat plots were used. A strip of *Phacelia* was sown in mid April along the longest edge (300-400 m) of the four winter wheat IFS plots. A transect of fluorescent yellow water traps (19 cm diameter) were located at crop height in a line at 45° and distances of 10, 25, 50 ,75 and 100 m from each *Phacelia* strip. Another transect of traps was also located on the opposite side of the field but adjacent to the conventional plot. An additional trap was placed within each *Phacelia* strip and on one edge of each conventional plot. The traps were emptied weekly from the 13th June 1993 until the 16th July 1993. *Phacelia* started to flower on the 21st June 1993. The contents of each trap were stored separately in 70% alcohol until identification. The following groups only are discussed: Syrphidae, *Aphidius* spp. (Braconidae), other Braconidae, Ichneumonidea, Proctotrupoidea, Chalcidoidea, Cynipoidea, *Platypalpus* spp. (Diptera:Empididae). Total numbers caught in each group were analyzed separately using Analysis of Variance (ANOVA) with farming system and distance from field margin as factors.

The proportion of Syrphidae which had been feeding on the *Phacelia* was determined by dissecting each syrphid and recording the presence/absence of *Phacelia* pollen. To obtain an estimate of fertility the presence/absence of eggs was also noted. The proportion of males and females with pollen and proportion of females with eggs (arc sine transformed) were analyzed separately using ANOVA with farming system and distance as factors. Means were separated using LSD tests.

Two winter wheat IFS plots were sampled using a Diettrich vacuum insect sampler (D-vac). Five samples were taken (10 s suction, repeated five times at 2 m intervals) from the *Phacelia* strip, at 3 m from the crop edge (in the selectively sprayed Conservation Headland area) and at 10 m from the crop edge. Samples were frozen and the above groups identified. To determine differences between sampling positions individual groups were analyzed using one-way ANOVA.

To estimate the proportion of parasitised aphids, 10 ears of aphid infested wheat were collected in the vicinity of each water trap on three separate occasions. Ten grain aphids per ear were removed and reared on wheat leaves in glass vials in the laboratory until each aphid had reached adulthood and begun to produce offspring. Emergent parasitoids and hyperparasitoids were identified.

Aphid numbers, species and life-stage (adult or nymph) was assessed on 50 tillers in each winter wheat plot on five occasions. To determine whether species composition differed between the two farming systems the ratio of the two species was calculated. To determine whether the population age structure varied between the two farming systems the ratio of adults to nymphs was calculated.

RESULTS

The ANOVA of the water trap data revealed that there were significantly ($P<0.05$) more *Platypalpus* spp. in the IFS plots and numbers in the *Phacelia* strip and conventional field margin were significantly ($P<0.05$) less compared to the crop (Table 1). The numbers of Proctotrupoidea and Ichneumonidea were significantly ($P<0.05$) greater at the field margin, but only significantly higher in the *Phacelia* strip with the latter. There were trends towards higher numbers of Syrphidae in the *Phacelia* strip and more Proctotrupoidea and other Braconidae in the IFS plots, and greater numbers of Chalcidoidea in the CFP plots.

TABLE 1. Total number per trap (± pooled SE) of each invertebrate group captured in water traps in each plot and results of ANOVA. (ns=not significant, D=significant differences between sampling distances, S=significant difference between farming systems).

Invertebrate group	CFP	IFS	SE	ANOVA results
Aphidius spp.	16.6	3.4	1.29	ns
Other Braconidae	59.4	77.7	10.3	ns
Ichneumonidea	12.7	14.8	1.3	ns
Proctotrupoidea	81.9	94.4	10.7	D
Chalcidoidea	346.2	295.6	43.5	ns
Cynipoidea	11.9	10.5	1.3	ns
Syrphidae	16.5	20.5	2.2	ns
Platypalpus spp.	251.0	378.3	26.0	S,D

The ANOVA comparing the different D-vac suction sample positions indicated that there were no significant differences (P>0.05). This was expected because of the low between-subjects degrees of freedom (2). There were trends towards higher numbers of *Aphidius* spp., other Braconidae, Proctotrupoidea and Syrphidae in the *Phacelia* strip whilst Cynipoidea and *Platypalpus* spp. were more abundant in the crop (Table 2).

TABLE 2. Mean number (± SE) of each invertebrate group per D-vac sample in *Phacelia* strip, crop margin and at 10m from the crop edge.

Invertebrate group	*Phacelia*		Crop edge	10m into crop
Aphidius spp.	10.6	(±2.4)	5.2 (±0.7)	7.6 (±0.9)
Other Braconidae	186.9	(±80.1)	26.7 (±6.1)	16.5 (±2.2)
Ichneumonidea	1.6	(±0.8)	2.2 (±0.5)	2.6 (±0.3)
Proctotrupoidea	16.2	(±8.5)	4.2 (±1.5)	6.9 (±0.7)
Chalcidoidea	8.0	(±1.0)	3.8 (±0.5)	7.5 (±1.3)
Cynipoidea	1.8	(±0.6)	2.6 (±0.5)	3.0 (±0.7)
Syrphidae	1.9	(±0.6)	0.6 (±0.3)	0.4 (±0.2)
Platypalpus spp.	3.0	(±0.9)	6.3 (±1.6)	8.9 (±1.1)

The ANOVA comparing the proportion of Syrphidae containing *Phacelia* pollen indicated that significantly (P<0.001) more males and females had fed on *Phacelia* in the IFS plots, but this only varied significantly (P<0.01) with distance from the field margin for females. The Syrphidae were distributing throughout the IFS plot and occasionally further into the CFP plots (Table 3).

TABLE 3. Percentage of Syrphidae males and females containing *Phacelia* pollen, and the proportion of females with eggs in each farming system.

Distance from field margin	Males CFP	Males IFS	Females CFP	Females IFS	Eggs CFP	Eggs IFS
0m	3	43	8	66	24	35
10m	6	33	4	22	48	24
25m	4	34	6	22	19	34
50m	5	15	0	31	13	36
75m	6	18	13	0	28	65
100m	0	24	6	13	35	30
Pooled S.E	8		7		10	

No significant (P>0.05) effects were found for the distribution of Syrphidae containing eggs. Results were variable with peaks at 10m in the CFP plots and at 75m in the IFS plots (Table 3).

TABLE 4. Mean number of each aphid species, the species ratio and adult/nymph ratio per tiller in each farming system.

Farming system	Sampling date 8/6/93	17/6/93	23/6/93	29/6/93	7/7/93
	Mean number of grain aphids (GA)				
CFP	0.3	1.1	0.5	0.9	1.2
IFS	0.6	1.7	0.9	1.1	0.7
	Mean number of rose-grain aphids (RGA)				
CFP	0.1	0.3	0.2	0.3	0.4
IFS	0.1	0.3	0.2	0.4	0.7
	Ratio of GA:RGA				
CFP	0.5	0.9	0.6	0.4	0.3
IFS	0.4	0.2	0.3	0.6	1.4
	Ratio of GA adults:nymphs				
CFP	1.7	3.0	4.7	3.1	6.9
IFS	1.9	4.6	5.7	1.6	4.0
	Ratio of RGA adults:nymphs				
CFP	2.3	4.0	10.4	3.2	2.8
IFS	4.5	1.6	5.2	5.7	2.9

Aphid populations were variable and low in all plots (0-1.7 per tiller) and there appeared to be no consistent difference in the number of species, their ratio or adult to nymph ratio between the CFP and IFS plots (Table 4).

The percentage parasitism was very low in all plots and did not vary with distance from the field margin (Table 5). The higher numbers of parasitoids found in the *Phacelia* strip was not therefore reflected in the proportion of parasitised aphids. *Aphidius* spp. were the most common aphid parasitoids throughout the plots.

TABLE 5. The percentage of parasitised aphids in each farming system at different distances from the field margin.

Distance from field margin	CFP	IFS
0m	5.7	5.7
10m	7.1	10.0
25m	0	1.4
50m	8.6	5.7
75m	4.3	2.9
100m	7.1	7.1
Overall mean	5.47	5.47

DISCUSSION

The results of the D-vac suction samples and water traps although generally not conclusive did indicate some possible trends. Syrphidae were utilizing the pollen source provided by *Phacelia* as found by Harwood *et al*., (this volume) and were subsequently distributing into the adjacent crop. The abundance of Syrphidae in the crop was, however, not improved by the presence of the *Phacelia* strip. Whether Syrphidae fertility was being improved was not verified because of variable results from gut dissections and the absence of eggs in the crop. *Phacelia* was also acting as an attractant to a range of parasitoids and hyperparasitoids notably *Aphidius* spp., other Braconidae and Proctotrupoidea but again the abundance was not increased in the IFS compared to the CFP plots. This was also reflected in the relatively even abundance of aphids and the proportion parasitised in the IFS and CFP plots. *Phacelia* may only be attracting airborne invertebrates from the adjacent hedgerow, although the water trap samples (insufficient space to present complete data) did not indicate a massive influx of parasitoids when the *Phacelia* started to flower. The impact of providing such a massive pollen source on aphid-specific predator populations requires further investigation to determine whether populations are increasing or redistributing so lowering levels of aphid control in other areas. Incorporating Umbelliferae species into the *Phacelia* may also enhance parasitoid populations as these are visited by a wide range of parasitoids species (Jervis *et al*., 1993). The greater abundance of *Platypalpus* spp. in the IFS plots was probably a result of another component of the integrated system and require further investigation.

The scale required to investigate different farming systems often limits within-year replication and therefore the likelihood of detecting statistically significant differences. Further investigations over several years are required to confirm the findings from these initial studies. In these investigations the effect of polyphagous aphid predators could not be excluded and therefore more specific experiments are required to isolate the impact of the aphid specific predators reported here.

ACKNOWLEDGEMENTS

The authors thank the Manydown Company for allowing field work on their farm and Dr N.J. Aebischer of The Game Conservancy Trust for statistical guidance.

REFERENCES

Anon, 1993. Integrated Production, Principles and Technical Guidelines. IOBC/WPRS Bulletin, **16**(1).

Dean, G.J. (1974) Effects of parasites and predators on the cereal aphids *Metopolophium dirhodum* and *Macrosiphum avenae* (Hemiptera: Aphididae). *Bulletin of Entomological Research*, **20**, 209-224.

Harwood, R.W.J.; Hickman, J.M.; MacLeod, A.; Sherratt, T.N.; Wratten, S.D. (this volume) Managing field margins for hoverflies.

Hickman, J. (1994) *Phacelia* as a resource for hoverflies. IOBC/WPRS Working Group 'Integrated control in cereal crops.' Bulletin 1994/XV (In press).

Holland, J.M. (1994) A MAFF/LINK project - integrated farming systems. IOBC/WPRS Working Group 'Integrated control in cereal crops.' Bulletin 1994/XV (In press).

Holland, J.M.; Frampton, G.K.; Wratten, S.D.; Cilgi, T. (1994) Arable acronyms analyzed -a review of integrated farming systems research in Western Europe. *Annals of Applied Biology*. (In press).

Jervis, M.A.; Kidd, N.A.C. (1986) Host-feeding strategies in hymenopteran parasitoids. *Biological Reviews*, **61**, 395-434.

Jervis, M.A.; Kidd, N.A.C.; Fitton, M.G.; Huddleston, T.; Dawah, H.A. (1993) Flower-visiting by hymenopteran parasitoids. *Journal of Natural History*, **27**, 67-106.

MacLeod, A. (1992) Alternative crops as floral resources for beneficial hoverflies (Diptera: Syrphidae). *Brighton Crop protection Conference - Pests and Diseases 1992*, **2**, 997-1002.

Schreiber, F. (1948) Beitrag zur kenntnis der Generationsuerhaltnisse and Diapause rauberischer Schwebfliiegen. *Mitteilungen der Scheizerischen Entomologischen Gesellschaft*, **21**, 249-285.

EFFECTS OF UNSPRAYED CROP EDGES ON FARMLAND BIRDS

G.R. DE SNOO, R.T.J.M. DOBBELSTEIN, S. KOELEWIJN

Centre of Environmental Science, Leiden University
P.O. Box 9518, 2300 RA Leiden, The Netherlands

ABSTRACT

In 1992 and 1993 the abundance of three farmland bird species, *Motacilla flava flava, Alauda arvensis* and *Anthus pratensis* in sprayed and unsprayed edges of winter wheat fields was investigated. Compared with the sprayed crop edges, the number of visits of *Motacilla flava flava* was significantly 3 - 4.5 times higher on the unsprayed edges. There was no difference in the frequency of visits for the other two species. The number of territories of *Motacilla flava flava* per hectare correlates with the percentage area of winter wheat on the farms.

INTRODUCTION

Over the past few decades there has been a substantial decline in the number of farmland birds in the Netherlands (Kwak *et al.*, 1988). This may be due in part to the use of herbicides and insecticides on arable land. In addition to direct effects there may also be indirect effects, for example resulting from changes in food abundance (*cf.* de Snoo & Canters, 1990). It has been demonstrated that leaving the edges of cereal fields unsprayed benefits populations of gamebirds such as *Perdix perdix* (Partridge) and *Phasianus colchicus* (Pheasant), because the greater abundance of insects in the unsprayed crop edges benefits chick survival (Rands, 1985; 1996). In the present study it was investigated whether unsprayed crop edges also attract small songbirds such as *Motacilla flava flava* (Blue-headed wagtail), *Alauda arvensis* (Skylark) and *Anthus pratensis* (Meadow pipit). *Motacilla flava flava* and *Anthus pratensis* are both insectivores, while *Alauda arvensis* is also partly herbivorous eating weed seeds, seedling cotyledons and leaves from weeds and crop (Cramp, 1988; Green, 1978; 1980). The research is part of the Dutch Field Margin Project being undertaken in the Haarlemmermeerpolder. The aim of this project is to develop a management strategy for promoting nature conservation in arable fields and reducing pesticide drift to non-target areas. Effects on weeds, invertebrates, vertebrates, pesticide drift and costs to the farmer are being studied.

METHODS

The research was conducted in the Haarlemmermeerpolder (clay soil) in 1992 and 1993. In this polder most parcels of land are 1000 m long and 200 m wide and are bordered by ditches. The most common rotation on the farms was: winter wheat, followed by potatoes, a second winter wheat crop, and finally sugar beet. The study was carried out on 8 farms in 1992 and on 7 farms in 1993. In one winter wheat field at each farm a 6-metre wide crop edge bordering on a ditch was left unsprayed; these crop edges had a mean length of 424 $\pm$ 116 m. The total length of unsprayed crop edge was 3790 m in 1992 and 2560 m in 1993. No herbicides or insecticides were used on these margins. Spraying with fungicides was allowed, but no organophosphate fungicides were used. The unsprayed edges were compared with sprayed edges which were generally in the same field. On the unsprayed crop edges weed coverage increased substantially, as did the overall number of weed species. Dominant species include *Matricaria recutita, Polygonum aviculare* and *P. convolvulus*.

The number of farmland birds frequenting the crop edges was determined in a linear transect census (strip census; Hustings *et al.*, 1989), with all birds visiting a sprayed or unsprayed edge being recorded (ground contact). In 1993 a census was also made of the number of birds visiting an imaginary strip in the centre of each field and visiting the other side of the field. A census was additionally carried out of the number of birds sighted across the ditch on the adjacent field, where in most cases the crop was potatoes or sugar beet rather than winter wheat. All the strips recorded had the same area. In 1992 10 census sessions were performed on each farm between 08.30 h and 14.30 h in the period from 26 May to 7 July. In 1993 a weekly census was undertaken between 15 April and 15 July, with 5 sessions being held at each farm between 06.00 h and 09.00 h and 7 sessions between 09.00 h and 16.15 h Central European Time.

In 1993, at the same time as the strip census, for each species of bird the number of territories was determined in the entire winter wheat field as well as in the first 100 m of each adjacent field. The area recorded on each farm had an average size of 27.8 ± 6.9 ha. In preparing the territory maps, both territory-indicative and nest-indicative sightings were used (SOVON, 1985). A positive territory indication was recorded only for conclusive observations or concentrations of very widely distributed observations that fell within the valid date limits for the species in question. Finally, in 1993 a number of individuals of *Motacilla flava flava, Alauda arvensis* and *Anthus pratensis* was observed in order to study differences in feeding behaviour in sprayed and unsprayed crop edges. To this end, between 08.00 h and 14.45 h from 4 June to 16 July a total of 15 birds were observed for more than one hour on 6 farms from an observation post camouflages with a net. A record was made of the time spent by the birds on the following activities; flying, singing, feeding, resting and 'unseen field presence' (= bird present in the field but not visible).

RESULTS

Major differences were found in the presence of the three species on the crop edges. *Motacilla flava flava* was most frequently observed. In 1992 a total of 110 visits by this species to sprayed and unsprayed edges was recorded; in 1993 this figure was 60. *Alauda arvensis* was recorded 20 and 23 times in these two years, respectively, and *Anthus pratensis* only 4 and 6 times. In the two years of study the total number of visits by *Motacilla flava flava* (Figure 1) to the unsprayed crop edges was significantly higher than in the sprayed edges (significant difference $P = 0.007$ in 1992 and $P = 0.03$ in 1993; Wilcoxon paired sample test). The difference was greater by a factor of 3 to 4.5 in the unsprayed edges. In the two years the average number of visits by *Motacilla flava flava* per km crop edge was 2.37 and 1.46, respectively, in the unsprayed margins as compared with 0.53 and 0.49 in the sprayed margins. Throughout the census period of both years the number of visits by *Motacilla flava flava* was consistently higher in unsprayed edges, with the exception of one census session in 1993.

In the case of *Alauda arvensis* and *Anthus pratensis*, in neither year there was a significant difference in species abundance in sprayed and unsprayed edges. The greatest number of visits to crop edges was 0.39 per km for *Alauda arvensis* and 0.13 per km for *Anthus pratensis*. It is noteworthy that 3 of the 6 sightings of *Anthus pratensis* were during the first census session and that 4 birds were observed on just one farm.

FIGURE 1. Number of visits by bird species per km in sprayed and unsprayed winter wheat edges in 1992 and 1993 (* = $P < 0.05$; ** = $P < 0.01$; Wilcoxon paired sample test)

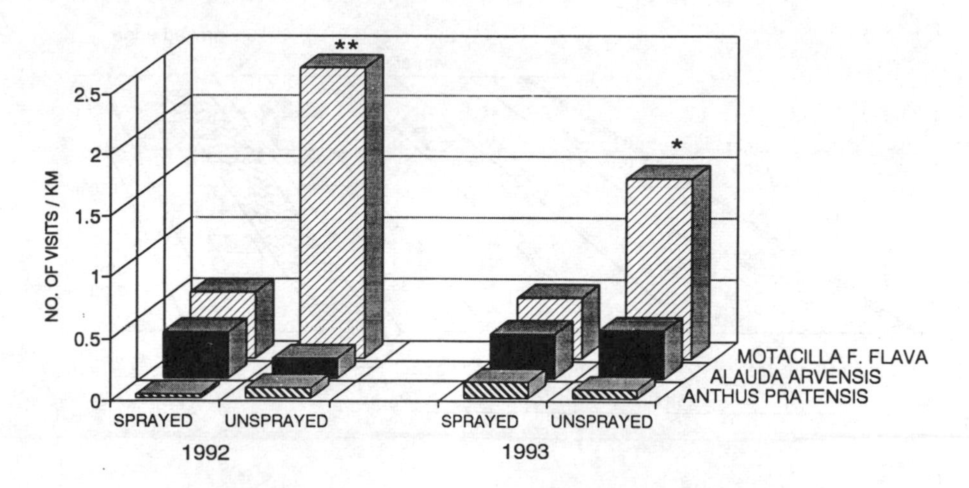

An analysis of species abundance over the entire plot indicated that the number of visits by *Motacilla flava flava* and *Alauda arvensis* to field edges was greater than that to the plot centre (Figure 2). This also seemed to be the case for *Anthus pratensis* but the number of observations was very small. The count was also higher for visits to sprayed edges bordering on a ditch than to sprayed edges directly adjacent to a second plot.

FIGURE 2. Total number of visits of bird species per km during the 1993 season in the various parts of the fields; 2A = *Motacilla flava flava*, 2B = *Alauda arvensis* and 2C = *Anthus pratensis*

2A: *Motacilla flava flava*

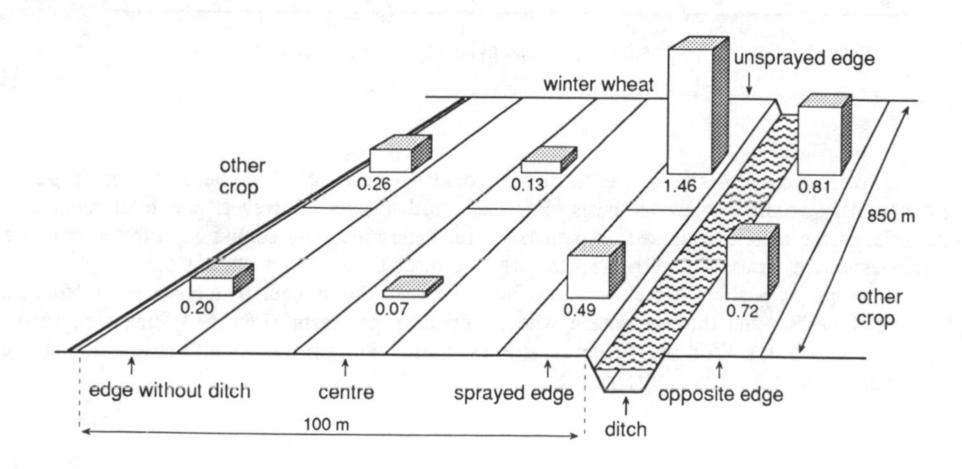

2B: *Alauda arvensis*

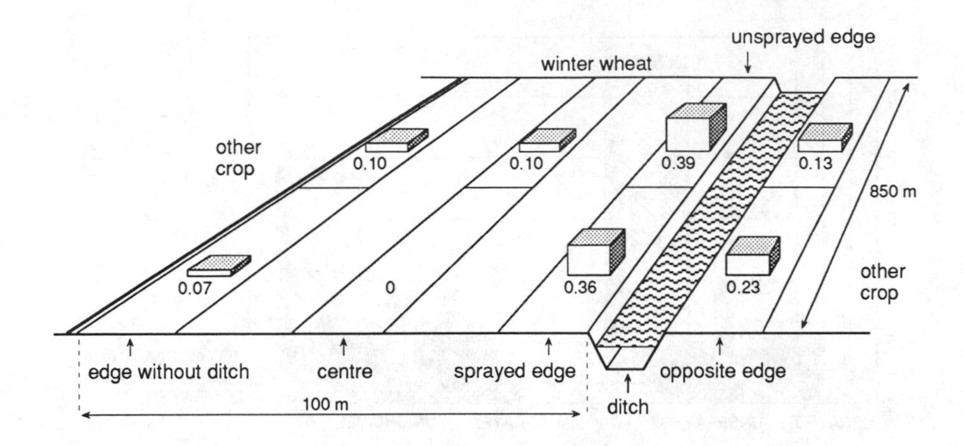

2C: *Anthus pratensis*

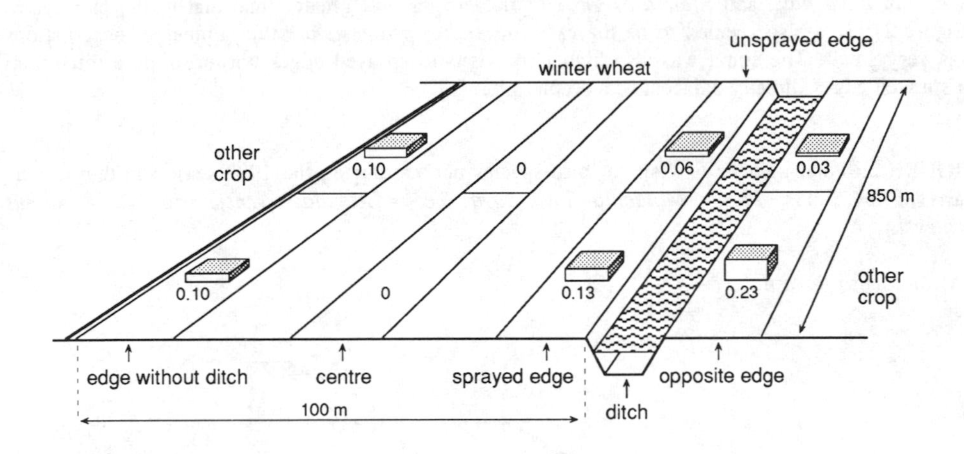

The exact shape and size of the territories could not be established, because there were too few boundary conflicts between birds. For each bird species, however, the total number of territories in the area investigated on each farm was determined and compared with the crop area on the respective farms. The territories of farmland birds generally encompass several plots, with different crops. A positive correlation was found between the number of territories of *Motacilla flava flava* per ha and the percentage wheat crop area per farm (Figure 3; linear regression coefficient r = 0.91). With the other two species, no relationship was found between the area of a given crop and the number of territories.

FIGURE 3. Number of territories of *Motacilla flava flava* in 1993 per ha compared to area of winter wheat (in %) in area studied on each farm

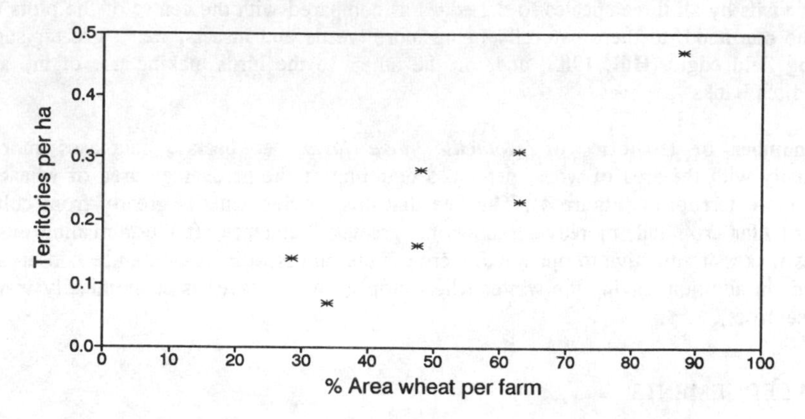

The results of the observations of individuals of *Motacilla flava flava* are shown in Table 1. For the larger part of the observation period this species had an 'unseen field presence', and consequently during this period no definite conclusions could be drawn as to activities. Feeding could be observed during a number of brief periods only. *Motacilla flava flava* was observed to feed both on the ground and on the crop itself. Because of the limited number of observations, no difference could be observed in the length of time spent feeding in sprayed versus unsprayed crop edges. Neither *Alauda arvensis* nor *Anthus pratensis* were observed to feed.

TABLE 1. Mean percentage of total time spent on each activity by *Motacilla flava flava*

Activity	Flying	Singing	Resting	Feeding	Unseen
% time	2.9 ± 1.3	4.7 ± 5.9	2.5 ± 4.2	1.2 ± 1.6	88.7 ± 8.1

DISCUSSION

Leaving a 6-metre wide strip along the edge of a crop unsprayed has pronounced positive effect on the number of visits by *Motacilla flava flava*. However, based on the small number of individual bird observations, it was not feasible to determine whether *Motacilla flava flava* indeed feeds more in unsprayed crop edges. Establishing unsprayed crop edges seemed to have no positive effect on the abundance of *Alauda arvensis*. A similar lack of effects of unsprayed crop edges on abundance of *Anthus pratensis* was indicated, but the number of observations of the latter species was extremely small. *Motacilla flava flava* is insectivorous, and insects such as Heteroptera, Chrysomelidae, Curculionidae and Carabidae are generally more abundant in unsprayed winter wheat edges (Rands, 1985; Storck-Weyhermüller & Welling, 1991 *etc.*). In our study area the most marked effects in the unsprayed edges were found on invertebrates which live on plants. The effects on soil invertebrates were only small (De Snoo, 1993; De Snoo *et al.*, 1994). The difference between *Motacilla flava flava* and *Alauda arvensis* may be explained since the latter species is partly herbivorous and a difference in feeding strategy: *Alauda arvensis* walks the ground feeding from the soil surface and the lower parts of plants, but never perches on plants to feed (Green, 1978, Cramp, 1988). However, *Motacilla flava flava* will also eat insects from te upper parts of the plants or even from the air (fly-catching) (Cramp, 1988). Moreover, *Alauda arvensis* may be avoiding the profuse weed growth in the unsprayed edges. The larger number of visits by all three species to the edges as compared with the centre of the plots may be due, on the one hand, to there naturally being more weeds and insects, i.e. a greater supply of food, along field edges (Hill, 1985) and, on the other, to the birds making use of the adjacent ditch and ditch banks.

The number of territories of *Motacilla flava flava* per hectare increases more than proportionally with the area of wheat per ha, a doubling of the percentage area of wheat giving 2.6 times more territories (Figure 4). The fact that this species benefits greatly from cultivation of this particular crop and, moreover, shows far greater abundance after establishing unsprayed crop edges makes it attractive to opt for this crop if the aim is to increase nature conservation of arable land. In addition leaving the winter wheat crop edge unsprayed is economically viable (De Snoo, these proc.).

ACKNOWLEDGEMENTS

The authors thank R. Cuperus for his help with the field work and with interpreting the results of the territory mapping.

REFERENCES

Cramp, S. (1988) *The birds of the Western Palaearctic*. Oxford: Oxford University Press, pp. 188-204, 413-433.

Green, R. E. (1978) Factors affecting the diet of farmland Skylarks, *Alauda arvensis*. *Journal of Animal Ecology* **47**, 913-928.

Green, R.E. (1980) Food selection by skylarks and grazing damage to sugar beet seedlings. *Journal of Applied Ecology* **17**, 613-630.

Hill, D.A. (1985) The feeding ecology and survival of pheasant chicks on arable farmland. J. Applied Ecology **22**, 645-654.

Hustings, M.; Kwak, R.; Opdam, P.; Reijnen, M. (1989) *Vogelinventarisatie. Achtergronden, richtlijnen en verslaggeving*. Nederlandse Vereniging tot Bescherming van Vogels Zeist, Wageningen: Pudoc, pp. 39-43.

Kwak, R.G.M.; Reyrink, L.; Opdam, P.; Vos, W. (1988) *Broedvogeldistricten van Nederland. Een ruimtelijke visie op de Nederlandse avifauna*. Wageningen: Pudoc, 143 pp.

Snoo, G.R. de (1993). Onbespoten akkerranden in de Haarlemmermeerpolder. *Landinrichting* **33** (4), 31-34.

Snoo, G.R. de; Canters, K.J. (1990) *Side-effects of pesticides on terrestrial vertebrates*. CML reports **35**, Centre of Environmental Science, Leiden University, pp. 147.

Snoo, G.R. de; Poll, R.J. van der; Leeuw, J. de (1994) Carabids in sprayed and unsprayed crop edges of winter wheat, sugar beet and potatoes. *Acta Jutlandica* in prep.

SOVON (1985) *Handleiding Broedvogel-monitoringproject*. Arnhem: Samenwerkende Organisaties Vogelonderzoek Nederland, pp. 73.

Rands, M.R.W. (1985) Pesticide use on cereals and the survival of grey partridge chicks: a field experiment. *J. Applied Ecology* **22**, 49-54.

Rands, M.R.W. (1986) The survival of gamebird (Galliformes) chicks in relation to pesticide use on cereals. *Ibis* **128**, 57-64.

Storck-Weyhermüller, S.; Welling, M. (1991) Regulationsmöglichkeiten von Schad- und Nutzarthropoden im Winterweizen durch Ackerschonstreifen. *Mitt. Biol. Bundesanst. Land-Forstwirtschaft. Berlin-Dahlem*. Heft **273** Berlin.

FACTORS INFLUENCING THE PLANT SPECIES COMPOSITION OF HEDGES - IMPLICATIONS FOR MANAGEMENT IN ENVIRONMENTALLY SENSITIVE AREAS.

C.A. HEGARTY, J.H. McADAM

Department of Applied Plant Science, The Queens University, Newforge Lane, Belfast BT9 5PX

A.COOPER

Department of Environmental Studies, University of Ulster at Coleraine, BT52 1SA.

ABSTRACT

A survey of hedges throughout Northern Ireland (NI), from 1990-92, revealed that most were gappy and that only 55% were managed (Hegarty, 1992). The aim of the survey was to assess the conservation status of hedges in NI and to determine the major factors affecting their plant species composition. Townland hedges, one of the oldest hedge types in Ireland, were found to have a greater tree and shrub species diversity and were associated with more woodland herbs than other field boundary hedges. Their greater structural diversity may account for this. Unmanaged hedges contained more *Prunus spinosa*, *Ilex aquifolium* and *Salix* spp. than managed hedges which contained more *Crataegus monogyna*, *Fraxinus excelsior* and *Rosa* spp. Stepwise multiple regression analysis showed that hedge width and soil total nitrogen content were the best predictors of hedge ground flora diversity. Hedge management and height had most influence on tree and shrub species diversity. Regional variation in species composition was strongly linked with associated land use. The majority of less intensively managed farmland and the greatest frequency of species-rich hedges were found in western NI, particularly Co. Fermanagh. Species-poor hedges, characterised by species such as *Urtica dioica* and *Galium aparine* were most common in the intensively farmed lowland areas of Counties Londonderry, Antrim and Armagh. The results indicate that when farm rationalisation involves hedge removal, the taller wider hedges should preferentially be left, if the objective is to maintain species diversity. They also emphasise the need to decrease the intensity of hedge management and to maintain a wide boundary strip. These findings should result in more detailed and practical hedge management guidelines and will be especially useful in Northern Ireland's three Environmentally Sensitive Areas, where enhanced grants are available for hedgerow regeneration and hedge planting.

INTRODUCTION

Hedges are an integral part of the Irish landscape, with an estimated length in Northern Ireland of 170,000km (Murray *et al*, 1992). Although in Northern Ireland widespread removal has not occurred on the same scale as in Great Britain, removal rates of 0.5% yr^{-1} since the 1960s are common (Murray *et al*, 1992). Regional variation in the plant species composition of hedges has been related to underlying differences in geology (Pollard *et al*, 1974) and other environmental variables. Hedge structure has been shown to affect species composition with the widest hedges supporting the most woodland herbs (Baudry, 1984; Baudry, 1988). Irish hedges are generally much younger than hedges in Great Britain, the majority having been planted between 1750 and 1850. The relationship between age of hedge and species diversity is further complicated by the fact that different tree species were frequently planted together (Robinson, 1977). Townland boundaries are associated with one of the oldest hedge types in Ireland. The townland was the first administrative division in Ireland to be mapped on the first Ordnance Survey of Ireland (1829-1836). All other divisions (parishes and counties) are made up of groups of townlands. These townland boundaries were often estate and farm boundaries and it may be assumed that they were the first to be planted with hedges to make them stockproof.

Northern Ireland contains only 1.8% semi-natural broadleaf woodland (Murray *et al*, 1992) and hedges are valuable as wildlife reservoirs. Hedge management is encouraged with financial grant aid for restoration, laying, replanting and guidelines have been published by the Department of Agriculture for Northern Ireland (DANI, 1992). In the three Environmentally Sensitive Areas (ESA), (The West Fermanagh and Erne Lakeland; The Mournes and Slieve Croob: The Antrim Coast, Glens and Rathlin ESA), enhanced grants are also available (80% standard cost).

The aim of the research described is to analyse the relative importance of factors such as hedge age, physical environment, structure and adjacent land use to hedge species composition, by studying the structure and botanical composition of hedges in six contrasting areas of Northern Ireland and relating this to their current management by farmers.

METHODS

Field Sampling Procedure

The Northern Ireland multivariate land classification (Cooper, 1986) was used to construct a stratified random hedge sampling program, based on 25ha Ordnance Survey Grid squares. Land classes considered to be representative of enclosed hedged landscapes in six study areas were sampled: 1) Fermanagh, 2) Sperrins, 3) Mournes, 4) Antrim, 5) Armagh, 6) Londonderry and Antrim.

To enable the full range of hedge types to be described, stratified sampling was necessary. This involved sampling hedges associated with differing boundary and land use types. Field records on the structure and botanical species composition of 418 hedges were recorded from linear plots (10m x 1m) located at the centre of each hedge.

Analytical procedures

Tree and shrub, ground flora, ditch and boundary strip plots were classified separately by TWINSPAN (Hill, 1979). Arbitrary stopping criteria for the classifications were based on minimum class size. A woodland flora characterised the species-rich hedges, whilst competitive grasses dominated the species-poor hedges. Pearson correlation coefficients, t-tests, analysis of variance and stepwise linear regression analysis were carried out with SPSS (Norusis 1990).

The chemical characteristics of hedge soils

To investigate the influence of soil properties on the hedge plant species composition, groups of samples from the eight major ground flora groups generated by TWINSPAN (Hill, 1979), were used to stratify a soil sampling program, representative of typical hedges in terms of their floristic composition. Twelve hedge samples were randomly chosen from each of the groups listed: a) Species-poor groups: *Galium aparine/Ranunculus repens, Lolium perenne/Phleum pratense, Lolium perenne/Urtica dioica, Anthoxanthum odoratum/Holcus lanatus;* b) Species-rich groups: *Rubus* spp./*Viola riviniana, Primula vulgaris/Hedera helix, Anthoxanthum odoratum/Digitalis purpurea, Juncus effusus/Rubus* spp.
From each of the chosen hedge samples, 5 soil samples were taken, 2 metres apart along the 10m linear plot, using a soil auger core (15cm deep x 3 cm diameter). The 5 samples were aggregated and when air-dried, sieved through a 2mm mesh. Chemical analyses were performed according to Allen *et al*(1986).

RESULTS

Regional Variation

Hedges in Co. Fermanagh contained significantly more tree and shrub species per plot than all other study areas, with the exception of County Armagh (Table 1). *Corylus avellana, Ilex aquifolium, Prunus spinosa* and *Fraxinus excelsior* were most abundant in Fermanagh. The upland Sperrins study area had the highest mean cover of *Sorbus aucuparia* and the Mournes contained the highest mean cover of *Ulex europaeus*. The lowland Londonderry/Antrim study area contained the greatest mean percentage cover of *Crataegus monogyna*. Fermanagh supported the highest abundances of woodland species such as *Primula vulgaris* and other species, for example *Filipendula ulmaria, Juncus effusus, Hedera helix, Taraxacum officinale* and *Thuidium tamarisicinum* were significantly more frequent in Fermanagh. Hedges in the Sperrins study area had the highest cover of *Polytrichum commune, Anthoxanthum odoratum* and *Pteridium aquilinum*. The mean cover value of *Digitalis purpurea* was significantly higher in the Mournes, than for any group except the other upland study area, the Sperrins. The mean cover values of the competitive-ruderal species *Galium aparine* and *Urtica dioica* were greatest in Armagh, as was *Anthriscus sylvestris*. The lowland Londonderry/Antrim study area contained the significantly highest mean cover value of *Cirsium vulgare* and the highest cover values of *Holcus lanatus* and *Veronica* spp.

Hedge structure

Complete hedges (with less than 10% gaps) had significantly more species in the tree and shrub plots and boundary strip plots, than gappy hedges. The ground flora of these stockproof hedges had the highest mean cover values of *Rubus* spp., *Anthriscus sylvestris, Galium aparine* and *Hedera helix*. The most gappy hedges (with more than 50% gaps) were found in the upland study areas of the Sperrins (46%) and the Mournes (51%). They supported species which are more characteristic of acidic soils such as *Ulex europaeus, Saxifraga* spp., *Vaccinium myrtillus* and *Anthoxanthum odoratum*. Taller and wider hedges were associated with a greater tree and shrub diversity ($r=0.04$, $p<0.05$). Hedge width was also significantly positively correlated with the mean number of boundary strip species ($r=0.01$, $p<0.01$).

The mean cover of *Acer pseudoplatanus*, *Corylus avellana*, *Salix cinerea*, and *Fraxinus excelsior* were positively correlated with hedge height. Woodland species such as *Anemone nemorosa*, *Primula vulgaris* and *Hyacinthoides non-scripta* were also significantly positively correlated with hedge height and width (Table 2).

TABLE 1. The mean number of species in the hedge plots in the 6 study areas.

	Mean number (+SE) of species in each study area						One-way ANOVA		
Hedge plot	1	2	3	4	5	6	F	df	p
Tree and shrub	5 (0.2)	3 (0.2)	3 (0.2)	3 (0.2)	4 (0.2)	3 (0.2)	26.3	5,413	***
Ground flora	11 (0.6)	9 (0.4)	10 (0.6)	11 (0.6)	9 (0.3)	8 (0.2)	9.06	5,831	***
Ditch	6 (0.3)	5 (0.3)	8 (0.8)	7 (0.4)	6 (0.3)	7 (0.5)	5.03	5,173	***
Boundary strip	7 (0.2)	7 (0.2)	7 (0.3)	8 (0.3)	7 (0.2)	7 (0.2)	3.29	5,831	***

Study area (n=number of plots) **p<0.01 ***p<0.001
1. Fermanagh (146) 2. Sperrins (140) 3. Mournes (106) 4. Antrim (136)
5. Armagh (160) 6. Londonderry/Antrim (148)

TABLE 2. Correlations between the number of plant species per hedge plot and species cover value with hedge height and width

	Pearson correlation coefficient r	
Plant species	Hedge height	Hedge width
Acer pseudoplatanus	0.14**	NS
Corylus avellana	0.21**	NS
Fraxinus excelsior	0.11 *	NS
Salix cinerea	0.20**	NS
Crataegus monogyna	-0.10 *	NS
Anemone nemorosa	0.07 *	NS
Anthoxanthum odoratum	-0.10**	-0.09**
Hyacinthoides non-scripta	0.09 *	0.24**
Primula vulgaris	0.09 *	NS
Oxalis acetosella	NS	0.08 *
Juncus effusus	0.07 *	0.10**
Ranunculus repens	0.10 *	0.10**

* p<0.05 **p<0.01

Hedge management in general did not have a significant influence on the mean number of species found in hedge plots. However management did effect the variation in abundance of plant species. Unmanaged hedges contained significantly higher mean cover values of *Prunus spinosa*, *Salix cinerea* and *Ulex europaeus* and ground flora species such as *Anthoxanthum odoratum* and *Thuidium tamarisicinum* (Table 3). Managed hedges contained more competitive-ruderal species such as *Galium aparine* and *Urtica dioica*. *Crataegus monogyna, Fraxinus excelsior* and *Rosa canina* were also more common in managed hedges.

Hedges cut by flail cutter had a significantly greater number of tree and shrub species (4.1±0.21; $F_{3,187}$=8.53, p<0.001) and boundary strip species (12.8±0.34; $F_{3,377}$=9.04, P<0.001) than hedges managed by any other method. Coppiced hedges (cut with a circular saw) had the greatest number of ground flora species (16.0±0.42; $F_{3,377}$=3.64, p<0.01) and ditch species (6.4±0.29; $F_{3,127}$=5.05, p<0.001). This agrees with recent findings from experimental work carried out on the effects of different hedge restoration techniques on hedges in NI (McAdam, Bell & Henry, 1992). They found that coppiced hedges had the significantly greatest plant diversity and related this to the initial increase in light caused by hedge coppicing.

TABLE 3. The mean cover of species which were significantly different between managed and unmanaged hedges

Plant species	Mean % cover (±SE) Managed hedges (n=380)	Unmanaged hedges (n=456)	t	df	2-tail p
Crataegus monogyna	52 (2.1)	30 (0.8)	-7.9	391	***
Fraxinus excelsior	6 (0.7)	4 (0.5)	-2.0	363	*
Prunus spinosa	5 (0.9)	9 (1.2)	2.9	334	**
Rosa spp.	3 (0.4)	1 (0.3)	2.7	400	***
Salix cinerea	1 (0.4)	4 (0.7)	4.6	334	**
Ulex europaeus	2 (0.5)	7 (1.0)	4.3	831	***
Anthoxanthum odoratum	5 (0.5)	8 (0.6)	4.3	831	***
Anthriscus sylvestris	4 (0.5)	2 (0.3)	-4.7	566	***
Dactylis glomerata	7 (0.6)	5 (0.4)	-2.7	734	**
Galium aparine	3 (0.4)	2 (0.2)	-3.6	604	***
Urtica dioica	8 (0.8)	6 (0.6)	-2.0	761	*
Thuidium tamariscinum	2 (0.4)	4 (0.6)	3.2	728	***

*p<0.05 **p<0.01 ***p<0.001

Land use and grazing

Hedges in the lowland study areas had the lowest plant species diversity and were associated with intensive agriculture. Intensively managed ryegrass pasture and arable crops were more usually adjacent to hedges with the significantly lowest species diversity. Hedges adjacent to rush-infested pasture and woodland had the greatest tree and shrub species diversity (4.6±0.24 species per hedge; $F_{8,410}$=5.81 p<0.001). Hedges adjacent to rush-infested pasture and woodland also had the highest mean number of ground flora species (10.5±0.54 species per rush-infested pasture hedge; 9.9±0.92 species per woodland hedge; $F_{8,828}$=2.76 p<0.01). Woodlands may have been the source of species-rich hedge floras, as *Anemone nemorosa* and *Corylus avellana* were more abundant in hedges next to woods or scrub. The highest mean number of boundary strip species was found in hedges next to roads (8.1±0.33; $F_{8,828}$=2.26 p<0.05). Woodland and other semi-natural land use types were most abundant in the Fermanagh study area and so may account for some of the diversity of Fermanagh hedges. Hedges alongside roads contained the highest mean cover of *Crataegus monogyna, Anthriscus sylvestris* and along the boundary strip species such as *Dactylis glomerata* and *Vicia sativa*.

Hedges which were adjacent to fields which showed no recent signs of grazing had significantly greater numbers of boundary strip species per plot than those with evidence of recent grazing (11.9±0.2 species per ungrazed hedge; t=4.4, df=820, p<0.01). The most grazed hedges were found in the Mournes and the Sperrins. These areas were associated with the largest numbers of sheep and it is widely reputed that sheep may reduce the species diversity of fields (Miles, 1988).

Hedge demarcation type

Riverside hedges contained the highest mean number of ground flora species (12.1±0.91; $F_{5,831}$=5.37, p<0.001) and boundary strip species (8.5 (±0.54); $F_{5,831}$=11.60 p<0.001). Townland boundaries had the highest mean number of tree and shrub species (4.0±0.59; $F_{5,413}$=0.65, p>0.05) and ditch plant species (9.0±1.40; $F_{5,173}$=3.79, p<0.01). Roadside hedges were more likely to be composed of *Crataegus monogyna* and *Fraxinus excelsior*. Townland boundary hedges were the most species-rich boundary type in terms of their trees and shrubs, containing higher frequencies of *Corylus avellana*, *Rosa* spp., *Salix* spp., *Prunus spinosa*, and *Ilex aquifolium* than other boundary hedges. They were more stockproof, taller, wider and more likely to be associated with a ditch than any other hedge boundary type. There was no significant difference in the mean number of ground flora species between townland hedges and other boundary hedges, but woodland species such as *Hyacinthoides non-scripta*, *Primula vulgaris* and *Hedera helix* were more common in townland hedges. The greater structural diversity of townland hedges may partly account for this, with perhaps the presence of a ditch and a wide, tall bank reducing the grazing pressure on the hedge flora. A ditch may also, help minimise fertiliser and other agricultural chemical drift into the hedge bottom.

Soil properties

Hedges with a low nutrient status and a high moisture content were significantly associated with semi-natural vegetation types, such as rush-infested pasture and woodland. Woodland species such as *Viola riviniana*, *Oxalis acetosella* and *Hedera helix* occurred on soils with a low nitrogen content and a low potassium and phosphorus content and low pH. Soils with the highest moisture content were found near hedges associated with rush-infested pastures and those with the greatest organic matter content were located near woods and scrub. The intensively farmed lowland study areas contained soils with the greatest nitrogen, potassium and phosphorus contents and highest pH values. The species-poor hedge ground flora groups all had significantly higher concentrations of soil nitrogen ($F_{7,89}$=2.97,p<0.01), phosphorus ($F_{7,89}$=2.25,p<0.05) and potassium ($F_{7,89}$=2.37,p<0.05) than the species-poor groups.

The relative importance of the factors influencing hedge species composition.

To investigate which environmental variables were the best predictors of hedge species diversity, a stepwise regression including hedge structural attributes, soil properties, hedge management, study area, associated land use and boundary demarcation type was performed (Table 4). Hedge height was the most significant predictor of tree and shrub species diversity, followed by hedge management and soil total nitrogen content. Hedge width was the most significant predictor of the diversity of hedge ground flora, with soil total nitrogen content the next most important predictor. Boundary strip species diversity was best predicted by adjacent land use. This agrees with the findings of a survey on lowland farms in Scotland by Shaw (1988), who found that the number of plant species present on any habitat type were most strongly and inversely related to the amount of arable land on the farm. Ditch width was significantly the best predictor of ditch species diversity, followed by hedge management.

TABLE 4. Stepwise multiple regression of the number of plant species from each hedge plot with hedge structural attributes and soil properties

Hedge plot	Hedge characteristic	r	F
ree and shrub	Hedge height	0.29	9.22 **
	Hedge management	0.42	9.77***
	Total soil nitrogen	0.47	8.80***
Ground flora	Hedge width	0.29	8.36 **
	Total soil nitrogen	0.47	8.80***
oundary strip	Adjacent land use	0.36	14.07**
Ditch	Ditch width	0.28	7.97 **
	Hedge management	0.44	7.50***

p<0.01;*p<0.001

DISCUSSION

The plant species composition of hedges is evidently determined by a wide range of inter-dependent variables. The structural diversity of a hedge may be regarded as one of the major influences on species diversity. Hedge width, hedge height and ditch width were all cited by stepwise-multiple regression as some of the best predictors of species diversity. Species-poor hedges are more associated with low elevation study areas on intensively managed farmlands. These hedges generally have a higher soil nutrient status and are structurally less diverse. In contrast, species-rich hedges are more associated with areas where less intensively managed farming is common on soils with low nutrient status. The high species diversity of Fermanagh hedges may be related to the higher proportion of adjacent semi-natural land use which occurs in Fermanagh. These hedges are more usually unmanaged and as a result are tall and wide. They are more likely to be associated with a ditch, which in turn increases species diversity. The wide diversity of hedge types and land use practices between different study areas, suggests that management guidelines should be more region specific. If hedge removal is necessary, the removal of tall, wide hedges should be avoided as these have the highest species diversity. Also townland boundaries may warrant special conservation attention due to their historical importance and the fact that they tend to have a higher wildlife value than other boundary hedges.

This work also suggests that management prescriptions should emphasise that a wide headland be left uncultivated and free from fertiliser and slurry application, if the wildlife value and species diversity of the hedge is to be maintained. In Northern Ireland it usually impracticable to leave headlands of more than 5 metres due to the small field size, especially in

Co. Fermanagh which has a particularly dense hedge network system. The ESA management prescriptions include restrictions on fertiliser inputs to habitats such as hay meadows in the West Fermanagh and Erne Lakeland ESA. These prescriptions should benefit not only the species diversity of grasslands but also the species diversity of their associated field boundaries.

ACKNOWLEDGEMENTS

This work was funded as part of a D.Phil. studentship award by the Department of Education for Northern Ireland and the Department of the Environment, Countryside and Wildlife Branch.

REFERENCES

Allen, S.E.; Grimshaw, H.M.; Rowland, A.P. 1986. Chemical Analysis. In: *Methods in Plant Ecology*. Ed. S.B. Chapman pp411-466. Blackwell Press, Oxford.

Baudry, J. 1984. Effects of landscape structure on biological communities: the case of hedgerow network landscapes. In: *Methodology in Landscape Ecological Research and Planning*. Ed. by J.Brandt and P.Agger. Vol **1** pp55-65.

Baudry, J. 1988. Hedgerows and hedgeroaw networks as wildlife habitat in agricultural landscapes. In: *Environmental Management in Agriculture*. Ed. J.P.Park pp111-124. Behalven Press, London & New York.

Cooper, A. 1986. *The Northern Ireland Land Classification*. Report to the Countryside and Wildlife Branch, Department of the Environment for Northern Ireland (DOENI), University of Ulster.

DANI (1992). Field boundaries, 1-7. A series of booklets produced as part of DANI Countryside management service to farmers, HMSO, Belfast.

Hegarty, C.A. 1992. *The ecology and management of hedges in Northern Ireland*. Unpublished DPhil Thesis, University of Ulster, Coleraine.

Hill, M.0. 1979. *TWINSPAN* - A FORTRAN program for arranging multivariate data on an ordered two-way table by classification of individual and attributes. 80pp. Cornell University, Ithaca, New York.

McAdam, J.H.; Bell, A.; Henry, T. 1992. The effects of restoration techniques on flora and microfauna of hawthorn dominated hedges. In: *Hedgerow Management and Nature Conservation*. Proceedings of a meeting of the British Ecological Society. Wye College, University of London 4-5 September, 1992.

Miles, J. 1988. Vegetation and soil changes in the uplands. In: *Ecological Change in the uplands*. Ed. M.B. Usher & D.B.A. Thompson. pp57-70. Blackwell Scientific Press, Oxford.

Murray, R.; McCann, T.; Cooper, A. 1992. *A Land Classification and Landscape Ecological study of Northern Ireland*. Report to the Environmental Science, DOENI. Department of Environmental Studies. University of Ulster.

Norusis, M.J. 1990. SPSS Bases system user guide. SPSS INC. Chicago USA.

Pollard, E.; Hooper, M.D.; Moore, N.W. 1974. *Hedges*. Collins, London.

Robinson, P. 1977. The Spread of hedged enclosure in Ulster. *Ulster Folklife* **23**, 57-69.

Shaw, P. 1988. *Factors affecting the numbers of breeding birds and vascular plants on lowland farmland*. Report to the Nature Conservancy Council Commission , No.838.

EFFECTS OF MECHANISED CUTTING ON THE SHORT TERM REGROWTH OF HAWTHORN HEDGEROWS

D.A. SEMPLE, J. DYSON, R.J. GODWIN

Department of Agricultural and Environmental Engineering, Silsoe College, Silsoe, Beds. MK45 4DT

ABSTRACT

This paper reports on an investigation into the short term effects of mechanical cutting on hedgerows. The subsequent regrowth of hawthorn was used as an indicator of the effect of alternative cutting methods.

The investigation was carried out over three years on farm hedgerows in mid-Bedfordshire. The hedgerows were predominantly comprised of Hawthorn and the study was confined to this species. The treatments were; three types of flail, a finger bar cutter, a circular saw, and an uncut control. A damage rating system was developed and used, together with other recorded data, to describe the type of severance and the effect on regrowth.

This study demonstrates that in general flails leave a more ragged cut than the other methods, and that as this apparent damage increases there are beneficial effects to regrowth in terms of position and number of new shoots.

INTRODUCTION

The traditional method of hedge management in this country has been 'hedge laying', however, as labour costs have increased the majority of managed hedges are now maintained using some form of mechanical cutter, often a version of the flail mower. Many observers have voiced opposition to the flail as a maintenance tool, because of perceived detrimental effects to the hedge due to the distressing appearance of a mature hedge that has been cut back in this way. Conversely, some users of flail trimmers have noted a beneficial effect on hedge growth.

METHODOLOGY

This investigation took place over three years with different hedges being cut each year. The hedges were cut during their dormant period, and the majority of the branches cut were less than three years old. The regrowth of the hedge was recorded during the subsequent growing season. The position of each treatment along the hedgerow was randomised. Each treatment was twenty metres long and three replicates of ten branches were taken. The hedges were all cut to a topped A shape, and the results of each treatment measured on the same side of the hedge.

Branches were tagged, and photographs taken at regular intervals during the growing season, as a permanent reference. The main assessment of the treatments was made after the initial regrowth of the hedge. A damage rating system was developed to record the type and amount of damage to the branch. The length and thickness of the branch, the number of new shoots and their position relative to the cut end were also recorded.

In order to give a repeatable assessment of the damage incurred by the branch the rating system was defined by the type of damage inflicted on the end of the branch and the length to which the damage extended. The main categories of damage are shown in Figure 1.

FIGURE 1. Types of damage inflicted by cutting the branches.

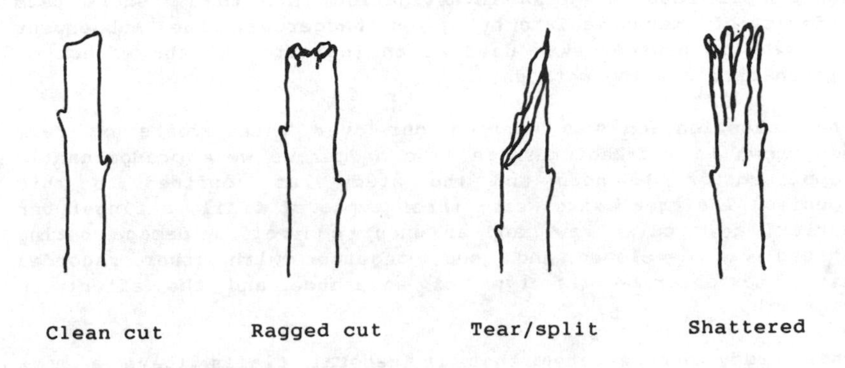

Clean cut | Ragged cut | Tear/split | Shattered

The damage scores awarded for different types of cut are shown in Table 1.

TABLE 1. Damage score awarded to types of cut

Type of cut	Damage score
Clean	0
Ragged	1
Tear < 5 x branch diameter in length	2
Shatter < 5 x branch diameter in length	3
Tear > 5 x branch diameter in length	4
Shatter > 5 x branch diameter in length	5

TREATMENTS

The cutting treatments were the heavy duty, standard and competition flails produced by Bomford Turner Ltd with the cutter rotation in both forward (rotating in the same direction as the tractor wheels) and reverse mode (rotating in the opposite direction as the tractor wheels), a finger bar cutter, a circular saw (0.75 m in diameter) and a control of no cut.

The heavy duty flail is recommended by the manufacturer for cutting thick material up to 100 mm in diameter or restoring an overgrown hedge.

The standard flail is recommended for cutting material up to 38 mm in diameter, typically one to four year old wood and is the flail normally supplied with this type of machine.

The competition flail is normally used for wood up to one year old and is designed to give a more even finish than the other flail types when used for annual trimming.

The finger bar cutter is normally used for wood up to two years old and cuts the branches by a shearing action, leaving full length trimmings.

The circular saw also leaves the trimming full length, but it can cut very thick wood in one pass.

RESULTS AND DISCUSSION

For the purpose of this paper the data collected from the three years of field trials has been amalgamated and the main results are shown in Figures 2 to 5. Not all of the treatments were used each year, so the number sets for each treatment vary. However, the overall trends shown in the graphs remain the same for the individual years as for the amalgamated data set.

Figure 2 shows the average damage score achieved by each cutting treatment. As can be seen from the graph, all of the flail treatments cause a similar level of branch damage. The saw has a lower average score than all the flails and the finger bar is very much lower than the flails.

These results may be explained by the type of cutting action employed by the four cutter groups.The flail has a relatively blunt cutting edge rotating at a high velocity and cuts by utilising the branch stiffness and inertia to generate a reaction force. Each branch is struck a number of times as the cutter head moves over it and this tends to leave a ragged severance, especially on thicker branches. The finger bar cutter, if properly maintained, has a sharp cutting edge and shears the branch cleanly between a pair of reciprocating blades. The circular saw has a sharp cutting edge and saws through the branch utilising the branch stiffness to react to the sawing action.

In Figure 3 the effect of damage to the branch at the time of cutting is demonstrated by the influence it has on the position of the new shoots relative to the cut end. As can be seen from the figure the shape of the cumulative percentage lines for each damage level remains fairly constant, but as the damage increases the position of the new shoots is displaced further away from the cut end. As most hedges are trimmed to the same size and shape in subsequent years, the implication is that by inflicting more damage to the branches the regrowth will be 'pushed' further into the hedge. Less regrowth will be removed in subsequent trimming, leaving a greater amount of new wood to promote the development of a thick hedge.

Figure 4 shows the effect of increasing cut damage on the number of shoots. There is an upward trend in the average number of shoot with damage score suggesting that increasing damage may have a beneficial effect on shoot numbers.

Figure 5 shows an upward trend between the diameter of the branch and the damage sustained by it when cut. This suggests, as would be expected, that the older and thicker the wood the more likely it is to be damaged when cut.

Due to the considerable variation in the data a much larger data set would be required to statistically prove the relationships shown in Figures 4 and 5.

CONCLUSIONS

Flail cutters inflict more damage on Hawthorn branches than the other mechanical cutting methods tested.

There is little difference in damage rating between flail types when cutting wood up to three years old.

Increasing the cut damage inflicted on a branch 'pushes' the position of new shoots further from the cut end.

Increasing the cut damage inflicted on a branch may slightly increase the number of new shoots.

Thicker branches appear to sustain higher levels of damage when cut by flails.

This study was confined to the short term effects on Hawthorn branches less than three years old. Further work will be required to determine the effect of the repeated mechanical cutting of hedgerows and the effects on older and thicker wood.

ACKNOWLEDGEMENTS

The authors thank Bomford Turner Ltd for providing equipment and technical support. D.A. Semple was funded by a MAFF CASE award and a DAFS studentship.

FIGURE 2. The effect of cutter type on damage score

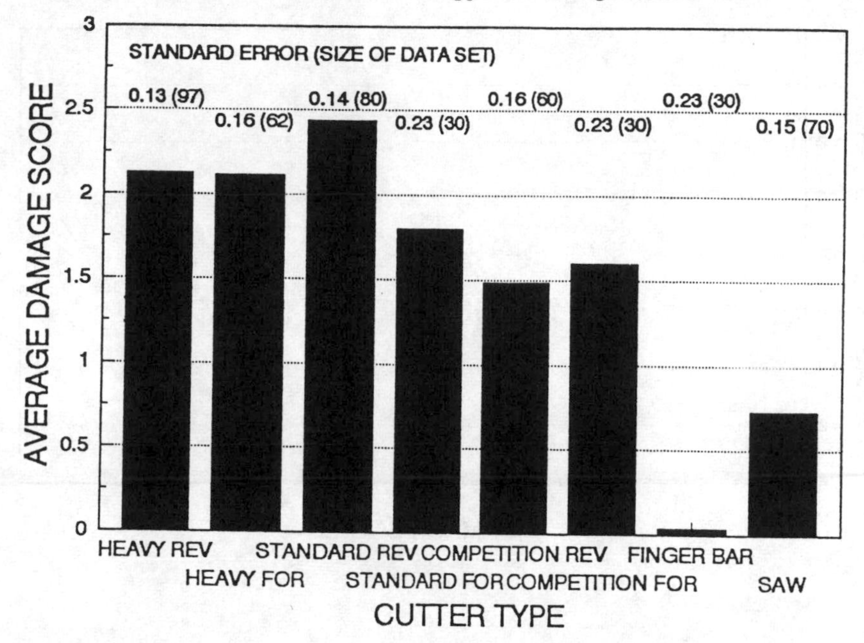

FIGURE 3. The effect of damage type on shoot position

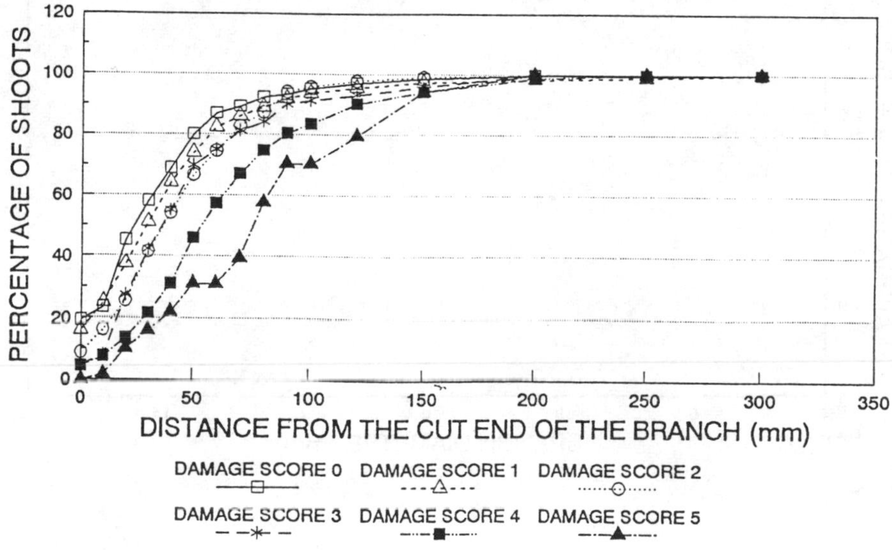

PERCENTAGE OF SHOOTS

DISTANCE FROM THE CUT END OF THE BRANCH (mm)

DAMAGE SCORE 0 DAMAGE SCORE 1 DAMAGE SCORE 2

DAMAGE SCORE 3 DAMAGE SCORE 4 DAMAGE SCORE 5

FIGURE 4. The effect of damage on shoot numbers

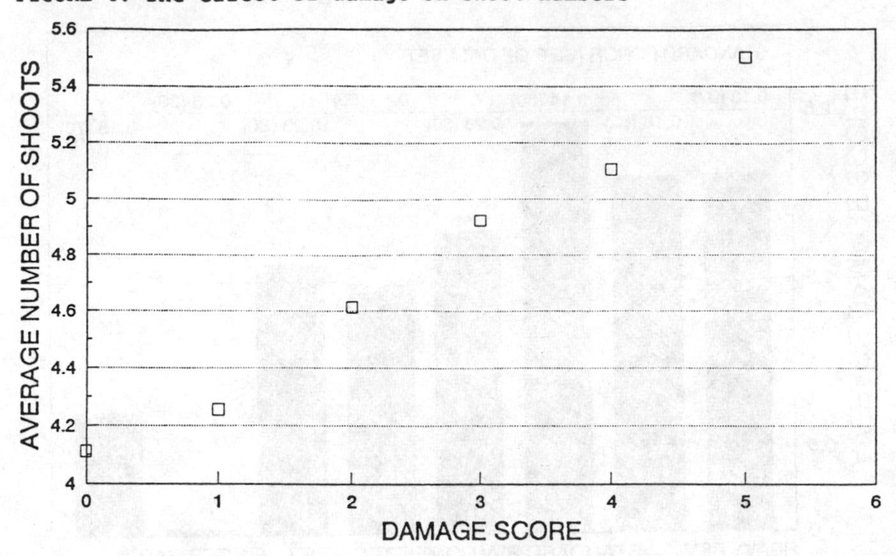

FIGURE 5. The effect of branch diameter on damage score

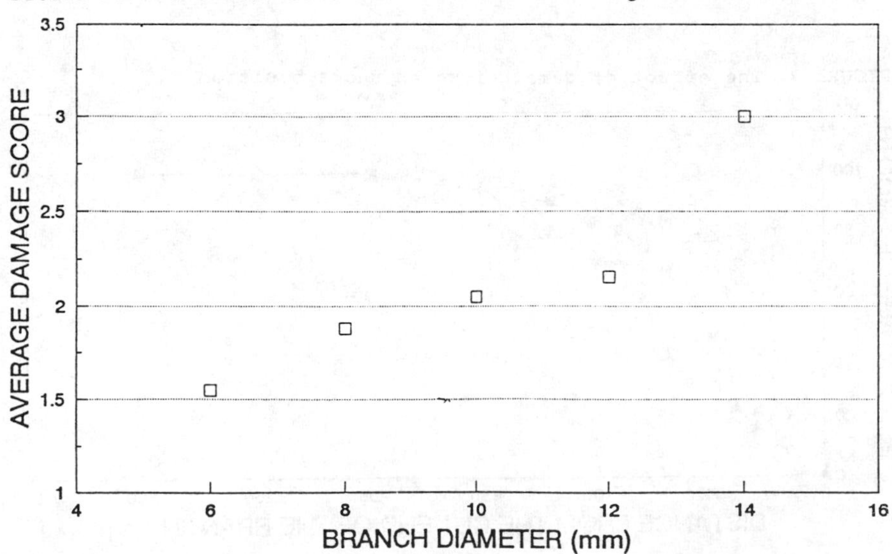

CLEMATIS VITALBA (OLD MAN'S BEARD) AS A COMPETITIVE "WEED" IN HEDGEROWS AND THE EFFECTS OF HEDGE CUTTING REGIMES ON ITS DEVELOPMENT

C P BRITT

ADAS Drayton, Alcester Road, Stratford-upon-Avon, Warwickshire, CV37 9RQ

ABSTRACT

Clematis vitalba is a woody, climbing plant commonly found on calcareous soils in the Midlands and south of England. Unless it is controlled by management practices, it is capable of smothering hedges. An experiment at Stratford-upon-Avon, Warwickshire examined the effects of four hedge management techniques on a *Clematis*-covered Hawthorn hedge. Cutting *C. vitalba* stems at their base and treating the stumps with glyphosate gave effective control of this species, with a consequent improvement in Hawthorn vigour. The method of mechanical hedge trimming had no clear short-term effect on *C. vitalba* abundance.

INTRODUCTION

Various national surveys have shown a continuing decline in the length of hedgerows in the UK over the past 20 years. Recent losses have been primarily due to general neglect, with many former hedges being reclassified as lines of trees or shrubs (Barr *et al*, 1991). Traditional hedge management techniques, involving regular laying of hedges, provided opportunities to cut out any species considered to be competitive "weeds". Several practical handbooks on hedges recommend removal of climbers before they damage the shrub species (eg Brooks, 1975). Modern farm management, however, often gives scant regard to hedges; hedge-laying, and consequently "weed" removal, are declining practices.

Clematis vitalba (Old Man's Beard) is one of a number of woody shrubs and climbers potentially damaging to hedgerows (Brooks, 1975). Found in several parts of the world, it is recognised as a problem species in the UK (Brooks, 1975), other parts of western Europe and in New Zealand (Popay, 1986 and Ryan, 1985). In the UK *C. vitalba* occurs in hedgerows and woodland-margins, chiefly on calcareous soils southwards from Denbigh, Stafford and South Yorkshire (Clapham *et al* 1962; Stace, 1991), although there is documented evidence of a northerly spread (Sinker *et al*, 1985). *C. vitalba* is often the dominant climbing species and many authors have referred to its smothering effect on hedge plants and trees up to 30 m in height (eg Brooks, 1975 and Ryan, 1985). Plants and seeds are long-lived (Salisbury, 1961 and Popay, 1986).

Clay (pers. comm.) reported that an extensive search had revealed no major study of the biology of this species. He stated that further research was required, studying the effect of different hedge management practices on the development and spread of climbers. This was the principal aim of the experiment described here.

MATERIALS AND METHODS

A hedge predominantly of *Crataegus monogyna* (Hawthorn) , with a uniform and fairly dense cover of *C. vitalba*, was selected for the experiment. The hedge, growing on heavy, calcareous, clay soil at ADAS Drayton, Stratford-upon-Avon, Warwickshire ran in a straight line, down a gentle slope, from west to east. It was situated between a Perennial Ryegrass ley, on the southern side, and a grassy road verge. Previous hedge management had been consistent, with annual mechanical cuts on both sides and the top, to produce a rectangular hedge, around 1.4 m high.

The hedge was subdivided into 16 plots, each 17 m long; of uniform height and density, and with similar species composition. These were used to study the effects of four experimental treatments (see Table 1) in a randomised complete block design.

On 20 December 1990 each plot was mechanically cut (flail cutter) according to the requirements of the experimental treatments. All *C. vitalba* and *Rubus fruticosus* agg. (Bramble) stems in treatment 4 plots were cut just above ground level using hand-held loppers, and the stumps painted with undiluted glyphosate herbicide (360 g l^{-1} , SL) on 14 May 1991. The hedge trimming treatments were repeated on 29 January 1992, 30 December 1992 and 20 October 1993. Hand cutting and glyphosate painting of *C. vitalba* and *R. fruticosus* stems, where required, were repeated on 10 June 1992 and 29 June 1993.

TABLE 1. Treatment list

Treatment	Frequency of Trimming	Shape	Cut	Target Height	Additional Treatments
1	Annual	Rectangular	Both sides/top	1.4 m	None
2	Annual	Rectangular	Alternate sides*/top	1.4 m	None
3	Annual	A-shape	Both sides	1.8 m	None
4	Annual	Rectangular	Both sides/top	1.4 m	Hand cut *Clematis* stems

* south side of hedge cut in 1990.

Three similar 50 m sections of *Clematis*-infested Hawthorn hedge, in an adjacent field, were used for related observations in 1993. *Clematis* and *Rubus* in one section were hand-cut, as described previously, on 15 March. The number of stems cut of both species, and the length of time taken, were noted. This exercise was repeated for the second 50 m section on 29 June. No stems were cut in the third section.

Assessments

The percentage *C. vitalba* cover was estimated using ten 0.1 m² quadrats per plot, held at 1.2 m height on the south side of the hedge. The number of *C. vitalba* stems per plot was counted. Hedge heights were measured at 13 regularly spaced points in each plot before and after trimming. 'Density', or more accurately a combination of density and thickness, was measured using 50 black diamond shapes arranged in 10 rows on a white board, 1.6 m tall and 0.45 m wide. Each diamond was 13 cm tall and 9 cm wide. This board was held vertically 2 m from, and perpendicular to, the mid-line of the hedge. An observer, positioned an equal distance on the opposite side of the hedge recorded the number of diamonds of which more than 50% of the total area was visible. Six counts were made per plot. Numbers of fruits were counted, using 0.1 m² quadrats, as for the *C. vitalba* cover assessments.

On the 21 September 1993 an assessment of percentage cover and relative abundance of *Clematis* and *Rubus* was made on the three separate observation plots.

All data were subjected to analysis of variance, using Genstat 5. Where necessary, data were transformed and re-analysed, and the back-transformed means and 95% confidence limits are shown in the results tables.

RESULTS AND DISCUSSION

TABLE 2 . Estimated % *C. vitalba* cover, at approximately 1.2 to 1.5 m height

Treatment	October 1991	September 1992	August 1993
1. Rectangular	30.5	31.3	28.9
2. Rectangular/alternate	36.3	33.9	34.7
3. A-shape	9.7	34.7	30.3
4. Rectangular + hand cut	10.7	5.0	0.0
SED	10.83	16.06	12.23
SE per plot	±15.32	±22.71	±17.29
df	9	9	9
CV%	70.2	86.6	73.6

Excellent control of *C. vitalba* was achieved by hand cutting stems (Table 2), with very few additional shoots being produced in 1992 and 1993. Differences between treatments were not, however, statistically significant at the 5% level of probability ($P = 0.08$, 0.26 and 0.07 for 1991, 1992 and 1993 data respectively).

There were significant treatment differences in mean numbers of *C. vitalba* stems in two shoot diameter groups when these were counted in March 1993 ($P < 0.001$) (Table 3). Hand-weeded plots had fewest *Clematis* stems; plots cut to an A-shape had most.

TABLE 3. Mean number of *C. vitalba* stems per plot in 5-20 mm (real values) and > 20 mm (back-transformed values) categories - 23 March 1993

Treatment	Shoot base diameter 5 - 20 mm	> 20 mm	L_l	L_u
1. Rectangular (control)	27.5	1.9	0.67	7.05
2. Rectangular/alternate	29.5	3.5	1.14	19.79
3. A-shape	37.8	6.2	1.75	86.85
4. Rectangular + hand cut	2.2	0.0	0.00	0.28
SED	5.35			
SE per plot	±7.57			
df	9			
CV%	31.2			

L_l = lower 95% confidence limit; L_u = upper 95% confidence limit

TABLE 4 . Mean hedge heights (cm)

Treatment	Nov 91	Mar 92	Sept 92	Feb 93	Sept 93
1. Rectangular (control)	188.1	138.0	194.7	137.6	201.0
2. Rectangular/alternate	186.5	142.1	202.9	140.6	199.3
3. A-shape	202.2	179.5	223.2	174.9	212.3
4. Rectangular + hand cut	208.3	142.3	205.8	142.6	220.0
SED	6.15	4.52	10.49	3.57	8.13
SE per plot	±8.70	±6.39	±14.84	±5.05	±11.49
df	9	9	9	9	9
CV%	4.4	4.2	7.2	3.4	5.5

Table 4 shows mean hedge heights measured at five dates. There were significant differences between treatments after hedge trimming (February/March assessments), as the A-shape treatment was deliberately left taller ($P<0.001$).

Although cut to a similar height to treatments 1 and 2 in December 1990, hand-weeded plots in treatment 4 were significantly taller by November 1991 ($P<0.05$). Height differences between treatments in September 1992 and September 1993 were not significant at the 5% level ($P = 0.116$ and 0.094 respectively). The greatest height increase, in each year, was shown by the hand-weeded treatment.

TABLE 5 . Hedge density scores (number of diamonds with ≥ 50% visibility)

Treatment	Feb 92	Feb 93
1. Rectangular	14.6	23.1
2. Rectangular/alternate	9.8	15.7
3. A-shape	8.4	18.5
4. Rectangular + hand cut	13.6	27.9
SED	1.57	3.33
SE per plot	±2.21	±4.70
df	9	9
CV%	19.1	22.1

* 0 = very thick or dense hedge; 50 = very thin or sparse hedge

Table 5 shows the mean scores of hedge density measured in winter, after trimming. All plots were fairly dense, without gaps. In both years, density scores were higher in treatments that had been trimmed to a rectangular shape (ie treatments 1 and 4) ($P<0.05$). This is probably simply because of the greater width of hedge sections receiving the rectangular/alternate and A-shape treatments, rather than any real differences in density.

Numbers of fruits were generally low in all three years (data not shown). The highest counts were recorded in October 1991, on plots trimmed to an A-shape.

TABLE 6. *C. vitalba* and *R. fruticosus* abundance scores and % cover - 21 September 1993

	Date of hand-weeding		
	Not Cut	15 March	29 June
C. vitalba			
Abundance score (0-5)*	3.0	0.0	0.6
% cover	36.4	0.0	2.0
R. fruticosus			
Abundance score (0-5)	0.7	0.1	0.0
% cover	8.6	0.7	0.0

* 0 = absent, 1 = rare, 2 = occasional, 3 = frequent, 4 = abundant, 5 = dominant

Table 6 presents the results of an assessment of *C. vitalba* and *R. fruticosus* in the hedge in the adjacent field, used for the supplementary hand-weeding observations. These results suggest that cutting *Clematis* and Brambles in March or June is effective.

Observations made on rates of hand-cutting showed them to be higher in March (217 *C. vitalba* stems cut in just 39 minutes) than in June, when tall nettles had to be strimmed before cutting was possible (total time taken was 60 minutes to cut 118 stems).

Annual botanical assessments along the southern (field) side of the main experiment hedge showed no discernible treatment effects (data not shown). Major short-term treatment effects on the flora were not expected. The mean number of plant species per plot (hedge and hedge-bottom) in 1993 was 20.1 (4.4 woody/climbing species, 5.2 grasses and 10.5 dicotyledonous herbs). In all, 51 species were recorded in 1993.

CONCLUSIONS

Cutting *C. vitalba* stems near the base, and treating the stump with glyphosate appears to be an effective method of control of this species. Preliminary observations suggest that winter treatment is as effective as summer cutting, and may in certain circumstances be easier and faster to perform. Hand-weeding by this method can improve hedge vigour and should be seriously considered where *C. vitalba* threatens to smother a hedge.

The method of hedge trimming had no clear short-term effect on *C. vitalba* abundance. A longer-term study might reveal changes in climbing weed frequency, and other gradual changes in the hedge and ground flora not observable within just three years.

ACKNOWLEDGEMENTS

Financial support for this work from the Ministry of Agriculture, Fisheries and Food (MAFF) is gratefully acknowledged. In addition I thank all colleagues at ADAS Drayton who have participated in this work.

REFERENCES

Barr, C.; Howard, D.; Bunce, R.; Gillespie, M.; Hallam, C. (1991). Changes in hedgerows in Britain between 1984 and 1990. NERC Contract Report to Department of Environment. Institute of Terrestrial Ecology. 13 pp.

Brooks, A. (1975). Hedging : A Practical Conservation Handbook. British Trust for Conservation Volunteers. 117 pp.

Clapham, A. R.; Tutin, T. G.; Warburg, E. F. (1962). Flora of the British Isles (2nd edition). Cambridge University Press. 1269 pp.

Popay, A. I. (1986). Old Man's Beard, *Clematis vitalba*, control measures. *Farm Production and Practice* 858, New Zealand MAF. 2 pp.

Ryan, C. (1985). Pests and problems - Old Man's Beard. *Soil and Water* 3. pp 13-17.

Salisbury, E. J. (1961). Weeds and Aliens. Collins. 384 pp.

Sinker, C. A.; Packham, J. R.; Trueman, I. C.; Oswald, P. H.; Perring, F. H.; Prestwood, W. V. (1985). Ecological Flora of the Shropshire Region. Shropshire Trust for Nature Conservation. pp 179-180.

Stace, C. (1991). New Flora of the British Isles. Cambridge University Press. 1226 pp.

MANAGEMENT OF DITCHES IN ARABLE FENLAND: MOWING REGIMES FOR DRAINAGE AND CONSERVATION

MILSOM, T.P.

Central Science Laboratory, MAFF, Tangley Place, Worplesdon, Surrey GU3 3LQ.

SHERWOOD, A.J.

ADAS Land and Environment, Block C, Government Buidlings, Brooklands Avenue, Cambridge, CB2 2DR.

ROSE, S.C.

ADAS Soil & Water Research Centre, Anstey Hall, Maris Lane, Trumpington, Cambridge CB2 2LF.

RUNHAM, S.R., TOWN, S.J.

ADAS Arthur Rickwood Research Centre, Mepal, Ely, Cambridgeshire, CB6 2BA.

ABSTRACT

In the peat Fens of Cambridgeshire, arable land is intensively managed for crop production. Field margins are potentially very important as wildlife refuges or corridors, and ditches are of particular interest for their relict aquatic flora and fauna. However, ditches in arable fenland are managed primarily for drainage and this requirement can conflict with those of wildlife management. The effects of altering bank mowing regimes on drainage and the emergent aquatic flora of ditches were investigated in a small replicated trial. Four mowing frequencies were chosen, from twice yearly to no cut over four years.

Two major gradients were identified in the species composition of the emergent aquatic plant associations in the ditches: the first reflected the transition from swamp to herbaceous fen; the second contrasted associations which were dominated by *Phragmites australis* and *Sparganium erectum* with associations which lacked those species. The species richness of all associations decreased during the course of the experiment, but the net loss of species was inversely correlated with mowing frequency. However, the effects of mowing on species composition differed between plant associations. Frequent winter mowing promoted the growth of *Phragmites* and adversely affected drainage efficiency.

INTRODUCTION

The peat Fens of Cambridgeshire contain some of the most intensively farmed arable land in western Europe. Consequently, field margins in this area have particular importance as wildlife refuges or corridors. Drainage ditches are the most widespread type of field boundary and, as the peat Fens were formerly a major wetland prior to drainage for agriculture, ditches are important as wildlife refuges because of their relict wetland plant and animal communities. However, the conservation management of ditches in arable fenland presents many problems (Milsom *et al.*, 1994).

Ditches in the peat Fens are managed primarily to maximise their drainage efficiency. This is done by periodic removal of accumulated plant debris and silt (slubbing), and by frequent cutting of vegetation on ditch banks and in the water (Newbold *et al.*, 1989). Some of these operations are thought to have an adverse effect on the flora and fauna of ditches but there are no empirical data from the peat Fens to determine whether current management regimes could be adjusted for the benefit of conservation without conflicting with drainage needs. Management for drainage is applied rigorously in the larger arterial drains by Internal Drainage Boards, but the regimes on smaller private ditches are more flexible and offer scope for adjustment.

This paper describes the emergent aquatic flora (as defined by Alcock and Palmer, 1985) of private ditches, about which little is known, and reports the effects of manipulating bank mowing regimes on the flora and on drainage efficiency. Further details of the experiment are given in Sherwood and Harris (1991) and in Milsom *et al* (1994).

METHODS

The experiment was carried out on two adjacent ditches at ADAS Arthur Rickwood Research Centre, in Mepal Fen, Cambridgeshire. Further details of the site are given in Sherwood and Harris (1991).

Three mowing treatments were compared against a control where vegetation on the ditch banks remained unmown for the duration of the experiment, which was four years. The treatments comprised: (i) a cut made twice a year in March and, again, in October; (ii) an annual cut in October and (iii) a cut made once every two years in October. October and March were chosen to avoid damage to crops during the main growing season and to minimise risk of destroying birds and their nests. Mowing was carried out using a Bomford 457 flail, following local practice. The flail was passed repeatedly over both ditch banks down to water level to create a low uniform sward. Each ditch was sectioned into 25m plots, with eight in one ditch and four in the other. Two replicates of treatments were assigned to the former ditch and one to the latter. Within ditches, the layout of the treatments was randomised.

Floristic assessments were made at six-week intervals, from April to September of each year. The middle 20 metres of each plot was split into 10 contiguous two metre sections and plant species lists were compiled for each section. These presence/absence data were used to calculate an abundance score on a ten point scale for each species in each plot.

The patterns of variation between plots in the species composition of the emergent aquatic vegetation were examined by using Detrended Correspondence Analysis (Hill, 1979). Detrended Correspondence Analysis (DECORANA) was used to ordinate plots in multi-dimensional space bounded by four axes, each of which represents a gradient in species composition. Data from separate years were entered individually so that changes in species composition over the course of the experiment could be evaluated using vector analysis (Hill *et al.*, 1993). Only the first two axes were considered because the proportion of the variation in the dataset that was explained by the remaining axes was very low, as indicated by the decrease of the eigenvalues from the first to the fourth axes: 0.1869, 0.0973, 0.0402, 0.0206. The DECORANA co-ordinates for each plot in 1989 were compared with their equivalents in 1992 and resultants were calculated trigonometrically from the differences between years on the first and second DCA axes. The angle between the resultant and Axis 2 was used as the measure of the direction of floristic change. Non-parametric analyses of variance (Siegel and Castellan, 1984) were used to test for differences between treatments in the resultants and angles of movement respectively.

RESULTS

An ordination of plots in the first and last years of the experiment is shown in Fig. 1. Axis 1 represents a gradient from swamp to herbaceous fen, which are denoted by low and high DECORANA scores respectively. Common Reed (*Phragmites australis*) was a characteristic species of the swamp associations but it was replaced by Reed Canary-grass (*Phalaris arundinacea*), Purple Loosetrife (*Lythrum salicaria*) and Hemp Agrimony (*Eupatorium cannabinum*) in the fen. The second axis contrasts associations, where either *Phragmites* or Branched Bur-reed (*Sparganium erectum*) were dominant (low scores), with those where *Phragmites* and *Sparganium* were absent and the Jointed Rush (*Juncus articulatus*) was locally abundant (high scores). A full species list given in Milsom *et al* (1994).

The effects of the mowing treatments were assessed against two criteria: (i) rates of change in species richness, and (ii) changes to the species composition of the emergent aquatic flora. Turnover of species occurred in all plots although losses tended to outnumber the gains. The loss of species was least in the plots that were mown twice yearly and greatest in those that were left unmown for four years. There was also a suggestion that number of gains per plot and mowing interval were inversely correlated but the test statistic was not quite significant. Overall, however, the net loss of species was significantly correlated with mowing interval (Fig. 2). These trends were investigated further using DECORANA (Fig. 1). The extent and direction of change in species composition between 1989 and 1992 varied considerably between plots, though in some, such as 5,6, 7 and 10, there was almost no change at all. Neither the extent of change nor its direction, as measured by the resultants and angle respectively, differed significantly between treatments (Kruskal-Wallis ANOVA: Chi-square 1.26, n=12, NS for resultants; Chi-square 1.67, n=12, NS for angle).

DISCUSSION

This experiment was the first attempt to describe the aquatic flora of arable fenland ditches in detail. Its main finding was that the emergent aquatic plant associations both within and between ditches were diverse which was unexpected. The floristic gradient, over 200 m, from swamp to herbaceous fen plant associations in one ditch (Plots 1-8), and the difference in swamp associations between adjacent ditches (Plots 1-8 vs 9-12), have implications for management prescriptions because of the variable effects of mowing.

Manipulation of the mowing frequency had an effect that was common to all emergent aquatic plant associations in the ditches. The reasons for this relationship are not clear from the results of the experiment. However, it is known that the species richness of fen plant communities is inversely correlated with their above-ground biomass (Wheeler and Giller, 1982), and mowing may well have affected the quantity of the vegetation in the ditches.

The effects of mowing on species composition were less uniform, however There was some evidence from the plots dominated by *Phragmites* (Plots 1,2 & 3) that the lack of mowing resulted in build up of litter and loss of species richness, whereas this process was slowed when cutting was done at least every two years. These differences between mowing treatments are reflected in Fig. 1. There was also some suggestion that the degree of change in the *Juncus* swamp association (Plots 9-12) may also have been inversely related to mowing frequency. However, in the herbaceous fen (Plots 4-8), there was no evidence of a simple correlation between mowing frequency and either the extent, or the direction, of change in species composition. These effects are being explored further in a follow-up experiment which has greater replication.

Ditch banks in the Fens are mown to minimise their resistance to water flow (hydraulic roughness) which may be critical during periods of high flow. This management is aimed particularly at *Phragmites* because of the significant contribution that the species makes to the hydraulic roughness of the banks and the water course itself. However, the evidence from this experiment was that frequent winter mowing promoted the growth of *Phragmites* on the banks, a finding which is in agreement with other studies (Haslam, 1968, Cowie *et al*., 1992). Reed growth also became more luxuriant in the water course itself, possibly due to lateral spread, where it impeded water flow and resulted in water backing up one of the ditches (Rose and Harris, 1993).

Apart from the hydraulic implications, pure stands of reed are undesirable from the conservation point of view. Though stands of reed are often the only nesting habitat for birds in ditches, they shade the water course and, thereby, reduce the species diversity of floating and submerged aquatic plant associations. Moreover, evidence from this experiment (Milsom *et al*, unpublished) suggests that swamp communities dominated by reeds support the least diverse species assemblages of dragonflies and butterflies. Summer cutting is a possible remedy to these problems because it is known to inhibit *Phragmites* growth (Haslam 1968) and to increase the diversity of herbaceous fen plant communities (Rowell *et al*., 1985). However, the practicality of summer cutting needs to be evaluated because maturing agricultural crops often hinder access to ditches and there are also implications for nesting birds.

ACKNOWLEDGEMENTS

The experiment was funded by the Land Use Conservation and Countryside policy group of MAFF. John Crow and Monica O'Donnell (ADAS) assisted in the field. Chris Feare, Roger Trout (CSL) and Carey Coombs (ADAS) co-ordinated the project. We are very grateful to them all. We wish to acknowledge the preliminary work on the ditches at Arthur Rickwood carried out by Kevin Hand and Nigel Critchley.

REFERENCES

Alcock, M.R.; Palmer, M.A. (1985). A Standard Method for the Survey of Ditch Vegetation. Nature Conservancy Chief Scientist Team Notes, **37**. Peterborough, J.N.C.C.

Cowie, N.R.; Sutherland, W.J.; Ditlhogo, K.M.; James, R. (1992). 'The effects of conservation management of reed beds. II. The flora and litter disappearance', Journal of Applied Ecology, **29**, 277-284.

Haslam, S.M. (1968). The biology of Reed *Phragmites communis* in relation to its control. Proceedings of 9th British Weed Control Conference, pp. 392-397.

Hill, D.; Rushton, S.P.; Clark, N.; Green, P.; Prys-Jones, R. (1993). Shorebird communities on British estuaries: factors affecting community composition. Journal of Applied Ecology, **30**, 220-234.

Hill, M.O. (1979). DECORANA - a FORTRAN program for Detrended Correspondence Analysis and Reciprocal Averaging. Section of Ecology and Systematics, Ithaca, Cornell University.

Milsom, T.P.; Runham, S.R.; Sherwood, A.J.M.; Rose, S.C.; Town, S.J. (1994). Management of fenland ditches for farming and wildlife: an experimental approach. Proceedings of Symposium on Nature Conservation and the Management of Drainage System Habitat. Loughborough, ICOLE, *in press*.

Newbold, C.; Honnor, J.; Buckley, K. (1989). Nature Conservation and Management of Drainage Channels. Peterborough, Nature Conservancy Council and Association of Drainage Authorities.

Rose, S.C.; Harris, G.L. (1993). The management of ditches in arable fenland for drainage and wildlife - Phase 1 hydrology report. Cambridge, ADAS Soil and Water Research Centre.

Rowell, T.A.; Guarino, L.; Harvey, H.J. (1985). The experimental management of vegetation at

Wicken Fen, Cambridgeshire. Journal of Applied Ecology, **22**, 217-227.
Sherwood, A.; Harris, G.L. (1991). The management of ditches in arable fenland for drainage and wildlife. Cambridge, ADAS Land and Environment.
Siegel, S.; Castellan, N.J. (1984). Nonparametric statistics for the behavioral sciences, New York, McGraw-Hill.
Wheeler,B.D.; Giller, K.E. (1982). Species richness of herbaceous fen vegetation in Broadland, Norfolk in relation to the quantity of above-ground plant material. Journal of Ecology, **70**, 179-200.

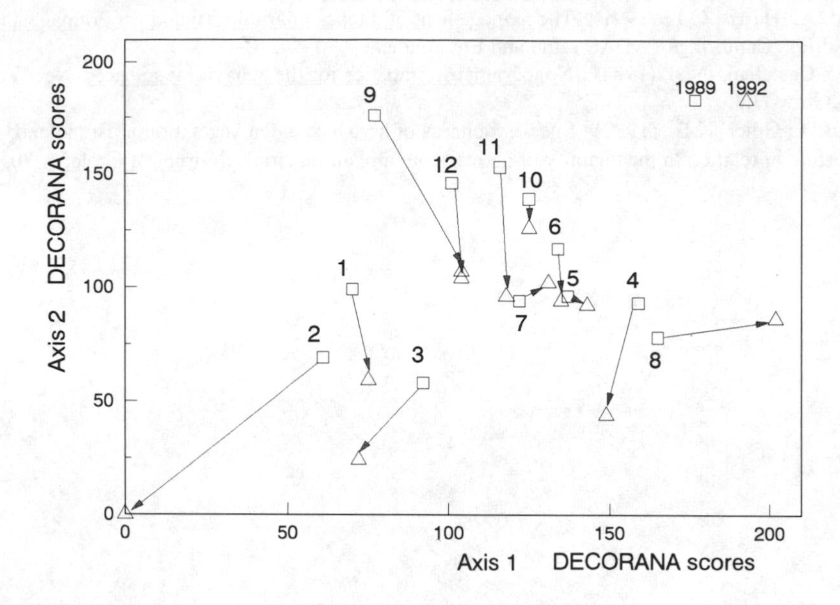

FIGURE 1. Ordination of plots with respect to first two axes from Detrended Correspondance Analysis showing DECORANA scores and changes within plots between 1989 and 1992. Plot numbers refer to those in the text. Plots 1,4 & 10 were mown twice a year; Plots 5, 8 and 11 were mown annually; Plots 3,6 and 12 were mown once every two years and Plots 2, 7 and 9 remained uncut over 4 years.

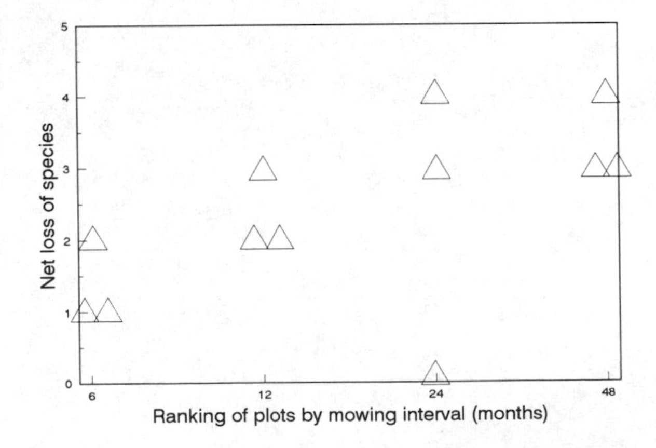

FIGURE 2. Correlation between net loss of species per plot and mowing interval. Kendall Rank Correlation coefficient: 0.495, P=0.024.

MANAGING FIELD MARGINS FOR THE CONSERVATION OF THE ARABLE FLORA.

Philip Wilson.

Farmland Ecology Unit, The Game Conservancy Trust, Fordingbridge, Hampshire, SP6 1EF.

ABSTRACT

The margins of arable fields are of great value for the conservation of endangered plants. Changes in farming practices during recent years have resulted in the severe decline of many species including some which were once very common, and some have even become extinct. There is therefore an urgent need for guidelines for the conservation management of these species and their communities. "The Wildflower Project" carried out between 1987 and 1990 showed that herbicide and nitrogen use were very important factors that could be easily manipulated within modern farming practice. The second phase of this project is designed to test these factors in practice and to provide some indication of the costs of management for the conservation of endangered arable plants.

INTRODUCTION

Since the revolution in arable farming since 1945, many species of arable weed such as *Scandix pecten-veneris* and *Ranunculus arvensis* which were once common, have become very rare. Others such as *Arnoseris minima* and *Caucalis platycarpos* have become extinct. In addition to the progressive restriction of many species to sites with conditions that are favorable to annual plants of mediterranean origin near the edge of their climatic range in Britain, most species are also becoming restricted to the extreme edges of arable fields where agricultural inputs are less efficient and crop yields are reduced (Wilson, in press).

Britain's arable field margins still contain populations of a number of threatened plant species. These include 25 that are classified as "Nationally Scarce" (recorded from fewer than 100 10km squares), and 24 "Red Data Book" species recorded from 15 or fewer 10km squares (Perring & Farrell, 1983). At least five further species are now of "Red Data Book" status. Of these "Red Data Book" species seven are now extinct, and ten others no longer occur in strictly arable habitats. Nine now receive full legal protection, although only two of these still occur at all in arable habitats (Wilson, 1994).

There is an urgent need for the conservation of these endangered species and the communities within which they occur, however until recently, little information has been available on which management can be based. It has also been difficult to persuade many conservationists that the arable habitat is of any importance.

THE WILDFLOWER PROJECT - PHASE 1.

A three-year research project was started in 1987 with the aim of investigating factors in the biology and ecology of a range of uncommon arable weeds which may be manipulated within the context of modern farming. Some of the findings are summarised below.

Germination periodicity and performance in relation to crop sowing time.

As a result of the germination periodicity of individual weed species, crops sown on different dates can support very different communities, even when the seed-bank is similar. Differences are greatest between autumn- and spring-sown crops, but can also be large between early and late autumn sowings and early and late spring sowings (Table 1). Changes in crop rotations and sowing times may have affected species with narrow germination periods.

TABLE 1. Mean numbers of plants in 4m² plots of cereal sown on seven dates. Significance levels: *** P<0.001, ** P<0.01, * P<0.05.

Crop	Winter barley			Winter wheat			Spring barley			
Date	29:9	13:10	2:11	13:10	2:11	19:11	16:2	9:3	28:3	P
Agrostemma githago	20.2	6.1	0	7.3	0	0	0	0	0	***
Petroselinum segetum	4.3	8.0	5.0	6.1	3.0	0	0	0	0	***
Torilis arvensis	8.1	12.4	14.4	8.0	12.2	8.0	0	0	0	***
Scandix pecten-veneris	1.4	5.4	5.2	5.6	5.0	4.2	0.4	0.1	0	***
Ranunculus arvensis	2.5	10.3	13.0	12.4	16.5	5.4	0	0.1	0	***
Buglossoides arvensis	3.7	14.6	16.6	17.4	17.3	9.6	2.2	9.5	8.7	***
Adonis annua	0.1	1.7	3.7	3.3	5.7	7.1	0.2	0.9	0	***
Papaver argemone	0.1	0.6	2.8	0.2	4.9	5.9	0.7	2.0	1.3	**
Valerianella rimosa	2.6	5.2	9.5	0.9	6.0	12.7	4.4	7.4	5.1	*
Papaver hybridum	0	0	2.4	0	2.7	13.5	2.1	4.8	3.5	***
Chrysanthemum segetum	3.4	0	0.2	1.0	0	0.4	5.7	9.7	8.9	***
Silene noctiflora	0	0	0	0	0	3.6	6.6	7.1	15.5	***
Misopates orontium	0	0	0	0	0	0	0	0.3	4.8	*

Effects of crop type.

It was also possible to compare fruit production in winter wheat and winter barley in the experiment described above. *Agrostemma githago, Buglossoides arvensis, Papaver hybridum, Ranunculus arvensis* and *Scandix pecten-veneris* all produced significantly more fruit per plot in winter wheat, while only *Valerianella rimosa* produced more fruit in winter barley.

Herbicides.

Many uncommon species are susceptible to a wide range of herbicides (Wilson, 1990), and it is likely that the introduction of modern herbicides from 1945 onwards has been a very important factor in the decline of many species. Species are however differentially susceptible to different herbicides, and even some uncommon species are non-susceptible to commonly used compounds (Table 2).

TABLE 2. The susceptibilities in relation to unsprayed control plants of seven uncommon annual plant species to four commonly used herbicides as determined in an experiment using pot-grown plants. Significance levels: *** P<0.001, ** P<0.01, * P<0.05.

	Chlor-toluron	Mecoprop	MCPA	Ioxynil/ Bromoxynil
Buglossoides arvensis	*	*	ns	**
Chrysanthemum segetum	*	ns	ns	*
Misopates orontium	-	***	*	*
Papaver hybridum	**	***	*	*
Ranunculus arvensis	*	**	-	-
Silene noctiflora	*	*	-	***
Scandix pecten-veneris	**	***	**	*

TABLE 3. Number of plants per m² present at harvest time in plots of cereals to which nitrogen was applied at three levels. Significance levels: *** P<0.001, ** P<0.01, * P<0.05. - = not significant.

	Nitrogen Level (Kg/ha²)			
	0	75	150	
Misopates orontium	0.9	0.02	0	***
Myosurus minimus	0.3	0	0	**
Arnoseris minima	0.2	0	0	***
Papaver hybridum	1.2	0.6	0.06	***
Filago pyramidata	3.2	0.9	0.3	***
Valerianella rimosa	3.0	1.7	0.6	***
Papaver argemone	0.7	0.3	0.2	*
Scandix pecten-veneris	1.8	1.1	0.8	-
Chrysanthemum segetum	1.2	0.9	0.6	-
Silene noctiflora	2.3	2.0	1.2	-
Ranunculus arvensis	3.0	1.8	1.6	**
Buglossoides arvensis	2.0	1.3	2.2	-

Competitive ability in relation to crops at different levels of nitrogen application.

Mean levels of nitrogen applied to arable crops increased by over 600% between 1943 and 1988 (Chalmers et al, 1990), and the cereal cultivars that are now grown tend to be highly responsive to these high levels (Fischbeck, 1990). Most of the weeds which have decreased most rapidly are relatively slow-growing annuals which might be expected to compete poorly with a fully fertilised crop. In experiments, levels of nitrogen similar to those applied to modern cereal crops suppressed the growth of many weeds almost as effectively as herbicides (Table 3).

THE WILDFLOWER PROJECT - PHASE 2.

Phase 1 of the Wildflower Project identified several potential reasons for the decline of some arable weeds, but did not investigate the effects of manipulating any of these factors on existing populations, or the costs of conservation management to the farmer. The second phase which started in 1992 was designed to test options for the management of endangered arable weeds and their communities within the context of modern farming.

Ten experiments were set up in both winter and spring cereals in the south of England during 1992 and 1993. In these the effects of omitting herbicide and nitrogen on the numbers and seed production of a range of common and uncommon arable weeds were studied. Preliminary results suggest that omission of both nitrogen and broad-leaved herbicides result in the greatest floristic diversity and number of uncommon species (Table 4).

TABLE 4. Mean numbers of species of arable weed per 2.5m² in plots of winter cereals grown under three regimes on three farms in Norfolk, Suffolk and Hampshire.

	Norfolk	Suffolk	Hampshire	Mean
Full nitrogen and herbicide	8.3	3.3	14.0	8.6
Full nitrogen, no herbicide	22.3	16.7	20.3	19.8
No nitrogen, no herbicide	24.3	19.0	25.7	23.0

A further series of experiments have been set up in 1993 and 1994. These have the aims of confirming the results of the previous year's investigations and also studying the effects of the use of selective graminicides on the growth and performance of uncommon broad-leaved species in fields where there are known to be large populations of highly competitive grass weeds including *Alopecurus myosuroides, Avena spp.* and *Lolium spp.*. It is possible that these grasses can have a detrimental competitive effect, especially where nitrogen applications are large. The effects of most of the modern selective graminicides on uncommon species are however not well known.

TOWARDS A MANAGEMENT STRATEGY FOR THREATENED ARABLE WEED COMMUNITIES.

Although the second phase of the "Wildflower Project" is at an early stage, it is possible to make some suggestions for the conservation management of endangered species and communities of arable land based on information gathered so far.

Conservation measures can, in most cases, be directed at the outermost four metres of fields, although there are exceptions where rare species grow outside this area. All broad-leaved herbicides must be omitted from conservation areas, although some selective graminicides may be permissible pending further work. Nitrogen should also be omitted as the competitive effect of a fertilised crop can suppress many uncompetitive species. It is important to drill crops at the correct time, although most species can withstand occasional years in crop rotations when conditions are not ideal. Practically this will mean either autumn or spring drilling, as the precise dates of farming operations are often determined by factors outside farmers' control. Winter wheat may allow better seed production than winter barley, and it is likely that modern crop varieties will be less competitive under low input conditions than older ones.

Some areas are already managed for their arable flora. Four sites have been scheduled as SSSIs by English Nature, although at only two of these are whole fields managed as part of a conventional arable farm. At one of these two sites, fertilisers and herbicides are applied as normal, while at the other no inputs are permitted. The headland only at the third site is managed. The fourth SSSI has recently been purchased by the Somerset Wildlife Trust, who intend to manage the three small fields as a working arable unit with minimal agrochemical inputs. The National Trust in Cornwall are currently managing some of their arable land for its arable flora, a particularly valuable initiative, as they are one of Britain's largest private landowners. Other sites are managed by informal agreement with landowners. One problem which has become apparent at some sites has been the build up of perennial weeds. Research into the selective control of these species will be essential future work.

In some fields where crops are not drilled up to the base of the field boundary, a zone of occasional cultivation exists between the crop and the perennial vegetation of the boundary. On nutrient-poor calcareous or sandy soils this zone can not only be of immense value for the least competitive arable annuals, but is also a habitat for annuals with atypical life-cycles and pauciennials, many of which, including *Ajuga chamaepitys* and *Teucrium botrys*, are now extremely rare. Research into the ecology of this habitat is also essential if the flora of the whole field margin is to be effectively conserved.

CONCLUSIONS.

It is already possible to propose provisional guidelines for the management of endangered arable plant communities and species, and on completion of the "Wildflower Project - Phase 2", detailed guidelines and costings of conservation management will be available. The implementation of such guidelines on a large scale is not impracticable. The experience of The Game Conservancy Trust with Conservation Headlands (Sotherton, 1990) has shown that given an incentive and soundly researched information, many farmers are happy to manipulate farming practices for environmental benefits (Thompson, 1993). Guidelines for the conservation of the arable flora can easily be incorporated into the management of Set-Aside land (Firbank & Wilson, in press) or the Countryside Commission's Stewardship scheme. Such programmes are already funded over large areas of Germany by local government (Schumacher, 1987), and with the political will, the future of some of our most endangered plants could become considerably brighter.

REFERENCES.

Chalmers A., Kershaw C. & Leach P. (1990). Fertiliser use on farm crops in Great Britain: results from the survey of fertilising practice, 1969-88. Outlook on Agriculture **19**, 269-278.

Firbank L.G. & Wilson P.J. (in press). Arable weeds and set-aside, cause for conservation or cause for concern ? In: Insects, Plants and Set-aside, Symposium of the Botanical Society of the British Isles and the Royal Entomological Society, 1994.

Fischbeck G. (1990). The evolution of cereal crops. In: The Ecology of Temperate Cereal Fields, L.G. Firbank, N. Carter, J.F. Darbyshire & G.R. Potts (Eds). Blackwell, Oxford. 31-54.

Perring F.H. & Farrell L. (1983). British Red Data Books: 1. Vascular Plants. RSNC, Lincoln.

Schumacher W. (1987). Measures taken to preserve arable weeds and their associated communities in central Europe. In: Field Margins, J.M. Way & P.G. Grieg-Smith (Eds), British Crop Protection Council Monograph 35, 109-113.

Sotherton N.W. (1990). Conservation headlands: a practical combination of intensive cereal farming and conservation. In: The Ecology of Temperate Cereal Fields, L.G. Firbank, N. Carter, J.F. Darbyshire & G.R. Potts (Eds). Blackwell, Oxford. 373-398.

Thompson P.G.L. (1993). The use of conservation headlands in Britain in 1992. The Game Conservancy Review of 1992. **24**, 67-68.

Wilson P.J. (1990). The ecology and conservation of rare arable weeds and communities. Unpublished PhD thesis. University of Southampton.

Wilson P.J. (1994). Botanical diversity in arable field margins. In: Field Margins - Integrating Agriculture and Conservation (This volume).

Wilson P.J. (in press). The distribution of dicotyledonous arable weeds in relation to distance from the field edge. Journal of Applied Ecology.

COMPARISON OF FIVE DIFFERENT BOUNDARY STRIPS - INTERIM REPORT OF FIRST TWO YEARS' STUDY

M.J. MAY

Morley Research Centre, Morley, Wymondham, Norfolk, NR18 9DB,

CLARE EWIN

Wymondham College, Morley, Norfolk NR18 9SZ

J. MOTT

University of East Anglia, Norwich NR4 7TJ

R. PACK and CAROLINE RUSSELL

Anglia Polytechnic University, East Road, Cambridge CB1 1PT

ABSTRACT

An experiment was initiated in winter wheat during autumn 1991 to compare five different boundary strips. These were 2 m wide by 16.5 m long and comprised conventional cropping (up to the field boundary), a sterile strip (by rotary cultivation), sown grasses, a game conservancy style strip (conservation headlands with barrier + sterile boundary strips) and sown wild flowers. In the first two years of the experiment, changes in weed flora have been recorded. Some invasive weeds such as *Elymus repens*, *Cirsium arvense*, and *Urtica dioica* are starting to ingress from the field boundary into some boundary strips and, where competition is poor, *Galium aparine* and *Bromus sterilis* are also starting to invade the boundary strip. Large differences in the numbers and species of invertebrates were recorded in pitfall traps, but the reasons for the differences cannot currently be explained.

INTRODUCTION

There is an increasing demand from farmers for more information concerning the practical, biological and financial effects of various boundary strips on farming systems. There is some information regarding the effects of conservation headlands on crop yields (Boatman & Sotherton, 1988); in general there was no significant yield loss. However, there is little comparative information between various types of boundary strips, especially for arable situations as encountered in Eastern England. This paper is a preliminary report of the first two years of a long term experiment to compare different boundary strips and their effects on crop yield and insect and weed fauna.

METHOD

Five different boundary strips were established on the east side of a north south embankment adjacent to a public road. The experiment was on a sandy loam soil at Manor Farm, Morley, Norfolk and established on the 'non-turning headland' of the field. The bank itself was relatively uniform with only a few small (<2 m tall) trees. A winter wheat crop (cv Riband as a second wheat) was established on 24 October 1991 and the boundary strip treatments set up during the autumn and early spring. They comprised conventional cropping (up to the field boundary), a sterile strip (by rotary cultivation), sown grasses, a game conservancy style strip (a conservation headland with 1 m left next to the bank for the field boundary vegetation to expand and a 1 m sterile strip next to the crop) and sown wild flowers. These flowers were a modified Emorsgate EM1 mixture (Anon, 1992) sown on 11 November. The grass strips were sown on 2 April 1992 with an Emorsgate General Purpose Meadow mixture (Anon, 1992). The sterile strips were rotovated in autumn 1991 and again in May and September 1992 and 1993. The crop for the 1992 harvest year was winter beans (cv Punch) sown on 20 October. The plots were 2 m wide, 16 m long and replicated five times as a randomised block design. Only the cropped boundary strip received normal farm inputs, the rest received none, and the game conservancy style treatments did not receive herbicides or insecticides on the first 6 m of crop. The bean crop was not treated with herbicide.

Weeds were assessed by counts or scores. The numbers and frequency of weeds were assessed either by the use of five 0.1 m x 0.1 m quadrats in June and July 1993 or by ten 0.25 m x 0.25 quadrats/wildflower boundary and their presence recorded on a 1 m x 16 m transect along the plot. The wild flower boundary strip was assessed in detail in 1992, but other strips were assessed to record the major weed species present. Insects were monitored by pitfall (100 mm tall disposable plastic cups containing water and a few millilitres of washing up liquid) or specially constructed directional traps. Assessments on the game conservancy style boundary strips were meaned for the two separate sections unless otherwise stated. The crops were harvested using an adapted Claas Compact combine harvester; the wheat on 1 September 1992 and the beans on 1 September 1993. Yields from areas of the adjacent crop as assessed on seven 2 m wide strips each 14 m long. The yield of the conventional crop on the boundary strip was also recorded.

RESULTS

Plants

The two methods of assessment provided different results as to the number of species present. The transect system recorded fewer species than the two sets of five quadrats. The number of species and overall densities recorded in the quadrats is given in Table 1.

The most prolific vegetation along the bank was *Urtica dioica*, *Arrhenatherum elatius* and *Poa pratensis*. *Convolvulus arvensis*, *Dactylis glomerata*, *Rumex* spp, *Rubus fruticosus*, *Elymus repens*, *Cirsium arvense* and *Equisetum arvense* were also present, but to a much lesser degree. In 1992, *Bromus sterilis* and *Galium aparine* appeared to make use of the extra light available in the absence of crop or tall weeds and 'leant' over the boundary strip, where they could have shed seed.

In both years, each treatment gave different dominant vegetation in the boundary strip. The dominant vegetation in the cropped boundary strip was, of course, wheat in 1992 and beans in 1993. There were few weeds present in 1992, mainly *C. arvense* and a few *Achillea millefolium*, but in 1993 *A. elatius*, *Papaver rhoeas*, *Polygonum aviculare* and *Tripleurospermum inodorum* were recorded.

In 1992 there were few weeds in the sterile strips, mainly *Rumex obtusifolius* and *C. arvense* with a few annual weeds such as *Chenopodium album* and *Poa annua*. In 1993 *C. album* was the dominant weed, along with *A. elatius*, *Senecio vulgaris* and *Lamium purpureum*.

The grasses established reasonably well in 1992, but *P. aviculare*, *L. purpureum* and *Viola arvensis* were also present. In 1993, the sown-grass strips were dominated by *Cynosurus cristatus* (included in the sown mixture). The next most dominant were *Festuca rubra commutata, F. rubra purinosa* and *F. rubra rubra*) (collectively these made up 35% of the seedling mix). These were followed in cover by *Holcus mollis*, *D. glomerata*, *P. pratensis* and *A. elatius*. Invasive weeds included *C. arvense*, *E. repens* and *U. dioica*. *Leucanthemum vulgare*, *Achillea millefolium*, *Agrostemma githago*, *Plantago lanceolata* and *F. rubra*, each included in the wild flower mixture, were present on all wild flower plots.

Table 3 shows a comparison between 1992 and 1993 of the main species present on the wild flower boundary. In 1992 the most abundant species were *P. rhoeas*, *Agrostemma githago*, *C. cyanus* and *A. millefolium*. In 1993 the dominant species had changed to *L. vulgare*, *A. millefolium* and *P. lanceolata*. *P. rhoeas* was absent from all plots and density of *A. githago* was low. *L. vulgare* was very successful in 1993, but was apparently absent in 1992.

In 1992, the weed spectra were similar in the sterile area of the game conservancy style and sterile boundary strips. In 1993 the most dominant vegetation in the first metre transect (essentially an extension of the bank) was *Trisetum flavescens* and *P. pratensis*. The one metre sterile strip suffered an invasion of *C. album* which varied between a trace to over 75% cover. Other weeds included *P. rhoeas*, *Galium aparine*, *T. inodorum, Heracleum sphondylium* and *A. millefolium*.

There were very few weeds present in the crop in 1992 (including the 6 m untreated area adjacent to the game conservancy style boundary), those present were mainly *P. annua* with *P. aviculare* and *T. inodorum*. The weeds invading the field crop in 1993 differed slightly between treatment, but the most invasive and dominant species was *P. aviculare* followed by *C. album* and *T. inodorum*. There was a greater variety of weed species next to the game conservancy style strips, but these did not exceed those on the sprayed strips.

B. sterilis was the most abundant species on the game conservancy style grass strip whereas on the sown grass boundary it was *C. cristatus*, *C. album* on the sterile areas, *A. millefolium* on the wild flower boundary and *P. aviculare* on the cropped headland (Table 2).

TABLE 1. Number of species recorded and mean density/m² in each boundary strip in June and July 1993

	Crop	'Game' bank	'Game' sterile	Sterile	Grass	Wild flower	Total species
Number of species							
Perennial dicot.	10	15	10	10	19	14	25
Annual dicot.	18	12	15	11	11	7	25
Total grasses	7	12	8	6	12	11	17
Total species	35	39	33	27	42	32	67
Density/m²							
Perennial dicot.		92	94	64	44	40	112
Annual dicot.		126	20	172	188	12	28
Total grasses		64	152	28	40	220	120
Total		282	366	264	272	272	360

TABLE 2. Average percentage of the major species recorded on each boundary strip in June and July 1993

	Crop	'Game' bank	'Game' sterile	Sterile	Grass	Wild flower
Perennial dicotyledons						
Rumex obtusifolius	1.5	2.1	2.7	4.5	1.2	
Cirsium arvense	3.1	3.7	6.1	3.2	1.2	1.2
Equisetum arvense	2.3	5.8	3.3	0.6	0.2	0.9
Achillea millefolium	3.1	3.3	0.5	1.9	2.0	12.8
Annual and perennial grasses						
Poa annua	6.1	2.1	4.4	1.3	1.2	1.2
Bromus sterilis	4.6	16.5	0.5	2.6	7.7	3.1
Dactylis glomerata		4.1	1.1		4.2	3.4
Elymus repens	0.8	1.6	1.1	7.1		
Cynosurus cristatus					30.9	9.7
Arrhenatherum elatius	2.3	13.2	1.1	2.6	2.0	
Poa pratensis	0.8	12.3			12.3	4.4
Triticum aestivum	8.4	1.6	1.1	0.6	1.0	1.2
Annual dicotyledons						
Chenopodium album	3.8		46.7	56.5		
Galium aparine	8.4	2.9	2.2		0.5	0.9
Myosotis arvensis	0.8		0.5			
Viola arvensis	1.5	0.4			0.2	
Senecio vulgaris	1.5	0.4	4.4	5.2	0.5	
Sisymbrium officinale	1.5					
Polygonum aviculare	13.7	0.4	2.2	0.6		

TABLE 3. Changes in flora dynamics (plants/m^2) of wild flower boundary strip, 1992-93

	1992	1993
Achillea millefolium	22	95
Leucanthemum vulgare	0	58
Centaurea cyanus	26	1
Agrostemma githago	12	16
Plantago lanceolata	0	24
Tripleurospermum inodorum	18	12
Papaver rhoeas	32	0

Invertebrates

Table 4 contains data for the invertebrates found most frequently in pitfall traps in 1992 and 1993. Individual species were not identified, but the majority of beetles was *Amara aulica*. The figures for the 'Game' boundary are the mean of the two areas.

In July 1993 there was a trend for more invertebrates to be caught in directional traps set in the field boundary (21.8) and less in the field (15.5) but this was not statistically significant ($p=0.05$) for any individual or group of species (Table 5). The traps did not provide conclusive data as to whether the invertebrates were moving within or simply traversing the areas sampled.

TABLE 4. Invertebrates caught in pitfall traps 28 May and 7 July 1992 and 21 June 1993

	Wildflower	Grass	Sterile	Crop	'Game'	SED
28 May 1992						
'spiders'	20	19	10	17	14	6.7
'beetles'	23	38	52	11	33	5.9
'ladybird larvae'	15	8	3	2	6	1.4
'aphids'	0	0	0	0	0	
7 July 1992						
'spiders'	14	14	7	21	6	2.4
'beetles'	35	38	22	4	37	6.6
'ladybird larvae'	1	3	0	0	0	0.3
'aphids'	15	6	21	153	272	10.56
21 June 1993						
'spiders'	3	3	3	2	4	1.1
'beetles'	2	4	3	3	5	1.5
'ladybird larvae'	>1	0	0	0	>1	0.2
'aphids'	0	0	1	0	>1	0.5

TABLE 5. Invertebrates caught in directional traps in July 1993

	Wild flower	Grass	Sterile	Crop	Game	SED
Adalia bipunctata	0.37	0.06	0.75	0.00	0.31	0.278
Amara aulica	0.38	0.69	1.56	3.12	1.94	0.767
Anthocomus fasciatus	1.37	0.81	0.31	3.62	0.88	1.082
Spharite glabratus	0.37	1.62	0.12	2.37	0.25	0.857
'snails'	1.50	0.94	0.31	0.13	0.37	0.406

Crop yields

TABLE 6. Mean crop yields for all harvested strips (t/ha at 85% dm)

	Wild flower	Grass	Sterile	Crop	Game	SED
wheat 1992	8.2	8.3	7.9	8.6	8.2	0.64
beans 1993	3.6	3.8	3.5	4.4	3.4	0.42

Harvest was delayed in 1992 owing to wet weather and brackling increased from 17% (adjacent to the field boundary) to 78% (10 to 12 m into the crop), but there was no significant difference between treatments. Whilst there was a trend of slightly poorer wheat yields near the field boundary, there was no significant interaction between the boundary strip treatments and distance into the field. The lowest wheat yields occurred where the crop was grown with a sterile boundary and highest where wheat was grown on the boundary strip (Table 6). There was no significant difference between the other boundary treatments.

DISCUSSION

These data show the complexity of regrowth onto the headland strips. All the unsown species found can be expected to occur as opportunists in similar situations, and the generally low percentage incidence confirms the large number of species which will colonise in these circumstances. It is likely that this diverse flora is of considerable benefit. The experiment has been sown to winter wheat for 1994 and will be sugar beet in 1995.

ACKNOWLEDGEMENTS

The authors acknowledge the help and encouragement given by Dr I.D.S. Brodie, Mr R.J. Cook, Dr C. Hill, Dr G. Martin, Mr R. McMullen and Mr M. Nowakowski.

REFERENCES

Anon (1992) Price list (spring 1992) Emorsgate seeds pp8

Boatman, N.; Sotherton, N., (1988) The agronomic consequences and costs of managing field margins for game and wildlife conservation. *Aspects of Applied Biology, 17*, 47-56.

THE CASE FOR HEADLAND SET-ASIDE

D.L. SPARKES, R.K. SCOTT

Department of Agriculture and Horticulture, University of Nottingham, Sutton Bonington Campus, Loughborough, Leics LE12 5RD.

K.W. JAGGARD

Institute of Arable Crops Research, Broom's Barn Experimental Station, Higham, Bury St. Edmunds, Suffolk IP28 6NP.

ABSTRACT

Headland yields were recorded by farmers on commercial cereal and sugar beet crops. The yield of the headland averaged 26% lower in the sugar beet crops and 11% lower in the winter wheat. One spring barley field was also monitored, and in this field the headland actually outyielded the midfield by 20%. Experimental work at Sutton Bonington indicated that yields of winter barley and sugar beet increased with distance into the field. When boundary strips of different widths around the crop were planted with grass no evidence for migration of the 'headland effect' into the field was found.

INTRODUCTION

The first set-aside scheme began in 1988. This was a voluntary programme which allowed farmers to take 20% or more of their land out of production for five years: for doing this farmers received a payment per hectare set aside. In 1991 a further voluntary scheme was introduced which allowed set-aside for one year only. The lower limit for this was set at 15%.

In 1992 the Common Agricultural Policy (C.A.P.) reforms introduced a new set-aside scheme across the European Community (E.C). The system of intervention payments was dropped, and replaced by Arable Area Payments. In order to qualify for these payments, farmers with more than 16 hectares of land in eligible crops had to set-aside 15% of their acreage on a rotational basis. From 1993, a non-rotational option is also available, set at the higher rate of 18%, making headland set-aside possible.

Depending on the size and shape of a field, the crop margin can be difficult to manage, and often yields poorly in comparison to the rest of the field (Boatman & Sotherton 1988, Speller, Cleal & Runham 1992). From studies on sugar beet headland yields, Jaggard (pers. comm.) estimated that headlands take up 15% of the area, require 20% of the effort, but only produce 10% of the yield of an average field. Headland set-aside would allow fields to be 'squared-off', and for all turning of sprayers, spreaders, hoes and harvesters to take place out of the crop.

At Bedfordia Farms, Bedfordshire where headland set-aside has operated for the past five years, it was estimated that taking 20% of the land out of production reduced yields by just 11% but reduced machinery and labour costs by 35%. These figures, were they repeated elsewhere, would make headland set-aside very attractive to the farmer.

The potential benefits of headland set-aside are not just economic. A grass strip between crop and boundary could act as a 'buffer zone', preventing nitrate and phosphate leaching into water courses. It could also provide nesting areas for game, hunting areas for owls and kestrels and, if wild flower species are encouraged, the diversity of flora will attract a wider range of fauna.

On-farm studies have been conducted to look at crop margin yields in comparison to field centre yields in cereals and sugar beet, the latter of which is more susceptible to soil compaction (Brereton 1986). Experiments at Sutton Bonington have been used to look at the yield profile of barley and sugar beet in detail. Various widths of grass strip were used to determine whether the 'headland effect' would be removed if a boundary strip were fallowed, or if it would migrate into the field.

METHOD

Data collection from commercial crops

Several farmers were asked to record, for one or more fields, the headland yield separately from the rest of the field. These data are not always directly comparable as 'headland' can be interpreted either to mean the crop margin all round the field, or just the at the turning ends of the field.

Sugar beet

Two farmers in East Anglia collected data on headland yields of sugar beet fields. At Upton Suffolk Farms, the two 'turning headlands' were recorded separately. At Weasenham Farms yield data were collected for four fields on the all-round headland as compared to the mainfield. Beet harvested from the headland was kept in a separate heap, and delivered separately to the factory in order to distinguish it from that lifted from the rest of the field.

Cereals

The Datavision system was used to record cereal yield on two fields at Moat House Farm, Sufolk. Combine harvesters with the Datavision system have a small radioactive source on one side of the grain elevator which emits low energy gamma rays. These are detected by a receiver on the other side. Yield is monitored as the grain passes up the elevator, interrupting emissions from the radioactive source. The headland was defined as twice around the field with the combine harvester, which had a cutter bar width of 5 metres. The on-board computer recorded yield and area harvested for the headland, and the whole field. Therefore, centre field yield was calculated by subtraction.

Experimental Methodology

An experiment was designed to investigate the effect of fallowing different widths of boundary strip on the yield of commercial crops. Three widths of grass strip (seed mixture of *Holcus lanatus, Phleum pratense* and *Dactylis glomerata*) were chosen to fit in with the farm machinery, e.g. drill, sprayer and fertiliser spreader. Due to the differences in the machinery used for the cereal and sugar beet, the widths are slightly different in each case. The three widths of boundary strip were compared with cropping up to the field boundary. Therefore, there were four treatments, each of which were replicated four times. Each plot was 18 metres long.

For practical reasons associated with the commercial use of crop protection chemicals by farm staff within the experiments, the treatments were arranged in a systematic design, within a block. The widths of grass strip were arranged as a progression 0, 5.4, 10.8, 18m: in the adjacent block this progression was reversed to produce a castellated effect. Each experiment had four replicates, therefore two of these 'castles'. (The grass strip widths given above are for the sugar beet experiment, in the barley crop, because of differences in machinery the widths were 0, 3, 9, 18m).

Crop cover measurements were taken throughout the growing season of both crops using a spectral-ratio meter. Measurements were made at 3 metre intervals from the field boundary to 36 metres into the field.

A cone penetrometer was used to measure soil compaction, also in a transect from the field boundary to 36 metres into the field, in the barley crop on 25 February. Equipment failure, followed by adverse weather conditions, prevented this from being repeated in the sugar beet crop.

Sugar beet

Sugar beet seed (cv. Celt) was drilled on 19 April 1993, at a within row spacing of 16.5cm and between row spacing of 45cm. Crop cover measurements were taken on 26 May, 23 June and 14 July. The crop was harvested on 1 and 4 October. Areas harvested were 6 rows by 2 metres in the headland, and 4 rows by 3 metres in the mainfield (because of the change in direction of the rows), thus 12 metres of row were harvested from each plot. The roots were pulled by hand and topped in the field. The samples were put through the tarehouse at Broom's Barn Experimental Station to determine clean root weight and sugar concentration.

Barley

Winter barley (cv. Pastoral) was drilled on 13 and 14 October 1992. Crop cover measurements were taken on 23 February, 21 April and 26 May. The crop was harvested on 27 and 28 July with a Wintersteiger plot combine (cutter bar width 1.75 metres). Strips were cut from the plots at different distances from the field boundary. Moisture content of sub-samples was determined and the yield standardised to 85% dry matter. Once the samples had passed through a seed cleaner, 1000 grain weights were determined using a Decca Mastercount to count 200 grams of seed.

RESULTS

Studies in commercial crops

Sugar beet

Data collected from two farmers in East Anglia on headland beet yields showed significant reduction in yield compared with the field centres. On one field, 6.24 ha in size, only turning headlands were recorded separately, these yielded 34.6 t/ha, while the rest of the field averaged 63.4 t/ha. Four fields were measured with an all-round headland kept separately from the field centre. On these fields yield on the headland was 21.7 - 30.7 % lower than that on the centre. (Table 1.).

Table 1. Headland and centre field yields of sugar beet from four fields in 1992

Field name	Lamberts	Pottle	Bullrush	Chicory
Field area (ha)	31.13	21.01	15.87	16.32
Headland area (ha)	3.61	3.09	2.62	2.52
Centre yield (t/ha)	43.26	34.53	45.69	46.25
Headland yield (t/ha)	27.52	26.56	35.77	33.69

Cereal

Winter wheat (cv. Estica), harvested on 20 August from a 7.73 ha field, yielded 7.31 t/ha on the headland (1.43 ha) and 8.27 t/ha on the rest of the field. A 4.78 ha field of spring barley (cv. Alexis) harvested on 17 August, yielded 5.22 t/ha on the 0.9 ha headland and 4.08 t/ha on the field centre.

Replicated experiments

The crop cover measurements, made with the spectral-ratio meter, did not show any significant differences due to position in the field.

In general, the density of the soil in the barley crop, measured as the cone penetrometer resistance, was least near the surface and increased to a depth of about 25 cm (Fig. 1). The density was greatest 8 to 10m from the field boundary, corresponding with the positions of the tramlines. The resistance then decreased by about 0.5 MPa in the region 12 to 16m from the field boundary: this corresponds to the area where the sprayer turns around. The outer 4m of the field had a similar density, probably corresponding with the turning of cultivation machinery. Cone resistance 18 to 36m from the field boundary was consistently small. There was no evidence that the grass strip affected cone penetrometer resistance.

Figure 1. Cone resistance of soil in a transect from field boundary to 36m.

(i) Plot A, cropped to field boundary

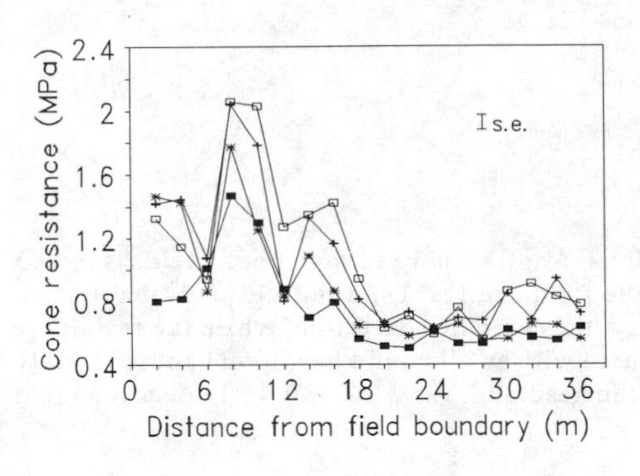

(ii) Plot B, 3m grass strip

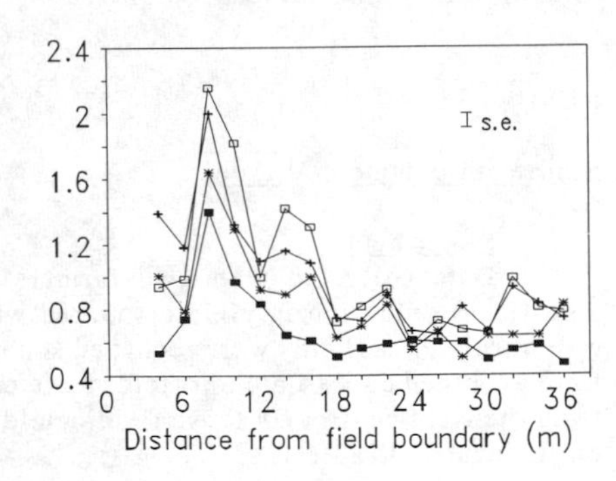

(iii) Plot C, 9m grass strip

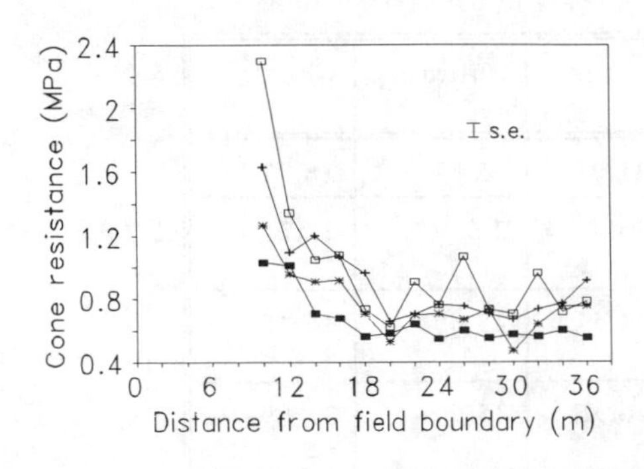

(iv) Plot D, 18m grass strip

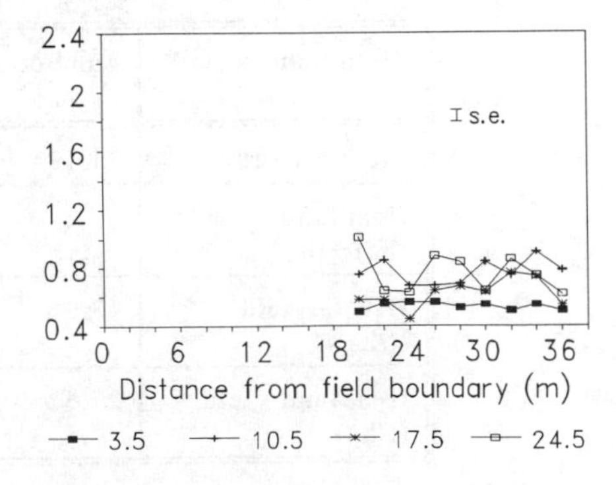

The yields of barley are shown in Fig. 2. The position of the tramlines can be seen clearly, corresponding to a serious yield reduction. There was little evidence that yield was influenced by width of grass strip. The thousand grain weight averaged over all plots, was 36g. There were significantly larger thousand grain weights at 6-9 and 9-12m from the field edge, i.e. the position of the tramlines.

Fig. 3 shows sugar yield of the plots without a grass margin. Yield was small at the very edge of the field, and at about 10m from the boundary. Again, this corresponds to the tramlines. Overall, yield at the crop margin was smaller than in the main field. Sugar percentage was not affected by position in the field, but there was a clear trend in root weight.

Figure 2. Yield of barley from the field boundary to 36 metres into the field.

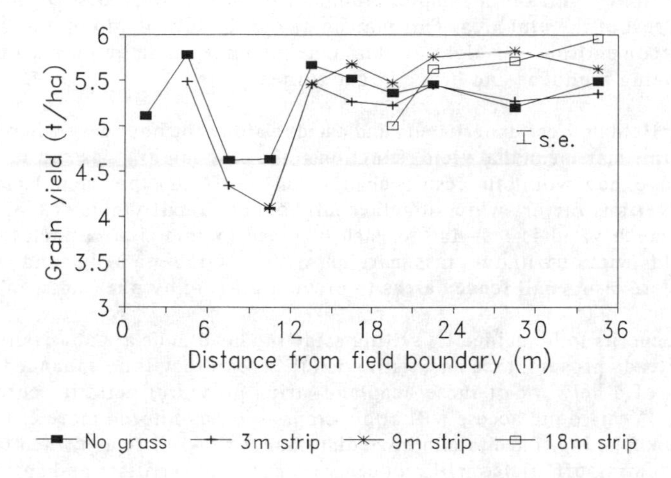

Figure 3. Yield of sugar from field boundary to 24m into the crop.

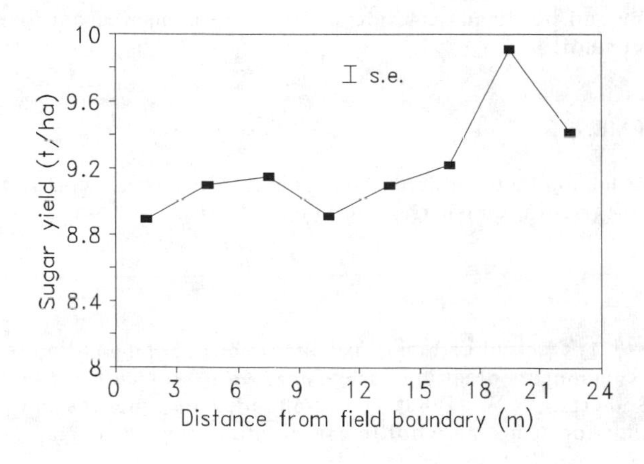

DISCUSSION

Studies in commercial crops, especially sugar beet, revealed the low yields of headlands. However, in one case, the spring barley field, the headland actually yielded more than the midfield. Boatman (1992), reported larger yields on headlands than the centre for two out of three spring barley fields. Boatman & Sotherton (1988), measured cereal headland yields over several years. Although, on average they yielded 18% less than the midfield, there was great variation, from 67% decrease to 24.9 % increase. All other on-farm measurements showed reduced yield on headlands, particularly evident in sugar beet crops.

As well as yielding poorly, the crop margin can be a difficult and time consuming area to manage. Therefore, even if the headlands were to yield as well as the midfield, it may be expensive to produce. The new set-aside rules provide an opportunity for farmers to create boundary strips around arable fields, which contribute towards their set-aside requirement. The experiments at Sutton Bonington showed no evidence for the headland effect moving into the field. However, the experiments so far have only tested one site for each crop in one year so firm conclusions cannot be drawn. The experiments will be repeated next year. It is interesting that thousand grain weights of the samples around the tramlines were significantly greater than those from the rest of the plot area. This may be due to the low plant population in this area, therefore less competition for light, water and nutrients, together contributing to more carbohydrate being produced and hence larger grains.

Neither field in these experiments had a hedge along the boundary where the trial plots were situated; this may mean that yield reduction at the crop margin was minimal. If there was an adjacent hedge, how would the results change? In 1993/4 an experiment has been designed to look at the various factors which together might cause small yields at crop margins, and attribute how much yield is lost due to each of these factors. It is a factorial experiment comparing yields with or without grass margins, with or without a hedge and with or without turning. There are also small fenced areas to prevent grazing by pheasants, rabbits etc.

Other benefits to be gained by setting aside the headlands are not easily quantifiable. If a hedge is already present, then an existing wildlife feature will be enhanced. There is also the possibility of a network of these headland strips providing wildlife corridors through arable England. Year-round access will allow crops to be monitored more easily, and hedge trimming / ditch cleaning timed to minimize disturbance to wildlife and when labour is not at a premium. 'Squaring off' fields will reduce overlapping of fertiliser and sprays.

The data accumulated from these experiments and studies will be used to calculate the cost of cropping the headlands, compared to setting them aside. The systems of whole field, rotational set-aside and headland set-aside will also be compared and their effects on the whole-farm budget studied.

ACKNOWLEDGEMENTS

Thanks are due to all the farmers who cooperated in this study. D.L. Sparkes is funded by the Ministry of Agriculture Fisheries and Food (MAFF).

REFERENCES

Boatman, N.D. (1992) Effects of herbicide use, fungicide use and position in field on the yield and yield components of spring barley. *Journal of Agricultural Science,* **118**, 17-28.

Boatman, N.D.; Sotherton, N.W. (1988) The agronomic consequences and costs of managing field margins for game and wildlife conservation. *Aspects of Applied Biology,* **17**, 47-56.

Brereton, J.C. (1986) *The sensitivity of barley, field beans and sugar beet to soil compaction.* PhD Thesis, University of Nottingham.

Speller, C.S.; Cleal, R.A.E.; Runham, S.R. (1992) A comparison of winter wheat yields from headlands with other positions in five fen peat fields. *Set-aside: British Crop Protection Council Monograph* No.**50**, 47-50

Session 4

Restoration and Creation of Field Margins

Session Organiser & Chairman	DR T A WATT
Session Organiser	DR H SMITH

HEDGE PLANTING: THE IMPLICATIONS OF USING ALIEN RATHER THAN NATIVE GENOTYPES

A.T. JONES

AFRC Institute of Grassland and Environmental Research, Department of Pasture Ecology, Plas Gogerddan, Aberystwyth, Dyfed SY23 3EB

ABSTRACT

Recent years have seen the widespread planting of hedges in Britain using alien (non-native) forms of hawthorn (*Crataegus monogyna*) and other species, especially from eastern Europe. There is evidence that alien forms may show poorer performance as hedging material than native ones probably because of poor adaptation to our climate. The planting of alien material does not follow the principles of the conservation of biodiversity. The implications of this need to be considered in terms of the genetic integrity of our native vegetation.

INTRODUCTION

Hedges are a quintessential feature of the British countryside, providing a network of woodland-edge type habitat for fauna and flora and delineating a landscape consisting of a pleasing patchwork of fields. It is well established that hedge-removal and decline has occurred on a large scale in post-war years. In areas where arable farming has increased, hedges have become redundant as a barrier and loss has mainly been through removal in order to increase field size for ease of soil cultivation with modern machinery. In pasture areas many hedges have become overgrown and eventually derelict with farmers having increasing reliance on fencing for stock proofing. Unfortunately hedge loss is still continuing with a sixth of our remaining hedges having disappeared between 1984 and 1990 (Barr *et al.*, 1990).

In recent years, there has been some replanting of hedges (Barr *et al.*, 1984) aided by grants from statutory bodies, e.g. Countryside Commission in England and the Countryside Council for Wales. Major planting of new hedges has also taken place along sections of newly engineered road and motorway (Dunball, 1982).

This paper addresses the problem of hedges being planted using material of non-native provenance and the consequences of this for both hedge establishment and the conservation of biodiversity in the British countryside. It also considers some guidelines for use in obtaining hedge planting material of native provenance.

PROVENANCES

There is substantial evidence to suggest that much of the material of a range of species that is planted in British hedges, both now (catalogues of the horticultural industry) and in the past two decades (Dunball, 1982) is of alien (non-native) provenance. Much of it appears to originate from eastern Europe where labour costs for the gathering of seed are low. For example, hawthorn (*Crataegus monogyna*) the most commonly planted species is often grown by British plant nurseries from imported Hungarian seed.

Implications of using non-native provenances

Though seed of alien material is often much cheaper, resulting in lower production costs for hedging plants, this advantage must be weighed up against the following implications.

Performance

The performance of newly planted hedges, encompassing their survival and persistence, growth, morphology, reproductive rate and disease resistance, may be poorer in alien than in native material. Jones and Evans (in press) found significant differences in growth rates, morphology and disease resistance between native-Welsh and Hungarian hawthorn, when planted together at an upland site in Dyfed, mid-Wales. Native material (Table 1) grew faster, had a more appropriate morphology, including greater bushiness and thorniness, and was less prone to hawthorn mildew, than the alien material.

Further work is needed to compare British hawthorn populations from other areas both with each other and with continental European hawthorn with respect to growth, morphology and disease resistance.

Continental material can often be identified, especially along new roads and motorways, by its early bud-burst which often occurs in February. It is also of note that newly planted hawthorn of continental provenance has been seen to develop extreme infestations of mildew in contrast to unaffected, native hawthorn nearby at low elevations in the West Midlands, England (pers. obs.) .

TABLE 1. A comparison of the growth, morphology and disease resistance of 2 year old native and Hungarian hawthorn (from Jones and Evans, in press).

	native (n = 129)		Hungarian (n = 63)	
	mean	s.d.	mean	s.d.
height (cm)	107.0	37.5	79.3	34.0
total stem (cm) length	235.0	134.7	129.0	84.0
branch number	3.1	1.7	1.8	1.4
thorns per plant	10.6	13.1	0.6	0.5
Mildew score	1.5	0.8	2.0	1.1

Mildew score: 1 = no disease and 4 = stems badly infected.
Anova, $P < 0.001$ for all variables.

Gene conservation

Virtually all species have some degree of variation within and between populations. It is important to conserve this variation and the most efficient way is *in situ* conservation, i.e. by planting material within its native range in those conditions where such variation evolved. It is desirable for hedge planting purposes that propagules of hedging species are collected from the nearest source populations to a planting site. This will increase the probability that genetic variation (including unique alleles and allele combinations) that may be related to adaptation to the local soils and climate is conserved.

Introgression within and between species

There is a high risk that introduced alien subspecies or ecotypes of a particular species will hybridise with conspecific natives, causing the formation of intraspecific but inter- subspecies or inter-ecotype hybrid swarms (Bonnemaison and Jones 1986). It is also possible that closely related species may be introduced with which hybridisation can take place. This can be seen as a form of genetic contamination of native populations.

Although hybrid material may be 'selected out' because of poor adaptation and consequently lower fitness, the time required for this to take place could be very great in the case of tree and shrub species. Selection pressures arising from plant competition may also not be great because of the disturbed and open nature of modern, intensively-managed agricultural environments.

Introduction of new species

There is the risk that seed obtained from the continent may be contaminated by alien species, especially by any which are closely related and morphologically similar to the target species.

Taxonomic confusion may increase the risk that alien species are inadvertently imported and planted. The taxonomy of *C. monogyna*, for example, has not been fully resolved. Tutin *et al*. (1969-1983) recognised six sub species of *C. monogyna* and 22 species within the *Crataegus* genus in Europe. At least four of these commonly hybridise with *C. monogyna* and are very similar in terms of their morphology. Recently, it has been suggested that some of the species are more likely to be hybrids (Holub, 1993).

In Czechoslovakia and Hungary where much seed is gathered, five very similar species and their hybrids are found. It would appear that there is a high risk of other *Crataegus* species being gathered along with *C. monogyna*. Such risks would not occur in collecting material from properly identified stands in Britain, where in many areas only *C. monogyna* is found.

Ethical arguments

Planting hedges of alien material, does not conserve local biodiversity and is it hardly in line with the treaty for maintaining biological diversity which Britain signed in Rio de Janeiro in 1992. Such habitats are of lower value in terms of their relevance to and association with other biota in the countryside and because of this they will be valued less in the future should they be identified as such and become threatened.

DISCUSSION

The British Isles has probably more exotic species planted for its area than anywhere else in the world and yet few of these species have 'escaped' from gardens and parks. Natural and semi-natural communities are still largely formed of native species and locally native populations of these species.

We should value the genetic integrity of our wild habitats and ensure that where trees or shrubs need to be replaced that we make use of propagules gathered from the most local populations or where possible rely on natural regeneration.

It may be impractical in every case of hedge-planting to obtain material of very local provenance, but at the very least it is conceivable that in Britain six hawthorn provenances could be made available by nurserymen for hedge-planting, representing an upland and lowland form in each of England, Scotland and Wales.

Guidelines for hedge and amenity plantings

Although the statutory organizations which provide hedging grants require that native material be planted for the provision of hedging grants, this needs to be more rigidly defined and enforced. The following guidelines should to be followed to ensure that native material is planted.

1. Managers of hedge planting programmes should specifically ask plant nurseries for native material and if possible material grown up from locally collected seed. Besides ensuring that the appropriate provenance is planted, this would ensure that sufficient demand for native material is created.

2. Conservation organisations should if possible establish shrub and tree nurseries using locally collected seed or cuttings.

3. Eventually, codes of conduct and even legislation for the genetic conservation of habitats should be formulated and introduced by the government funded conservation organisations. At the moment the only standards with regard to the provenance of material are those maintained by the horticultural industry.

ACKNOWLEDGEMENTS

This research programme was funded by the Ministry of Agriculture, Fisheries and Food as part of an open contract.

REFERENCES

Barr, C.J.; Benefield, C.; Bunce, R.G.H.; Risdale, H.; Whittaker, M.(1984) *Landscape changes in Britain*. Huntingdon: Institute of Terrestrial Ecology, Monks Wood Experimental Station, Monks Wood. Natural Environment Research Council, pp. 15-18.

Barr, C.J.; Bunce, R.G.H.; Clarke, R.T.; Fuller, R.M.; Furse, M.T.; Gillespie, M.K.; Groom, G.B.; Hallam, C.J.; Hornung, M.; Howard, D.C.; Ness, M.J. (1990) *Countryside Survey*. Merlewood: Institute of Terrestrial Ecology; Department of the Environment, pp. 96-102.

Bonnemaison, F.; Jones, D.A. (1986). Variation in alien *Lotus corniculatus* L. 1. Morphological differences between aliens and native British plants. *Heredity*, **56**, 129-138.

Dunball, A.P. (1982) The management and planting of motorway verges. In: *The Flora of a Changing Britain*, Perring, F. (Ed.) London: Botanical Society of the British Isles, pp. 84-89.

Holub, J. (1993) *Flora of the Czech Republic, 3rd Volume*. Prague: Academy Publishing House.

Jones, A.T.; Evans, P.R. (in press) A comparison of the growth and morphology of native and commercially obtained continental European *Crataegus monogyna* Jacq. (hawthorn) at an upland site. *Watsonia*.

Tutin, T.G. *et al*. (Eds.) (1969-83) *Flora Europaea*. Cambridge: University Press.

THE DEVELOPMENT AND TESTING OF HEGS, A METHODOLOGY FOR THE EVALUATION AND GRADING OF HEDGEROWS

R.J. TOFTS, D.K. CLEMENTS

Countryside Planning and Management, Knight's Gate, Quenington, Cirencester, Gloucestershire, GL7 5BN.

ABSTRACT

The development and testing of a system for the evaluation and grading of hedges (HEGS) is described. A flow-chart grading system was originally favoured over a simple additive system following a study of the literature, and desk-based experimentation. Field testing suggests that the flow-chart method does have advantages, but further testing is required before it is known whether ecologists agree sufficiently on hedgerow evaluation to make this method widely acceptable. A brief analysis of completed survey forms is provided, giving sample distributions of scores for 4 groups of hedgerow features. An alternative and less prescriptive approach to hedgerow evaluation is also discussed. Such an approach may be preferable if there is a significant element of disagreement on hedgerow evaluation amongst ecologists.

INTRODUCTION

A draft methodology for the evaluation and grading of hedges (Hedgerow Evaluation and Grading System (HEGS)) has been produced (Clements and Tofts, 1992). In the present paper we describe developments on this project and outline some areas which need future research.

Initial experimentation and a survey of the literature (e.g. Margules, 1986) suggested that a simple additive scoring system did not possess the flexibility to enable evaluation of hedges in a way which satisfactorily reflected subjective judgement. From both published sources and discussion, the view was formed that surveyors evaluate different aspects of a habitat before arriving at an overall judgement (e.g. Ratcliffe, 1977). As a consequence, a different and more versatile method of scoring was developed, using a flow-chart. At one extreme, a flow-chart may produce identical results to an additive scheme, whilst at the other extreme it may invert some of the grades that would have been awarded by simple additive methods, depending on the design adopted. The flow-chart adopted in the test draft of HEGS was of this latter kind, and was based on the judgement of the authors. Due to the possible implementation of the Hedgerow Protection Bill around the time when the draft version of HEGS was produced, it was considered preferable to publish the document at the earliest opportunity, and then to continue testing the methodology at the same time as receiving comments from other ecologists.

In order to grade a hedge using HEGS, a field record card is completed. This allows the surveyor to record 12 hedgerow features (such as hedgerow height) in a categorical fashion, using up to five classes to which scores are attached according to their perceived wildlife value. Class limits were initially determined by pragmatic reasons (e.g. width classified in units of 1 metre), although an attempt was made to ensure that each of the classes would be encountered with moderate frequency in the field. The 12 hedgerow features are divided into four groups (structure, connectivity, diversity and associated features) and a total score for each group is derived.

At the time of writing, approximately 500 hedgerows have been surveyed using the HEGS methodology, with valuable data being supplied by surveyors from as far afield as Cornwall, Ireland and the Scottish Borders. A summary of these data is given below, following an analysis of the HEGS grading methodology.

ANALYSIS OF GRADING SYSTEM

Comparison of flow-chart and subjective grading system

During the ecological survey of a number of sites, we have collected information on hedgerow characteristics and have graded the hedgerows subjectively on a 12 point scale running from 1+ (highest value) to 4- (lowest value). Using the data obtained, it was also possible to grade the hedgerows using the HEGS methodology which likewise employs a 12 point scale. In total, 87 hedgerows were surveyed and then graded using the two techniques. A comparison of the results derived by the two grading methods is given in Figure 1.

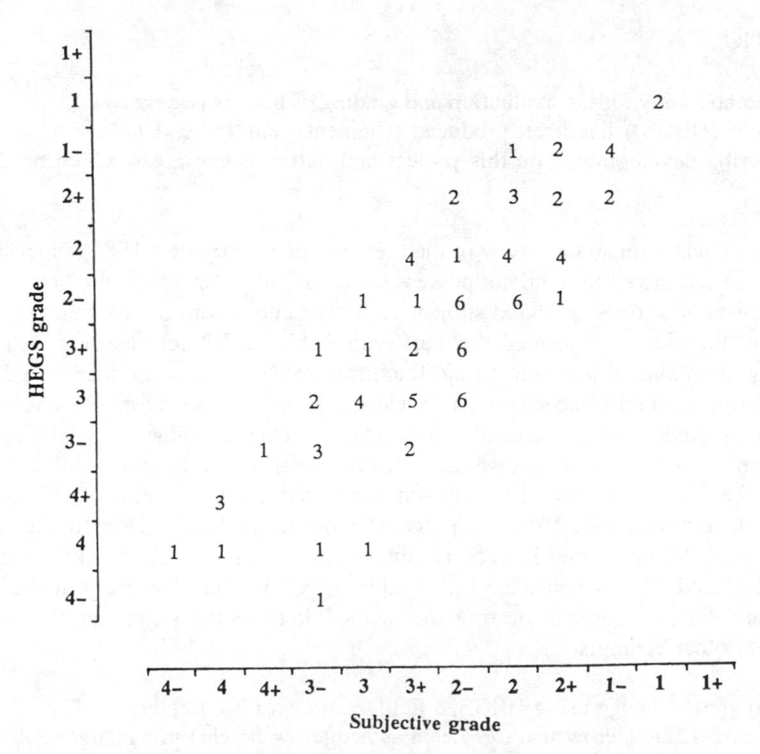

FIGURE 1. Comparison of HEGS grades and subjective grades. The numbers represent the number of observations of each kind.

Overall, the level of agreement is relatively good (using Spearman's rank correlation coefficient and allowing for tied values, r_s=0.83, P<0.005), although the number of 'mis-classifications' suggests that the present flow chart is not sufficiently sophisticated to assign grades reliably to a 12 point system. In the present example, 28 hedges were correctly classified, 24 were given higher grades than considered appropriate whilst the remaining 35 were under-graded. Analysis of these data using the sign test does not suggest that the flow-chart produces unduly biased grades in comparison with the estimated grades (P>0.10), however, and in all but two cases the hedges were graded to within two points of their subjective grade.

Comparison with additive scoring methods

Although designed for use with the flow-chart grading system, completed HEGS forms may be used to obtain one overall additive score for each hedge by ignoring the four subdivisions of structure, connectivity, diversity and associated features, and simply producing a total score. This procedure was followed for the 87 hedges analysed above, giving a range of total scores from 11 to 35. These scores were partitioned in various ways, in an attempt to produce a 12 point grading system which had the highest positive correlation with the subjective grades (details not presented). The 'best fit' that could be obtained (r_s=0.79, at P<0.005) was found to have a slightly lower correlation with the subjective grades than did the HEGS grades. It is instructive to note that in this case the additive system is being fitted retrospectively, whilst the flow-chart system was being used in a predictive fashion. In this respect, the comparison is biased towards favouring the additive system, and it is therefore of interest that the flow-chart method appears to be slightly superior.

For any ecological grading system to gain widespread acceptance, it is vital that it should produce results which are regarded as reasonable by the majority of people likely to use it. Useful information on acceptability may be obtained by running experiments involving field testing, and also by methods such as sending out questionnaires requesting the views of users. The former method has the advantage of greater statistical rigour, but is only practical for a restricted sample size, since field meetings can be difficult to arrange and are only likely to attract volunteers from a relatively small area. The latter approach can involve a much larger sample, but the results may be more open to interpretation. Due to their different strengths and weaknesses, it was decided to adopt both approaches in the testing of HEGS. It is hoped to organise a formal experiment during 1994, and a questionnaire survey is currently underway.

Most of the feedback received so far is eith from questionnaire respondents, or others who have expressed their views in general conversation or letter. Most respondents have indicated that HEGS gives reasonable results, or would do so with slight modifications, although a full analysis awaits further questionnaire returns. A few respondents have, however, indicated that their judgement is substantially at variance with HEGS in at least some cases.

A study by a student at Southampton University (C. Walker, 1993) compared HEGS with a grading system of her own devising. On a sample of 20 hedges, the correlation between grades produced by the two systems was found to be positive but relatively weak (r_s=0.47, 0.025>P>0.01).

Discussion

The present evidence indicates that the flow-chart grading system does assign grades which we feel to be reasonable, based upon our knowledge of hedgerow ecology. The decision to adopt a flow-chart grading system rather than a simple additive system also appears to have been appropriate, given the results of the comparative study of the two methods. What is less certain, however, is the general applicability of HEGS, and further field testing by a range of ecologists is required. The comments we have received from other HEGS users have, however, been generally very encouraging, and further development of the system certainly seems worthwhile.

It remains possible, however, that there is insufficient agreement between ecologists to make HEGS (or any other method which produces grades along a single scale) widely acceptable. The implications of this are discussed in our concluding section.

ALTERNATIVE APPROACHES TO HEDGEROW GRADING

Figure 2 shows the sample distributions for 454 hedgerows (the number of fully completed forms received) with respect to scores for the four groups of hedgerow features.

The results obtained so far suggest that a flow-chart grading system does have a flexibility which is lacking in simple additive schemes, and it may be used to 'model' the subjective response of an individual, or perhaps a limited number of surveyors. What is not known is the extent to which practising ecologists agree or disagree about assigning values to hedges. A widespread lack of agreement would mean that approaches such as HEGS would lack general approval. We hope to investigate this problem through further field work, but are also researching alternative approaches to hedgerow grading. One of these is outlined below.

HEGS identifies certain hedgerow features (eg dense growth, high species diversity) as being generally of high wildlife value, and produces a score for each of four groups of features. There does appear to be general agreement that this approach is satisfactory. If, however, there is widespread disagreement about how a single overall grade is subsequently to be produced from these separate scores, a more modest method which simply places hedges into a context (e.g. national or county) may be preferable to one which attempts to model the response of ecologists.

In the present case, one might argue that for any particular group of features, a hedge which is better than 'average' has some conservation merit, and a hedge which is better than average in several respects is generally more worthy of conservation than one which is better in one respect alone. Using the data shown in Figure 2, it is possible to assign (provisional) median values for each of the 4 categories of features currently recognised by HEGS. It is therefore possible to assign grade A, for example, to a hedge which is better than average in all 4 groups of features, and to continue the grading down to grade E hedges which are not better than average in any features (this approach can be made more sophisticated to reflect greater and lesser departures from the median value). Each grade would then have a clearly defined meaning, and would act as a general indicator of value to inform the judgement of the evaluator. Detailed value judgements may then be made on a less 'mechanistic' basis, by considering data for other hedges in the vicinity (if such data are available). A better than

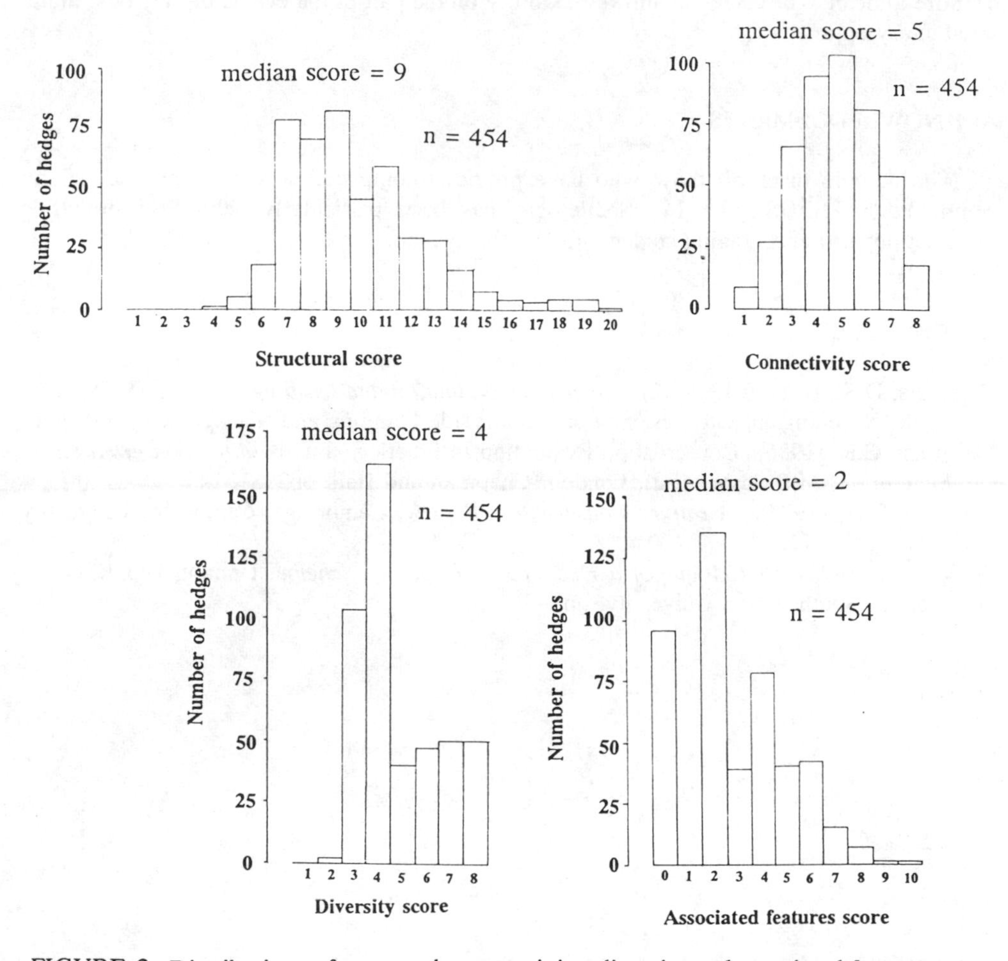

FIGURE 2. Distributions of structural, connectivity, diversity and associated features scores.

average diversity (in whatever way it is defined) may, for example, be valued more highly than a better than average structure, if species diversity in hedgerows is generally poor in the vicinity and structure is not. This type of comparison is extremely difficult to build into a simple grading system, but is certainly a factor which can influence the judgement of hedgerow ecologists. A record card of the HEGS type would aid comparisons of this kind, by providing scores for individual groups of features.

CONCLUSION

At its present stage of development, the grading system adopted by HEGS does appear to have some advantages over a simple additive grading system, but it is not yet known whether there is sufficient agreement between ecologists undertaking evaluation to make a simple 'modelling' approach widely acceptable. If experienced ecologists do not agree sufficiently about the values to be assigned to particular hedges, a system which provides one or more indicators of value but allows flexibility on the part of the evaluator may be a more satisfactory approach.

ACKNOWLEDGEMENTS

The authors thank all those who have provided completed survey forms and helpful comments on HEGS. Dr I.F. Spellerberg has been particularly helpful in providing information, and encouraging students to test the system.

REFERENCES

Clements, D.K; Tofts, R.J. (1992). *Hedgerow Evaluation and Grading System (HEGS)*, Test Draft. Quenington, Gloucestershire: Countryside Planning and Management, pp1–61.

Margules, C.R. (1986) Conservation Evaluation in Practice. In: *Wildlife Conservation Evaluation*, M.B. Usher (Ed), London: Chapman and Hall, pp298–314.

Ratcliffe, D.A. (1977). *A Nature Conservation Review*, Cambridge: Cambridge University Press, pp10–11.

Walker, C. (1993). *A Critique of a Hedgerow Evaluation Scheme*. Unpublished BSc dissertation, Southampton University, pp1–36.

RELATIONSHIPS BETWEEN VEGETATION AND SITE FACTORS IN UNCROPPED WILDLIFE STRIPS IN BRECKLAND ENVIRONMENTALLY SENSITIVE AREA

C.N.R. CRITCHLEY

School of Environmental Sciences, University of East Anglia, Norwich, NR4 7TJ and ADAS Newcastle, Kenton Bar, Newcastle Upon Tyne, NE1 2YA

ABSTRACT

One option for management of arable field margins in Breckland Environmentally Sensitive Area is the creation and maintenance of uncropped, cultivated field boundary strips. The aim is to conserve plant communities containing species adapted to the special conditions of Breckland. From a sample of strips on a typical Breckland farm, the relationship between vegetation and soil variables, field boundary types and cropping history has been investigated by multivariate analysis of plant community data and by analysis of selected species. Soil pH, overhanging trees, broadleaved shelterbelts and previous cropping with sugar beet were the most important factors affecting the plant community composition. Relationships were detected between some individual species and site factors. The results are discussed in the context of the ESA objectives.

INTRODUCTION

The area of Breckland in East Anglia is characterised by its light soils, relatively continental climate and past land use, which comprised extensive grazing of heathland and shifting patterns of short-term cultivation. Land use changes and intensification of agriculture have caused declines in plant communities once typical of the area. These communities contain species adapted to the special conditions of Breckland, which might be defined in an ecological context by soil surface disturbance, summer drought and moderate or low nutrient availability. One aim of the Breckland Environmentally Sensitive Area (ESA) is to re-establish such plant communities by the creation and maintenance of cultivated field boundary strips known as "Uncropped Wildlife Strips" (UWS). Farmers electing to join this part of the scheme are required to cultivate 6m wide strips between August and December. Cultivations must be done 3-5 times in a 5-year period, but only once in any 12-month period. Most other inputs are restricted or prohibited (MAFF, 1988a).

Vegetation data from the environmental monitoring programme of the ESA show considerable variation between UWS sites (MAFF, 1993). To assist farmers and their advisers to target suitable sites for this management, the investigation reported here was carried out to relate UWS vegetation to soil variables, field boundary types and recent cropping history on a Breckland farm.

METHODS

The survey was carried out on a farm which was typical of Breckland in terms of its soils (sands, loamy sands and sandy loams) and cropping, and where management had been consistent between UWS sites. Consistency of management was particularly important because the survey was done in June 1992, at which time the UWS were in their fourth year following establishment. A site is defined as a UWS along one side of a field. From the 192 sites on the farm, a random sample of 32 was selected. One sampling location was randomly selected per site, at which three 25 x 25cm quadrats were positioned 2m, 3m and 4m from the field boundary respectively. Presence of plant species rooted in each quadrat was recorded and local rooted frequencies for each site calculated. Soil samples collected at each sampling location were hand textured and analysed for pH, P, K, Mg and total N. Texture classes were converted to available water estimation values (MAFF, 1988b). Presences of seven field boundary categories (grass/tall herb verge, hedge, broadleaved shelterbelt, coniferous shelterbelt, mixed shelterbelt, unmetalled track, metalled surface) as well as overhanging trees, were recorded separately for the whole site and for the sampling location at each site. Orientation of the site was recorded and a measure of "north-facing" obtained by conversion to absolute degrees from due south. Slope was minimal at all sites and was not recorded. Cropping, fertiliser, herbicide and cultivation data were collected for each site for the three years prior to establishment of UWS, but only previous crop type (spring barley, spring wheat, sugar beet, winter barley, winter wheat) was used in the analysis due to correlations between crop type and crop management factors. Two sites from one field were omitted from the crop analyses to remove the confounding effect of the field, as it was the only one under continuous barley for the three year period. Four separate aspects of cropping were analysed: (i) crop rotation, (ii) crop type in the year prior to establishment, (iii) number of consecutive years the previous crop existed prior to establishment and (iv) an index of occurrence of each crop type in the 3 years prior to establishment. The formula for the index *I* was

$$I = \sum_{i=1}^{3} (x / y)$$

where y is the number of years prior to establishment of the UWS (1, 2 or 3) for each *x* occurrence (0 or 1) of the crop type. To investigate the variation between different parts of the farm, sites were allocated to one of 4 blocks of fields. Long-established roads were used to define the block boundaries, as these were most likely to correspond with past management history.

Canonical correspondence analysis (CCA; Ter Braak, 1987) with downweighting of rare species, forward selection of environmental variables and Monte Carlo permutation tests was carried out using field block, soil, field boundary and crop variables separately. Partial canonical correspondence analysis (PCCA; Ter Braak, 1988) using field block as a covariable was carried out on soil variables. Intercorrelations between environmental variables were checked using Spearman rank correlation coefficients, Kruskal-Wallis tests or chi-square tests as appropriate for the measurement scales. To enable interpretation of results in the context of the ESA objectives, a list was compiled from the literature of indicator species which are adapted to some or all of the factors (i) disturbance, (ii) low summer soil moisture and (iii) moderate or low nutrients. In addition, species were allocated to three classes according to their established strategies (Grime *et al*, 1988), (i) species tolerant of disturbance and stress induced by low resource availability (SR or R/SR strategists), (ii) species which tolerate disturbance (R strategists) and (iii) other species. CCA biplots of species and environmental variables were examined for patterns relating to indicator species and established strategies.

To elucidate further the possible relationships identified in the multivariate analyses, individual species were also analysed against selected environmental variables. Species chosen for analysis were the five commonest indicator species (*Anthriscus caucalis, Apera spica-venti, Arenaria serpyllifolia, Descurainia sophia, Veronica arvensis*), the five commonest serious agricultural weeds (*Bromus sterilis, Cirsium arvense, Elymus repens, Galium aparine, Poa trivialis*) and the five commonest remaining species (*Viola arvensis, Poa annua, Polygonum aviculare,*

Tripleurospermum inodorum, Fallopia convolvulus). Individual species data (arcsine(square root) transformed) were analysed against ratio scaled and nominal environmental variables using simple linear regression and one-way analysis of variance respectively.

RESULTS AND DISCUSSION

Of 58 species recorded, 20 (34.5%) were indicator species and nine (15.5%) were SR or R/SR strategists (Table 1).

TABLE 1. Indicator species and SR or R/SR strategists (*)

Amsinckia micrantha	*Conyza canadensis*	**Medicago minima*
Anchusa arvensis	*Descurainia sophia*	**Myosotis arvensis*
**Anthemis arvensis*	*Echium vulgare*	*Reseda lutea*
Anthriscus caucalis	**Erodium cicutarium*	*Silene latifolia*
Apera spica-venti	**Geranium pusillum*	*Sisymbrium altissimum*
**Aphanes arvensis*	*Lamium amplexicaule*	**Veronica arvensis*
**Arenaria serpyllifolia*	**Legousia hybrida*	

Multivariate analyses

The CCA analysis showed that (Table 2) one field block (no. 4) had a significant effect on species composition. Of the soil variables, only pH (range 6.8-8.1) had a significant effect on species. However, when the effect of field block was removed using PCCA, the effect of total N and K became significant (P<0.05) but pH

TABLE 2. Main relationships between vegetation and site factors from CAA.

Analysis	Total variance in species data	% of total variance explained by all environmental variables	Variables with significant relationship with species (P, Monte Carlo test)
Field block	2.027	13.8	block 4 (0.01)
Soil	2.027	21.9	pH (0.03)
Field boundary (sampling point)	2.027	26.7	overhanging trees (0.02) broadleaved shelterbelt (0.04)
Cropping (index of occurrence)	1.919*	17.3	sugar beet (0.03)

* 30 sites only - see text.

was not significant. Soil pH differed significantly between field blocks (Kruskal-Wallis H = 9.14, P<0.05) with pH lowest in block 4. Site maps showed that block 4 was formerly heathland, and that a railway bisected former field boundaries. These results suggest that the effect of field block 4 on species is at least partly due to edaphic factors, and that any effect of management probably pre-dates construction of the railway. Overhanging trees occurred more often at south facing sites (Kruskal-Wallis H = 4.83, *P*<0.05), so orientation was omitted from the analysis. None of the field boundary categories for the whole site showed a significant relationship with species. Of field boundary categories at the sampling point, overhanging trees and broadleaved shelterbelt had a significant effect on species; these variables were positively correlated ($\chi^2 = 4.94$, $P<0.05$). This indicates a localised effect of trees. Crop rotation failed to produce significant results. The other three expressions of previous cropping produced consistent results with only sugar beet having a significant effect on species.

Biplots showed no obvious patterns for either indicator species or established strategies in the analyses of field block, soil or field boundaries. Crop analyses showed that SR or R/SR strategists occurred mainly away from sugar beet, with the exception of *Anthemis arvensis*, which only occurred at one site. No patterns were evident for indicator species.

Analysis of individual species

Of the species showing a significant difference in frequency between field blocks, none were related to soil pH, suggesting that other factors also affected species composition between different parts of the farm (Table 3). Two weed species (*Galium aparine* and *Poa trivialis*) were associated with higher soil K levels, and the latter was also associated with broadleaved shelterbelts. Soil K and Mg were positively related to broadleaved shelterbelt (Kruskal-Wallis H = 4.89, *P*<0.05; H = 3.87, *P*<0.05), and soil P was just outside the limits of significance (*P*=0.05). This suggests a possible causative effect of broadleaved trees on soil nutrients, for example from leaf litter (Rode, 1993) or rainwater throughfall (e.g. Velthorst & Van Breemen, 1989). The fact that the relationship between vegetation and these field boundary types is stronger at individual sampling points than over sites as a whole, lends support to the possibility that trees may have a local effect on soil nutrients. *Anthriscus caucalis*, typically a hedgebank species, showed a positive relationship with overhanging trees.

TABLE 3. Summary of individual species analyses. I - indicator species, W - weed species, C - commonest species. Species/environmental variable combinations not shown are not significant.

Soil (r^2, *P*, +/- association)

Species	K	Mg	Total N	Available water
Veronica arvensis I	n.s.	n.s.	20.6, 0.009,+	18.1, 0.015,+
Galium aparine W	20.0, 0.01,+	n.s.	n.s.	n.s.
Poa trivialis W	19.3, 0.012,+	n.s.	n.s.	n.s.
Viola arvensis C	12.8, 0.045,-	n.s.	n.s.	n.s.
Fallopia convolvulus C	n.s.	12.6, 0.046,+	n.s.	n.s.

TABLE 3 (Continued)

Field block, field boundary and crop (*P*, +/- association for field boundary types)

Species	Field block	*Previous year's crop	Overhanging trees	Broadleaved shelterbelt
Anthriscus caucalis I	n.s.	n.s.	0.045,+	n.s.
Apera spica-venti I	0.003	n.s.	n.s.	n.s.
Veronica arvensis I	n.s.	n.s.	0.012,-	n.s.
Bromus sterilis W	0.033	n.s.	n.s.	n.s.
Cirsium arvense W	0.002	0.004	n.s.	n.s.
Elymus repens W	n.s.	0.001	n.s.	n.s.
Poa trivialis W	n.s.	n.s.	n.s.	0.004,+
Tripleurospermum inodorum C	0.01	n.s.	n.s.	n.s.

*30 sites only - see text

Veronica arvensis showed a negative relationship with overhanging trees, and a positive relationship with total N and available water estimation value. Two weed species (*Cirsium arvense* and *Elymus repens*) were most frequent at sites following sugar beet. It is notable that sugar beet cropping is characterised by applications of Mg, Na and K, although release of mineralised N from ploughed in sugar beet tops is known to have a greater effect on soil nutrients (B. Chambers, pers. comm.). This might also explain the tendency of SR or R/SR species not to occur at sites following sugar beet.

CONCLUSIONS

The results showed significant relationships between species and site factors, and suggested that the effects of field block, trees and previous crop may be operating partly via soil characteristics. As is often the case in ecological studies a large proportion of variance in the species data is unaccounted for, and although the study was confined to one farm, the findings may be of use in helping to decide which sites are likely to most successful in meeting the ESA objectives. The site factors analysed here are readily measured in the field or by standard laboratory techniques. Other major factors affecting species composition of UWS are likely to include chance events such as the introduction of seeds with organic fertiliser. Since the majority of species in UWS have a persistent seed bank, an assessment of the arable weed community at the edge of a cropped field may give further indication of the suitability of a site for this management. Although the lightest soils might be expected to support more indicator species and R/SR strategists on account of lower moisture and nutrient availability, there appears to be no advantage in targeting the lightest soils within the range of soil textures in the sample. Soil pH affected plant community composition but with indicator species and R or R/SR strategists present throughout the range of pH values in the sample. To achieve a diversity of communities therefore, sites with a range of pH values might be selected.

The effect of overhanging or broadleaved trees was localised. Sites with the occasional tree need not be discounted, particularly since the indicator species

Anthriscus caucalis was more frequent where overhanging trees were present. However, sites with an abundance of overhanging or broadleaved trees might be best avoided. Of the four aspects of previous cropping history used, similar results were obtained for all except the most generalised (crop rotation). In practice therefore, there was no advantage in using cropping history data for more than the year prior to establishment of the UWS.

Within the ESA management prescription for UWS, there is some flexibility in how farmers might manage their sites. This study has concentrated on the effects of site variables on species composition under similar management conditions. Variation in management of UWS sites will affect how their vegetation develops after initial establishment. The study also concentrated on sites under typical Breckland conditions. Some rare arable weed species also occur on heavier soils (Wilson, 1991), and this field margin management prescription may have applications elsewhere.

To meet the ESA objectives of encouraging plant communities containing species adapted to the conditions of Breckland therefore, sites with a range of soil pH values and sandy or loamy soils can be equally targeted. Within this range of soil textures, a small advantage might be gained by avoiding sites with overhanging or broadleaved trees and those where sugar beet has been cropped in the previous year. These factors may be related to soil nutrient availability.

ACKNOWLEDGEMENTS

Financial support from the Ministry of Agriculture, Fisheries and Food is gratefully acknowledged for this work, carried out as part of the wider environmental monitoring programme of ESAs. I am also grateful to the farmer for allowing access to sites; Ms A.J. Sherwood for collecting cropping data; Dr B. Davies for advice on soils; Dr J.P. Barkham, Dr S.P. Rushton and Dr R.A. Sanderson for comments on a draft version.

REFERENCES

Grime, P.J.; Hodgson, J.G.; Hunt, R. (1988) *Comparative Plant Ecology*, London: Unwin Hyman Ltd. 742pp.

MAFF (1988a) *The Breckland Environmentally Sensitive Area. Guidelines for Farmers BD/ESA/4)*, London: MAFF. 9pp.

MAFF (1988b) *Agricultural Land Classification of England and Wales. Revised Guidelines and Criteria for Grading the Quality of Agricultural Land*, Alnwick: MAFF. pp41-46.

MAFF (1993) *Breckland Environmentally Sensitive Area Report of Monitoring 1992*, London: MAFF. 56pp.

Rode, M.W. (1993) Leaf-nutrient accumulation and turnover at three stages of succession from heathland to forest. *Journal of Vegetation Science,* **4**, 263-268.

Ter Braak, C.J.F. (1987) The analysis of vegetation-environment relationships by canonical correspondence analysis. *Vegetatio,* **69**, 69-77.

Ter Braak, C.J.F. (1988) Partial canonical correspondence analysis. In: *Classification and Related Methods of Data Analysis*, H.H. Bock (Ed.), Amsterdam: North Holland, pp. 551-558.

Velthorst, E.J.; Van Breemen, N. (1989) Changes in the composition of rainwater upon passage through the canopies of trees and of ground vegetation in a Dutch oak-birch forest. *Plant and Soil,* **119**, 81-85.

Wilson, P.J. (1991) *The Ecology and Conservation of Rare Arable Weed Species and Communities*, Unpublished PhD. thesis, University of Southampton, 335pp.

THE ROLE OF WILD FLOWER SEED MIXTURES IN FIELD MARGIN RESTORATION

HELEN SMITH, RUTH E. FEBER, DAVID W. MACDONALD

Wildlife Conservation Research Unit, Department of Zoology, University of Oxford, South Parks Road, Oxford OX1 3PS

ABSTRACT

Experimental data are used to evaluate the relative benefits to agriculture and conservation of sward re-establishment on boundary strips by natural regeneration and by sowing a wild flower seed mixture. We show that wild flower-seeded swards can be richer in plant and invertebrate species, and produce more rapid and effective weed control for equivalent management effort, than naturally-regenerated swards. These benefits can substantially offset the higher establishment costs of seed mixtures. Benefits to amenity, invertebrates and weed control, however, are likely to depend on only a small number of plant species in the mixture. We suggest that simple seed mixtures, comprising species which confer these benefits and which thrive under the intended management regime, are likely to be most cost-effective. Wild flower seed mixtures are effective in excluding desirable as well as weedy elements in the local flora and should therefore be used only where the flora in the immediate vicinity of the field margin is depauperate and weed control is a problem.

INTRODUCTION

The severe degradation of many arable field margins in post-war years, by close-ploughing, indiscriminate fertiliser application and either deliberate or accidental herbicide application, has resulted in both severe weed management problems and loss of wildlife and amenity interest (Way & Greig-Smith, 1987; Smith *et al.*, 1993). We postulated in 1987 that the conservation interest of boundary strips could be restored, and weed control problems simultaneously combatted, by reversing these deleterious trends. The re-establishment, on expanded-width boundary strips protected from agrochemical drift, of a more diverse flora, dominated by non-aggressive perennial grasses, seemed a likely means of achieving this objective. We tested this idea using a large-scale experiment to examine the re-establishment of swards on fallowed extensions to degraded and weedy boundary strips (Smith & Macdonald, 1989).

We established perennial vegetation on the fallowed extensions to the strips both by allowing natural regeneration and by sowing a wild grass and forb mixture. Considerable prejudice surrounds the use of such seed mixtures in farmland habitat restoration. From the farming perspective they are often viewed as inherently expensive, inferior to conventional agricultural leys and difficult to establish. Many conservationists have reservations about their effects on the genetic integrity of the local flora.

In this paper we use data from the boundary strips experiment to evaluate the role of wild flower seed mixtures as management tools in the restoration of degraded arable field margins. In particular, we compare the performance of seed mixtures with that of naturally regenerating swards in manipulating weed control, plant species richness and butterfly species richness and abundance. We discuss the choice of species for inclusion in mixtures and the situations in which sowing seed mixtures is likely to be both an appropriate and a more cost-effective option than natural regeneration.

METHODS

We created 2 m wide boundary strips around six arable fields at the Oxford University Farm, Wytham, in autumn 1987. The strips comprised the original boundary strip (the 'old' zone), which was usually about 0.5 m wide, and a fallowed extension of about 1.5 m onto cultivated land (the 'new' zone). Ten treatments were each imposed on 50 m lengths of boundary strip in a randomised complete block design with eight blocks. Each block was located around a single field. Eight of the treatments formed a 2x4 factorial structure: four were sown with a mixture of wild grasses and forbs ('sown' plots) and four were allowed to regenerate naturally ('unsown' plots). They then received one of the four following cutting regimes: uncut, or cut (with cuttings removed) in (a) summer only (b) spring and summer or (c) spring and autumn. The plots were first cut in June 1988 and in subsequent years in the last weeks of April, June and September ('spring', 'summer' and 'autumn' respectively). The new margins were rotavated in March 1988 just before the seed mixture was sown. It contained six 'non-aggressive' species of grass and 17 forbs, in a 4:1 ratio, and was sown at 30 kg/ha (Smith *et al.*, 1993).

The relative frequencies of plant species on the old and new zones of the boundary strip and in the crop edge were monitored five times a year until 1990, by recording presence/absence in eight sub-cells in three 50 x 100 cm permanent quadrats in each zone of each plot, in six blocks of the experiment. The frequencies of some key weed species were also monitored in either 1992 or 1993. The relative abundance of butterflies was measured using transect recording methods on eight blocks of the experiment. Counts of all butterfly species on each plot were made at least once a week during the summer months from 1989 to 1991.

In this paper we concentrate on botanical data, collected in June each year from the new zones of the boundary strips, and on annual measures of butterfly abundance and species richness. Data from the eight treatments that allow factorial comparison of sown and unsown plots are analyzed by analysis of variance, following appropriate transformation to achieve homogeneity of variance. We present significance levels for the main effect of sowing together with adjusted means for sown and unsown plots.

RESULTS

Plant species richness

The effect that wild flower seeding has on the development of species richness is at least partly dependent on the species richness of the sown mixture. The numbers of species sown are modified both by the success with which they establish and the numbers of unsown species that are accommodated. Our relatively complex mixture resulted in species richness values that were consistently higher in sown than in unsown plots in 1988, 1989 and 1990 (Table 1).

The numbers of naturally regenerating species colonising the sown and unsown plots did not differ in June 1988, two months after sowing, but the numbers of unsown species that had established by June 1990 were very significantly lower in sown than in unsown plots (Table 2). Sown swards thus effectively excluded natural colonists.

TABLE 1. Mean numbers of species in sown and unsown plots. Numbers in parentheses are mean numbers of sown species in sown plots. Analyses were performed on log-transformed data. Means presented are back-transformed.

Date	Mean no. species /0.5m^2			$F_{[1,35]}$	P
	Sown plots		Unsown plots		
June 1988	18.6	(5.89)	11.97	66.79	<0.001
June 1989	15.8	(10.95)	8.95	97.93	<0.001
June 1990	13.4	(11.06)	10.46	30.89	<0.001

TABLE 2. Mean numbers of unsown species in sown and unsown plots. Analyses were performed on log-transformed data. Means presented are back-transformed.

Date	Mean no. unsown species /0.5m^2		$F_{[1,35]}$	P
	Sown plots	Unsown plots		
June 1988	12.52	12.95	0.34	>0.05
June 1989	6.74	9.74	33.32	<0.001
June 1990	5.12	11.75	213.53	<0.001

Although in both 1989 and 1990, all sown treatments were more species rich than all unsown treatments, species richness could also be substantially modified by mowing (Smith & Macdonald, 1992; Smith *et al.*, 1993). Of the regimes applied, the dominant effect on both sward types was of mowing in spring and autumn, which increased species richness.

Weed control

Sown swards were extremely effective in controlling pernicious weeds. The adjusted mean frequencies of six of the commonest pernicious annual and perennial weed species on the field margins are presented for June 1989, June 1990, and for either 1992 or 1993, for all species except *Elymus repens* (Table 3). In June 1988, two months after the mixture was sown, sown and unsown plots did not differ significantly in the mean frequencies of any common weed species. By the following June, however, frequencies of the three annual grass weeds, and of *E. repens* and *Urtica dioica*, were significantly lower in sown swards. By 1990 differences for all six species were highly significant, with *Avena* species and *A. myosuroides* being virtually eliminated from sown swards. By 1992 both *Avena* species (*A. fatua* and *A. sterilis* subsp. *ludoviciana*) and *Alopecurus myosuroides* had also declined to low frequencies in unsown swards (there were too few data for *A. myosuroides* for formal analysis: Table 3) but the differences for the remaining species remained large and highly significant. The mean frequencies of these species appeared to be relatively stable in both sward types between 1990 and 1992/93.

The effectiveness of weed control in both sward types could be substantially modified by mowing but, in general, even the least successful of the mowing regimes that we applied to sown swards resulted in much more rapid and effective weed control than any mowing regimes applied to unsown swards over this time period (Smith & Macdonald, 1992; Smith *et al.*, 1993).

TABLE 3. Relative frequencies of weed species in sown and unsown plots in June. Analyses were performed on arcsine square-root transformed data. Means presented are back-transformed.

Species	Year	Mean % frequency		$F_{[1,35]}$	P
		Sown	Unsown		
Avena species	1989	1.43	10.75	9.73	<0.01
	1990	0.33	9.96	25.21	<0.001
	1992	0.12	0.42	1.92	>0.05
Alopecurus myosuroides	1989	0.13	5.06	8.75	<0.01
	1990	0.03	4.22	11.94	<0.01
	1992	0.00	0.07	-	-
Bromus sterilis	1989	86.67	20.51	104.18	<0.001
	1990	6.17	85.77	155.88	<0.001
	1992	5.56	58.12	34.21	<0.001
Cirsium arvense	1989	4.03	11.57	3.06	>0.05
	1990	2.32	14.53	12.68	<0.001
	1993	2.67	12.94	11.33	<0.01
Elymus repens	1989	39.91	69.80	8.11	<0.01
	1990	41.80	88.83	20.01	<0.001
Urtica dioica	1989	2.76	16.18	11.56	<0.01
	1990	0.67	14.31	28.35	<0.001
	1993	0.77	14.74	18.91	<0.001

TABLE 4. The mean annual abundance of butterflies in sown and unsown plots

Year	Mean no. per 50m plot		$F_{[1,49]}$	P
	Sown	Unsown		
1989	28.3	25.15	0.70	>0.05
1990	55.0	19.11	45.43	<0.001
1991	33.0	21.20	10.60	<0.01

Butterfly abundance

Sowing resulted in highly significant increases in the abundance of adult butterflies in 1990 and 1991 (Table 4). Species richness was also higher on sown plots but the difference was significant only in 1990 ($F_{[1,35]}=4.55$, $P<0.05$). Summer mowing, however, also had profound effects on butterfly distribution. The removal of important nectar sources, predominantly *Leucanthemum vulgare*, from the half of the sown plots that were mown, resulted in the concentration of butterflies on the remaining sown plots (Feber *et al.*, 1994). Later in the summer, butterflies made more use of unsown plots, feeding particularly on increasingly abundant flowers of *Cirsium* and *Carduus* species (Smith *et al.*, 1993). As a consequence of these changes, sowing had highly significant effects on abundance prior to, but not after, the summer cut in both 1989 and 1990 ($F_{[1,49]}=13.36$, $P<0.001$; $F_{[1,49]}=138.2$,

$P<0.001$ respectively). In 1991, however, the main flight period of the dominant species and the main flowering period of *L. vulgare* were later. Under these circumstances sowing resulted in a significant increase in butterfly numbers after, rather than before, the summer cut ($F_{[1,49]}=12.5$, $P<0.01$).

DISCUSSION

Our results show that wild flower seed mixtures can potentially be powerful tools in field margin restoration. When sown on newly-fallowed boundary strips they can be used to manipulate plant species richness and both annual and perennial weed populations with minimal management. They can also result in increased abundance and species richness of butterflies. However, whilst these assets can offset the high initial outlay on seed mixtures, they are unlikely to be an intrinsic property of such mixtures. Rather, they depend on the careful selection and subsequent establishment of the component species. Species that do not contribute to the attributes required of a mixture, or which do not thrive under the proposed management regimes, reduce its cost-effectiveness as a management option.

Thus, in our experiment, the very rapid and effective control of annual and perennial weeds that was achieved, even with minimum subsequent management, was likely to have depended largely on the dominant grassy components of the mixture, including *Festuca rubra* subspp. *littoralis* and *commutata*, *Phleum pratense* subsp. *bertolonii*, *Poa pratensis* and *Trisetum flavescens*, which rapidly formed a very dense sward base. Most broad-leaved species were sown, and established, at much lower frequencies and had relatively little effect on weed control. Thus, where weed control is the only aim of field margin restoration, grass-only mixtures are the most cost-effective solution. The effectiveness of fine-leaved grasses has not been rigorously tested against that of rye-grass dominated mixtures but the denser sward base that they produce, together with their lower productivity on fertile agricultural soils (Smith *et al.*, 1993), suggests that they may be a superior option.

Inclusion of broad-leaved species, however, has enormous benefits for the amenity value of boundary strips and for nectar-feeding invertebrates. The latter include beneficial species such as hoverflies and bees, as well as attractive and declining species of butterflies (Feber & Smith, in press). Most of these benefits also depended largely on a very few plant species, which together provided an abundant and continuous nectar supply throughout the summer (Feber *et al.*, 1994).

The invertebrate community can be further manipulated by the composition of the seed mixture. For example, we found that the overall abundance of invertebrates, caught by Deitrick-Vacuum suction sampling, was significantly increased by sowing (Smith *et al.*, 1993). Much of this increase was attributable to groups, such as spiders, which benefitted from the structural diversity contributed by species in the seed mixture. Similarly, inclusion of appropriate larval foodplants in mixtures can attract more specialist feeders (Feber & Smith, in press), while that of tussock-forming grasses, such as *Dactylis glomerata* and *Holcus lanatus*, benefits overwintering populations of polyphagous predators (Thomas *et al.*, 1991).

Since only a small number of plant species is likely to contribute to the weed control properties and attractiveness to invertebrates of sown swards, seed mixtures must be restricted to these species to be cost-effective in achieving these objectives. Plant species richness is likely to be sacrificed as a consequence. The relatively few species that were effective in excluding pernicious weeds were also likely to be most effective in excluding other naturally regenerating colonists from the swards. Simple seed mixtures should thus be as effective as more complex mixtures in excluding local colonists, and are therefore likely to remain species poor. However, where sown swards comprise a balance of three to five grass species and include two or more broad-leaved species, they can still be attractive assets. Where objectives and finance permit, inclusion of more broad-leaved species can result in better imitations of semi-natural grassland.

The species composition of cost-effective seed mixtures must also be tailored to suit the management regime intended. For example, in 1990 we found that *Cynosurus cristatus*, which was the most abundant grass species in all sown treatments in 1988, had not changed in abundance in treatments cut twice a year but had declined by 70% in treatments which were left uncut, and by 27% in those cut once annually. By contrast, *Centaurea nigra* increased in frequency during the experiment although the increase was significantly smaller in plots cut in summer than in those left uncut (Smith *et al.*, 1993).

Where a reasonably diverse and attractive flora remains in the immediate vicinity of newly-created boundary strips, sowing wild flower seed mixtures is likely to prevent its successful colonisation. Under these circumstances, and particularly where pernicious weed populations are small, natural regeneration, with mowing management, is likely to be both more cost-effective, and a more desirable option for nature conservation, than sowing. However, in many intensively-farmed areas of lowland England, the potential for natural establishment of swards that are both relatively species rich and acceptable to farmers is low. The enormous potential of carefully designed seed mixtures for creating swards that both control weeds and are attractive to wildlife in these situations should be seen in the light of our long history of sowing wild flower species in agricultural grass leys.

ACKNOWLEDGEMENTS

This work was funded by English Nature, with additional support from the Ernest Cook Trust, the People's Trust for Endangered Species and the Co-Op Bank. We are grateful to Drs. Stephen Baillie and Trudy Watt for helpful comments on the manuscript.

REFERENCES

Feber, R.E. & Smith, H. (in press) Butterfly conservation on arable farmland. In: *Ecology and Conservation of Butterflies*, A.S. Pullin (Ed.). London: Chapman and Hall.

Feber, R.E., Smith, H. & Macdonald, D.W. (1994) The effects of the restoration of boundary strip vegetation on the conservation of the meadow brown butterfly (*Maniola jurtina*). In: *Field Margins - Integrating Agriculture and Conservation,* N.D. Boatman (Ed.), *BCPC Monograph No. 58*, Farnham: BCPC Publications.

Smith, H., Feber, R.E., Johnson, P., McCallum, K., Plesner Jensen, S., Younes, M., & Macdonald, D.W. (1993) *The conservation management of arable field margins*. English Nature Science No. 18. Peterborough: English Nature.

Smith, H. & Macdonald, D.W. (1989) Secondary succession on extended arable field margins: its manipulation for wildlife benefit and weed control. *Brighton Crop Protection Conference - Weeds 1989*, **3**, 1063-1068.

Smith, H. & Macdonald, D.W. (1992) The impacts of mowing and sowing on weed populations and species richness in field margin set-aside. In: *Set-aside,* J. Clarke (Ed.), pp. 117-122. *BCPC Monograph No. 50*. Farnham: BCPC Publications.

Thomas, M.B., Wratten, S.D. & Sotherton, N.W. (1991) Creation of 'island' habitats in farmland to manipulate populations of beneficial arthropods: predator densities and emigration. *Journal of Applied Ecology*, **28**, 906-917.

Way, J.M. & Greig-Smith, P.W. (1987) (Eds) *Field Margins*, *BCPC Monograph No. 35*. Thornton Heath: BCPC Publications.

THE EFFECTS OF FIELD MARGIN RESTORATION ON THE MEADOW BROWN BUTTERFLY (*MANIOLA JURTINA*)

RUTH E. FEBER, HELEN SMITH, DAVID W. MACDONALD

Wildlife Conservation Research Unit, Department of Zoology, University of Oxford, South Parks Road, Oxford, South Parks Road, Oxford OX1 3PS.

ABSTRACT

We examine the extent to which simple changes in management of boundary strips can increase the value of farmland as a habitat for the meadow brown butterfly (*Maniola jurtina*). Butterfly transects were conducted on a farm with conventionally managed field boundaries, and on experimental, extended-width field margins which were subject to ten contrasting management regimes. More butterflies were associated with grassy boundary strips than with other field edge habitats. Plots which were sown with a wild flower seed mixture and left uncut during the summer attracted the highest numbers of *M. jurtina*. Nectar sources which were preferred by *M. jurtina* were most abundant on these plots. Plots sprayed with glyphosate became progressively less attractive to butterflies over a three year period.

INTRODUCTION

Agricultural intensification has not only resulted in the loss of semi-natural habitats, but has also had deleterious effects on smaller, interstitial areas of uncultivated land within the arable landscape. Arable farmland in lowland Britain is currently considered to support an impoverished butterfly fauna, consisting mainly of mobile species such as the large white (*Pieris brassicae*; Thomas, 1984). In other European countries, the declining abundance of several species has been linked to agricultural intensification (e.g. van Swaay, 1990).

Intensification has affected habitat suitability for butterflies on farmland in several ways. Field boundaries, which remain the primary uncropped habitats available to butterflies within arable farmland, have traditionally been associated with high floral and faunal diversity, but have been degraded by modern farming methods. Herbicide drift and the application of herbicides directly to the hedge base, together with the high nutrient status of field boundary soils caused by inadvertent fertiliser application, ensures the perpetuation of species-poor communities dominated by pernicious weeds (Smith & Macdonald, 1989; Smith *et al.*, this volume). These practices are likely to have reduced substantially the quality and quantity of both adult and larval food resources on farmland (Feber & Smith, in press; Dover, this volume). Habitat fragmentation and unpredictable patterns of resource supply further militate against the persistence of populations of less mobile species.

Simple methods for re-creating and managing field boundary swards could potentially result in radical improvements in the availability of larval and adult resources for common grassland and hedgerow butterflies. We describe the effects of contrasting methods of creating and managing boundary strips on the meadow brown butterfly (*Maniola jurtina*). Although the larvae are grass feeders, they require permanent swards in which to overwinter and this, combined with relatively low adult mobility, makes *M. jurtina* vulnerable on farmland. However, because its habitat requirements are fairly simple, it is a good target species for conservation within farmland and the high densities which can occur in established colonies make it ideal for comparative studies of habitat management.

METHODS

In autumn 1987 we created 2 m wide boundary strips around arable fields at the University of Oxford's farm at Wytham. These comprised the original boundary strip, about 0.5 m wide, and a fallowed extension of about 1.5 m on to cultivated land. Swards were established on the fallowed strips either by allowing natural regeneration ('unsown' swards) or by sowing a mixture of wild grasses and forbs ('sown' swards: see Smith *et al.*, this volume). 50 m long plots were established on both sward types and subjected to the following management regimes: uncut, or cut (with cuttings removed) in (a) summer only (b) spring and summer or (c) spring and autumn. Two further treatments were imposed on unsown plots only: (a) cut in spring and summer with hay left lying and (b) uncut but sprayed with glyphosate in late June or early July. The plots were cut in the last weeks of April, June and September ('spring', 'summer' and 'autumn' respectively). Glyphosate (3 l/ha Roundup in 175 litres water) was first sprayed in 1989. The treatments were randomised in eight blocks, each occupying a single field. From the time that they were fallowed, the field margins were protected from fertiliser and spray drift.

Butterfly transects (see Pollard, 1977) were conducted on the experimental margins weekly from April to September between 1989 and 1991. In 1991, transects were also conducted on a neighbouring farm with narrow, unmanaged boundary strips, which received agrochemical drift and direct herbicide applications. Mark-release-recapture studies, together with behavioural observations, were also conducted on *M. jurtina* on the experimental margins (Feber, 1993). The larvae were sampled by sweep-netting and visual searching in spring 1991. The abundance of flowers of all broad-leaved species on the experimental margins was estimated monthly on a six-point scale in the same year (Feber, 1993).

RESULTS AND DISCUSSION

Effects of boundary strip creation on the abundance of adult *Maniola jurtina*

Maniola jurtina individuals were significantly more abundant per kilometre on the extended-width experimental margins at the University Farm than on the neighbouring farm (Wilcoxon 2-tailed test, $P<0.05$). At the University Farm, significantly more butterflies per unit area were associated with the grassy boundary strips than with the hedges, ditches, crops, tracks, or sterile strips ($\chi^2=1164.26$, $P<0.001$). Grassy boundary strips are thus important field margin components for butterfly conservation.

Effects of boundary strip management on abundance of adult *Maniola jurtina*

Although butterflies were more abundant on the expanded experimental margins than on narrow conventional margins, transect results showed that experimental plots at the University Farm were not equally attractive to butterflies. There were highly significant differences in the abundances of *M. jurtina* between experimental treatments in all three years of the study (Table 1).

In all years there was a significant effect of mowing on *M. jurtina* abundance. Butterfly abundance was highest on treatments which were left uncut, or which were cut in spring and autumn. Although butterflies utilised all plots before the summer cut, *M.*

TABLE 1. The effects of management on the abundance of *Maniola jurtina* on the field margins 1989-1991. Analyses performed on log-transformed means. Means presented are back-transformed. Significance levels: *** $P<0.001$, ** $P<0.01$, * $P<0.05$, ns not significant.

(a) Mean numbers of *M. jurtina* individuals per 50 m length of field margin.

Treatment	Mean number of butterflies per 50m length of field margin[1]		
	1989	1990	1991
Sown, cut spring+autumn	13.5	52.2	19.9
Sown, uncut	18.9	39.2	13.3
Sown, cut spring+summer	15.0	21.4	7.8
Unsown, cut spring+autumn	14.6	8.2	7.3
Unsown, cut spring+summer, hay left	7.4	4.8	4.6
Sown, cut summer only	11.9	22.1	4.4
Unsown, uncut	12.7	9.9	4.3
Unsown, cut summer only	6.7	5.4	4.1
Unsown, cut spring+summer	9.6	5.6	3.9
Unsown, sprayed	14.5	4.1	2.6

[1] Minimum Significant Differences (Tukey's Studentized Range Test; $P=0.05$): 1989, 0.94; 1990, 1.26; 1991, 1.07.

(b) Significance of planned comparisons between treatments. Main effects from 3-way ANOVA using the eight treatments that form a 4x2 factorial design (Smith *et al.* 1993).

	d.f.	1989	1990	1991
Main effects				
Block	7,49	*	***	ns
Sowing	1,49	*	***	***
Cutting	3,49	*	*	***
Sow x cut	3,49	ns	ns	ns
Planned comparisons between means				
Cut *vs* uncut	1,49	ns	ns	ns
Cut in summer *vs* not cut in summer	1,49	*	**	***
Cut spring+autumn *vs* uncut	1,49	ns	ns	*

jurtina responded to the summer cut by immediately moving to uncut plots (Figure 1). This change in distribution was typical of the majority of butterfly species whose flight period spanned the summer cut (Feber, 1993). The main advantage to *M. jurtina* of plots which were left uncut, or which were cut in spring and autumn, lay in their continuity of provision of nectar supply throughout the flight period. In many arable areas nectar sources are likely to be patchily distributed, both spatially and temporally, and may limit the potential for this, and other less mobile species, to establish populations. Of the two treatments, spring and autumn cutting is a more desirable management strategy than leaving swards uncut. As well as preventing invasion by woody species, our experiments have shown that plant species richness was significantly lower on uncut than on spring and autumn cut swards (Smith & Macdonald, 1992). As these underlying differences increase they are likely to have increasing impacts on the availability of nectar sources and larval foodplants. By 1991, *M. jurtina* was significantly less abundant on uncut swards than on spring and autumn cut swards (Table 1).

Sown swards attracted more butterflies than unsown swards and, in 1990 and 1991, sprayed plots attracted significantly fewer butterflies than other plots (Table 1). These differences were attributable to the abundance of a small number of preferred nectar sources rather than to the gross abundance of flowers. Flower abundance in July and August on sown and unsown plots did not differ significantly (Friedman's Test, P=0.132 and P=0.483 respectively). Sown plots, however, contained several important nectar sources including oxeye daisy (*Leucanthemum vulgare*), field scabious (*Knautia arvensis*) and common and greater knapweeds (*Centaurea nigra* and *C. scabiosa*), which were heavily utilised by *M. jurtina* (Table 2). The best single predictor of *M. jurtina* abundance in July was the abundance of *L. vulgare* (R^2=0.212, P<0.001). Sown plots also contained more perennial plant species in flower than unsown plots in July and August (3-way ANOVA, $F_{(1,49)}$=13.47, P<0.001 and $F_{(1,49)}$=14.11, P<0.001 respectively; Figure 2). There is some evidence to suggest that perennials may provide higher nectar rewards than annuals (Feber & Smith, in press).

TABLE 2. Nectar source utilisation by *Maniola jurtina* in July 1991. All broad-leaved plant species were ranked in order of the mean abundance of their flowers. Data are percentages of total observations of feeding butterflies (300 systematic observations).

Plant species	rank abundance	percentage of visits
Cirsium/Carduus spp.	13	38.0
Leucanthemum vulgare	4	33.0
Centaurea spp.	45	12.0
Knautia arvensis	19	7.7
Tripleurospermum inodorum	6	4.3
Trifolium spp.	12	2.0
Convolvulus spp.	24	1.3
Ranunculus spp.	25	0.7
Rubus spp.	23	0.3
Senecio jacobaea	59	0.3

Effects of the creation and management of boundary strips on *Maniola jurtina* larvae

Larval abundance did not differ significantly between treatments and larvae were found on all treatment types. A mean of 2.73 larvae was recorded per plot. They were recorded feeding on *Elymus repens*, *Lolium perenne*, *Trisetum flavescens*, *Arrhenatherum elatius*, *Bromus sterilis*, *Poa trivialis* and *Dactylis glomerata*. Several of these species are either common components of agricultural leys or are agricultural weeds. *Maniola jurtina* larvae overwinter in the base of grassy tussocks and there was no evidence that they were affected by spring or autumn mowing. Although more grass species were present in sown than unsown swards, these results suggest that the extension in width and exclusion of agrochemicals from the margin were sufficient to meet the requirements of *M. jurtina* larvae.

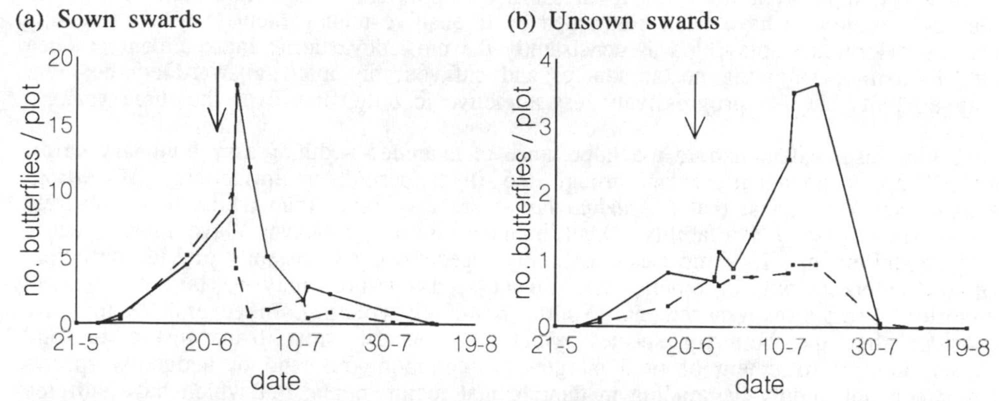

FIGURE 1. Changes in abundance of *M. jurtina.* Solid lines: abundance on swards not cut in summer. Dashed lines: abundance on swards cut in summer. Arrows indicate the date of cut. Note different vertical scales in (a) and (b).

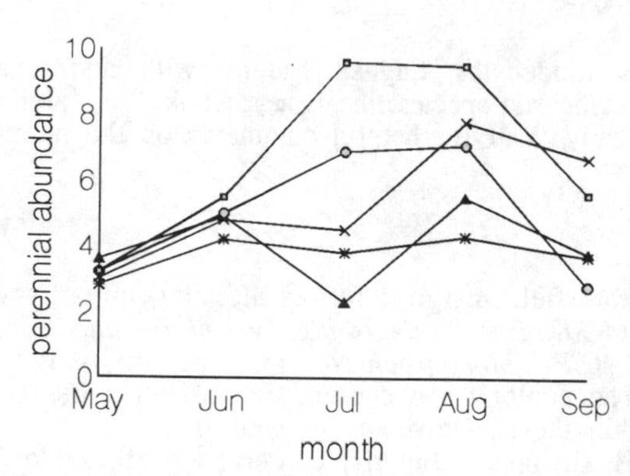

FIGURE 2. Differences in mean abundance of perennial species in flower, between plots sown and uncut in summer (-▫-), unsown and uncut in summer (-○-), sown and cut in summer (-×-), unsown and cut in summer (-▲-) and sprayed with glyphosate (-✳-).

CONCLUSIONS

Maniola jurtina is considered to be a generalist in its habitat requirements but it nonetheless showed clear and significant responses to the creation and management of boundary strips. Simple management changes that were central to the establishment of the experimental field margins - extension in width and exclusion of agrochemicals - increased its abundance.

Of the different management regimes, summer mowing had the most profound effects on *M. jurtina* abundance. This removed nectar sources at a critical time during the butterfly's life cycle. Summer mowing may also remove larval foodplants of other butterfly species. It had consistently detrimental effects on the overall abundance and species richness of butterflies on the experimental margins (Feber, 1993). Sowing with a wild flower seed mixture, by contrast, significantly increased *M. jurtina* abundance.

On many conventionally managed arable farms, the existing flora is unlikely to provide nectar resources adequate to support butterfly populations. Seed mixtures are a particularly appropriate tool for reddressing this problem in situations where suitable sources of colonists have been eliminated by insensitive management (Smith *et al.*, this volume). Herbicide spraying was consistently the most devastating management practice for *M. jurtina*, removing nectar sources and unfavourably altering sward composition. Sprayed plots became progressively less attractive to butterflies over the three years.

Our results demonstrate the importance of extended-width, grassy boundary strips, which supply abundant nectar through the flight period, in influencing *M. jurtina* abundance. We suggest that *M. jurtina* larvae are less likely than adults to be severely constrained by food availability. Other butterfly species, however, have more specific larval requirements. In some cases, naturally regenerated swards may provide sufficient larval foodplants but, in others, wild flower seed mixtures may be tailored to meet specific conservation requirements (Smith *et al.*, this volume). In general, though, we consider that the butterfly species which can benefit most from restoration and conservation management of field margins are common grassland or hedgerow species which are not unduly demanding in their habitat requirements, but which have suffered as a result of agricultural intensification and poor field margin management practices.

ACKNOWLEDGEMENTS

This work was funded by English Nature, with additional support from the People's Trust for Endangered species, the Ernest Cook Trust and the Co-Op Bank. We are grateful to Dr. Trudy Watt for helpful comments on the manuscript.

REFERENCES

Dover, J. (1994) Arable field margins: factors affecting butterfly distribution and abundance. *Field Margins - Integrating Agriculture and Conservation*, N.D. Boatman (Ed.), *BCPC Monograph No. 58*, Farnham: BCPC Publications.

Feber, R.E. (1993) The ecology and conservation of butterflies on lowland arable farmland. D. Phil. thesis, University of Oxford.

Feber, R.E.; Smith, H. (in press) Butterfly conservation on arable farmland. In: *Ecology and Conservation of Butterflies*, A.S. Pullin (Ed.). London: Chapman and Hall.

Pollard, E. (1977) A method for assessing changes in the abundance of butterflies. *Biological Conservation*, **12**, 115-134.

Smith, H.; Macdonald, D.W. (1989) Secondary succession on extended arable field margins: its manipulation for wildlife benefit and weed control. *Brighton Crop Protection Conference - Weeds - 1989*, pp. 1063-1068. Farnham: BCPC Publications.

Smith, H.; Feber, R.E.; Macdonald, D.W. (1994) The role of wild flower seed mixtures in field margin restoration. *Field Margins - Integrating Agriculture and Conservation*, N.D. Boatman (Ed.), *BCPC Monograph No. 58*, Farnham: BCPC Publications.

Smith, H.; Feber, R.E.; Johnson, P.; McCallum, K.; Plesner Jensen, S; Younes, M; Macdonald, D.W. (1993) *The Conservation Management of Arable Field Margins. English Nature Science No. 18*. Peterborough: English Nature.

Thomas, J.A. (1984) The conservation of butterflies in temperate countries: past efforts and lessons for the future. In: *The Biology of Butterflies*, R.I Vane-Wright and P.R. Ackery (Eds.), Royal Entomological Society, London, pp. 333-353.

van Swaay, C.A.M. (1990) An assessment of the changes in butterfly abundance in the Netherlands during the 20th century. *Biological Conservation*, **52**, 287-302.

COMPARISON OF DIFFERENT MANAGEMENT TECHNIQUES FOR CROP MARGINS IN RELATION TO WILD PLANTS (WEEDS) AND ARTHROPODS

A.B. HALD

National Environmental Research Institute, Frederiksborgvej 399, DK-4000 Roskilde, Denmark

ABSTRACT

A pilot experiment was carried out in 1988 to evaluate the effect on wild plants (weeds) and arthropods of four unsprayed management techniques in crop margins of fields in rotation. The techniques were: spring barley; spring barley undersown with a mixture of grasses, legumes and herbs; sown mixture of grasses, legumes and herbs, and no crop. The experiment was carried out at nil, half and full dose of nitrogen for spring barley. The density and biomass of the wild plants and of arthropods were measured. In interpretating the results, the plant and arthropod species were divided into groups according to density or to taxonomic group.
Sowing had a greater effect on the analyzed groups of plants and arthropods than did nitrogen. Nitrogen mostly affected the sown species, the biomass of arthropods, and - in no crop - the biomass of a common wild plant species. The wild plants benefitted mostly from no crop and sown mixture of grasses, legumes and herbs. An arthropod group comprising all species occurring less frequently was found at highest densities in techniques involving sown species.

INTRODUCTION

The EU has launched programmes to reduce its surplus production of cereals and other crops, for example by set-aside. At the same time, there is a concern about environmental problems in the landscape deriving from agricultural production. These include impoverishment of the wild flora and fauna. However, the problems of surplus and of environmental degradation may both be considered within the regulations. It is, therefore, essential to know more about the beneficial effects of different management techniques on the occurrence and abundance of wild plants and animals and to evaluate them bearing in mind eventually future weed problems. In particular, attention has to be paid to management of crop margins, where the species potential is relatively high.

This paper gives the results of a pilot experiment carried out in 1988 with different management techniques in crop margins of a field in rotation. It describes the relative abundance of wild plant species (weeds) in crop margins which were grown with spring barley or grown with a mixture of grasses, legumes and herbs in combination with spring barley or alone, or left without a crop. All sowing treatments were carried out at three rates of nitrogen. The influence on the arthropod fauna is also assessed.

METHODS

The experimental area was situated in a 6 m crop margin on the south side of a 3 m high, uniform and managed hedgerow in a spring barley field in the eastern part of Denmark. The area was divided into 12 plots each divided into an inner and an outer subplot (Figure 1). Plots 1-6 were sown with spring barley. The inner subplot of plots 4-9 were sown with a mixture of grasses, legumes and herbs. The remaining plots were left untouched after rolling the whole experimental area. The sown mixture (Dæhnfeldt no. B5) consisted of *Lolium perenne, Festuca pratensis, Phleum pratense, Trifolium repens* and *T. pratense* supplemented by *Vicia sativa, Fagopyrum esculentum, Medicago lupulina, Phacelia tenacetifolia,* and *Brassica napus*. The sowing treatments are abbreviated SB, SBGLH, NCGLH and NC (Figure 1). Together SB and SBGLH are named "spring-barley-plots". Similarly,

NC and NCGLH are named "fallow-plots". The plots were treated with 0, 1/2, and 1/1 N-dose where 1/1 N-dose was 110 kg N ha^{-1}. Pesticides were not used.

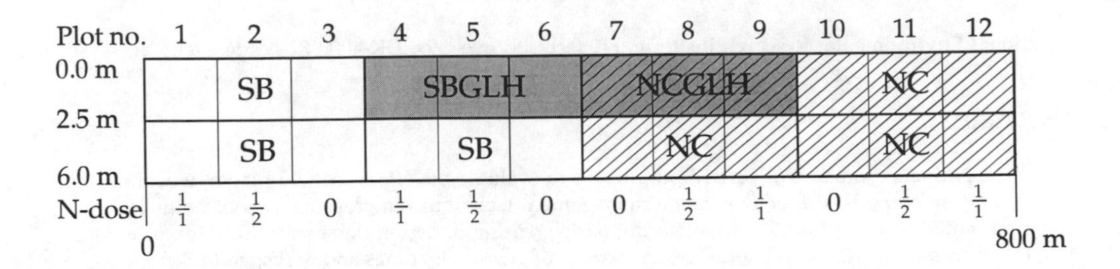

FIGURE 1. Experimental design in crop margin indicating distance from boundary, sowing treatments (SB: Spring Barley; NC: No Crop; SBGLH: Spring Barley undersown with Grasses, Legumes and Herbs; NCGLH: No Crop with Grasses, Legumes and Herbs), and N treatment (0, 1/2, 1/1 N-dose) in inner and outer subplots.

In each subplot, the vegetation was sampled 6½ and 10½ weeks after sowing using five circles of 0.1 m^2. The plants were identified and counted. At the second date the above ground plant biomass was harvested and sorted into: a) sown species (spring barley and mixture of grasses, legumes and herbs), b) *Chenopodium album*, c) other wild plant species. The abundance of the wild plant species that occurred in at least 10 % of the circles was analyzed at a species level.

The arthropod fauna of the inner plots only was sampled on 29th June with three D-vac samples per plot. Each sample comprised five suctions giving a total area of 0.5 m^2 per sample. The fauna in the samples (excluding Collembolans and Mites) was counted, and the total biomass (ww) excluding Aphids was measured. The arthropod fauna was divided into taxonomic groups: Aphids, Flies, Pollen Beetles, Rove Beetles, and the remaining species, "Other A-species", which among others included bird food items (cf. Rands, 1985) such as Ground Beetles, Cicadas, Weevils, Ants and Plant Bugs.

Sowings were allotted to the plots bearing in mind to minimize interactions of the arthropod fauna between treatments. The N treatments were allotted randomly to plots 1-3, replicated in plots 4-6, and reversed in plots 7-12. Statistical analyses were made seperatedly for the inner and outer subplots. The area seemed homogenous and had for long been treated uniformly. Therefore, the circles and D-vac samples have been considered as independent and random samples of treatments. Statistical analyses were performed with SAS/STAT PROC NPAR1WAY (SAS, 1988) using Mann-Whitney U test (sowing treatment in outer subplots, only) or Kruskal-Wallis test followed by Multiple Comparison Between Treatments (Siegel & Castellan, 1988). All tests were performed at P=0.05. A Wilcoxon Signed Ranks Test (Siegel & Castellan, 1988) of the counts of plants at first and second analysis was made for all subplots. As a difference was found in one case only (P=0.07), the mean of the two observations within each circle has been used.

RESULTS AND DISCUSSION

The biodiversity (total number of species) was higher in "fallow-plots" than in "spring-barley-plots" (Figure 2). In total, 36 wild plant species were recorded. The density of plants was twice as high in NC than in the other sowing treatments (Table 1). In all, 13 species occurred in ≥10 % of the circles, and together they made up 95 % of all the wild plants. *Chamomilla recutita* was the dominant species (Figure 3). The distribution of seven of the 13 species was affected by sowing treatments.

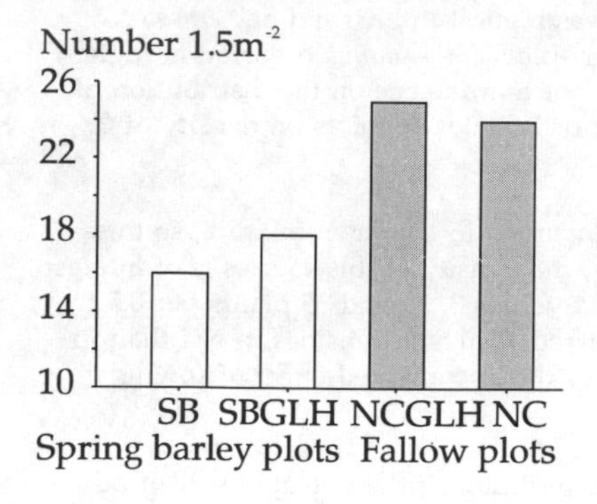

FIGURE 2. Total number of wild plant species in sowing treatments (all samples in three subplots = 1.5 m^2).

TABLE 1. Mean density (plants 0.1 m^{-2}) of wild plants in sowing treatments in inner and outer subplots. Means with the same letter within a subplot are not significantly different.

Subplot	Inner	Outer
SB	13 b	8.6
SBGLH	13 b	-
NCGLH	17 b	-
NC	28 a	17
P-value	<0.001	<0.001

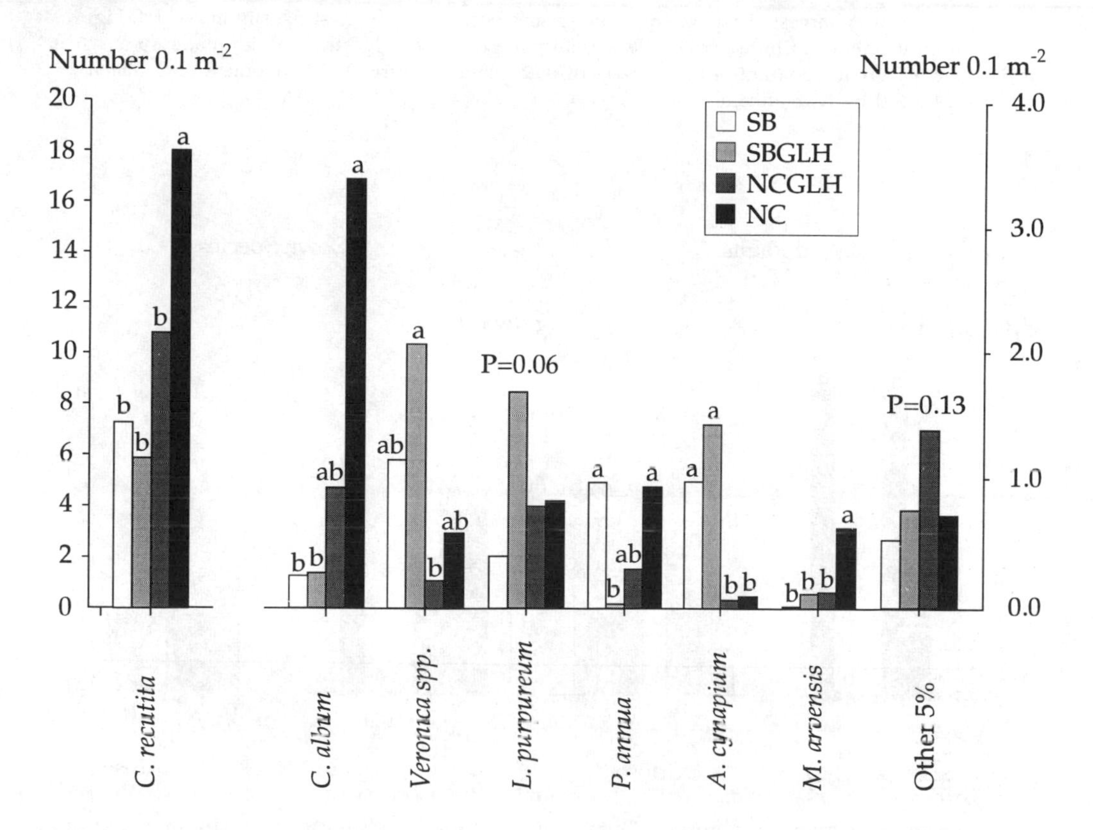

FIGURE 3. Mean density (plants m^{-2}) in inner subplot of the seven species among the 13 common species (making up 95 % of all plants) that were affected by sowing treatment in inner or outer subplot. Similarly for the remaining species making up other 5 % of all plants. Means with the same letter within species/group are not significantly different.
Similar analyses from outer subplots were in agreement with the results from inner subplots.

Density of *C. recutita*, *C. album* and *Myosotis arvensis* was highest in NC, and of *Aetusa cynapium* and *Veronica agrestis/persica* highest in "spring-barley-plots". *Poa annua* occurred at highest densities in plots without sown mixture. Sowing was found not to influence on the distribution of "Other 5 % species". No effect of N-dose was found in inner or in outer subplots on density of *C. recutita* or the remaining species together.

B. napus occurred naturally in the area, but was also included in the sown mixture, so this species was not included in the statistical analysis. However, the density of this species is of interest for the occurrence of especially Pollen Beetles and was respectively 0, 3, 31 and 35 plants per 0.5 m^2 in SB, SBGLH, NCGLH, and NC. The fact that *B. napus* occurred at higher densities in NC than in plots where this species was sown, namely SBGLH, underlines the documented effect of sowing treatment on the occurrence of wild plant species.

The biomass of the wild plants is of interest as a valid indicator of the potential seed production as well as of living space for the associated fauna. The biomass of wild plants was highest in NC (Figure 4a1). Compared to the "spring-barley-plots", the biomass of wild plants in NC was 40 times higher and to NCGLH four times higher. There was no effect of N treatment on the biomass of wild plants (Figure 4a2). The biomass of the sown species was highest in "spring-barley-plots" (Figure 4b1) and in plots with N application (1/2 or 1/1 N-dose): (Figure 4b2).

C. album was harvested separately. This species occurred at highest density in NC (cf. Figure 3). Although it only contributed very little to total plant density in NC this species made up a proportionately greater amount of the biomass of wild plants (Figure 5). The biomass of *C. album* only was affected by N treatment in NC.

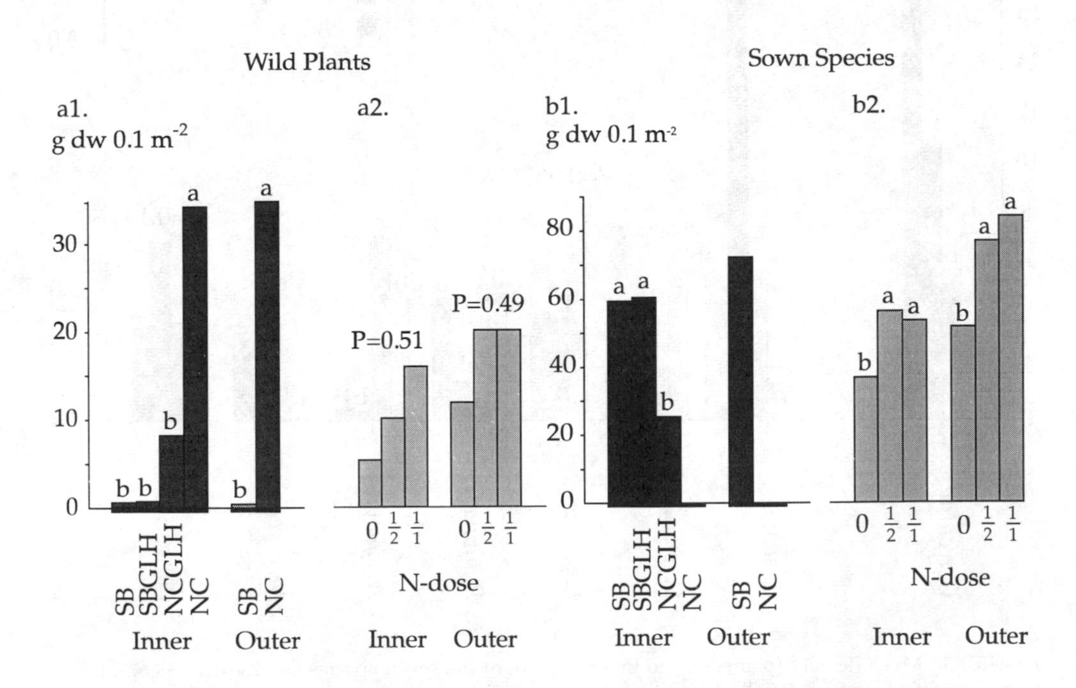

FIGURE 4. Mean biomass (g dw 0.1 m^{-2}) in inner and outer subplot of **a**) all wild plants **b**) sown species. Means with the same letter within plant group and subplot are not significantly different. **a1** & **b1**. Sowing treatments. **a2** & **b2**. Nitrogen treatments.

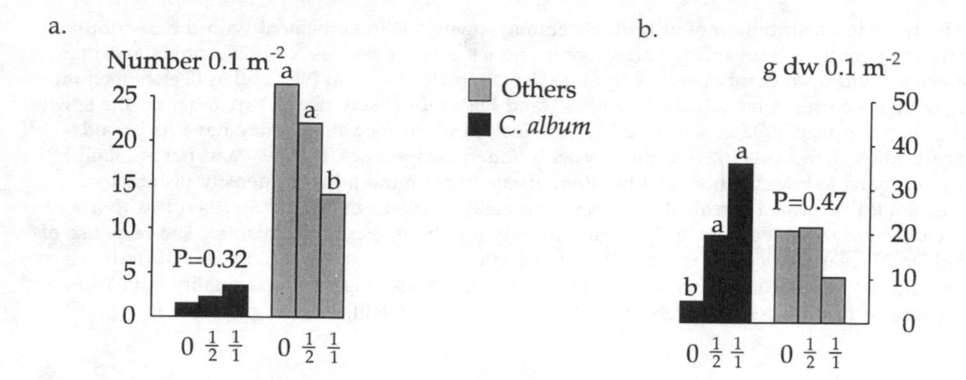

FIGURE 5. Performance of *C. album* and the remaining wild plant species (Others) in sowing treatment NC (No Crop) at different N-dose. **a.** Mean density (plants 0.1 m^{-2}). **b.** Mean biomass (g dw 0.1 m^{-2}). Means with the same letter within plant group are not significantly different. Only the 26 circles in which *C. album* was observed are included.

A total of 2640 specimens of arthropods excluding Aphids was collected in the 36 D-vac samples. The density in each of the five groups were affected by sowing (Figure 6). Aphids were found at highest densities i "spring-barley-plots". Rove Beetles and "Other A-species" were found at highest densities in plots with sown species, namely "spring-barley-plots" and NCGLH. Pollen Beetles and Flies were found at highest densities in "fallow-plots" - an especially high density of Flies was found in NCGLH. Only Flies and "Other A-species" occurred at higher densities at 1/1 and 1/2 N-dose than at zero N-dose (P<5 %). A similar tendency was seen for Aphids (P=0.08). The biomass of arthropods was not affected by sowing (P=0.43), but was affected by N (P=0.002). Biomass (0.5 m^{-2}) at 1/2 and 1/1 N-dose (0.15 and 0.14 g respectively) was different from the biomass at zero N-dose (0.052 g).

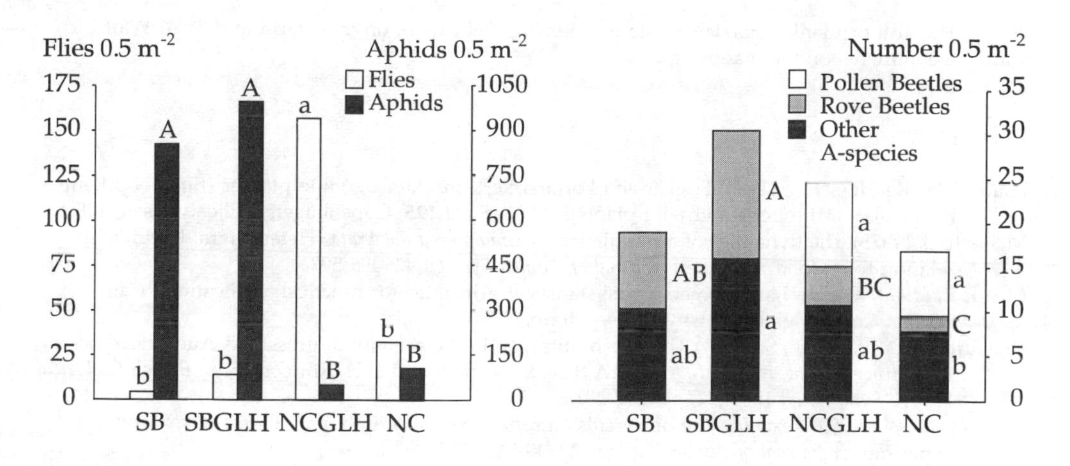

FIGURE 6. Density of arthropods (0.5 m^{-2}) in sowing treatments. Means with the same letter within arthropod group are not significantly different.

The recorded distribution of arthropods among sowings is in agreement with the distribution of their hosts: Aphids feed on spring barley, some species of Rove Beetles feed on Aphids and many of the species avoid open plant stands with more dry microclimate as in NC, Pollen Beetles feed on *B. napus*, which occurred mostly in "fallow-plots", and Flies most likely used *C. recutita* and the sown herbs as pollen supplier. "Other A-species", which includes bird food items, may have had good chances of finding their hosts in spring barley with undersown grasses, legumes and herbs (SBGLH). Thus in a comparable investigation, Hald & Reddersen (1990) found a higher density of bird food items in ecologically farmed cereals than in conventionally farmed cereals and ascribed this to a higher incidence of sown legumes and a higher density and biomass of wild plants. The response of arthropods to N treatment could be a result from selection pressure for high N in agricultural landscape and better food quality in the fertilized plots. Such responses to food quality have been described for sap feeding species (McNeill, 1973; Prestidge & McNeill, 1983).

CONCLUSION

Sowing had a greater effect on the analyzed groups of flora and fauna in the crop margin than did nitrogen. The number of wild plant species (biodiversity), the density of plants and arthropods, the distribution of plant species and arthropod taxa, as well as the biomass of wild plants, were all affected by sowing treatment. Compared to treatments including sown spring barley, the biomass of wild plants in No Crop was 40 times higher and compared to mixture of Grasses, Legumes and Herbs four times higher. Only the biomass of arthropods was found not to be affected by sowing treatment. The N treatment mostly affected the sown species (biomass), density and the biomass of arthropods. In No Crop treatment, application of fertilizer increased the biomass of *C. album*, while the pooled biomass of the other species of wild plants was not affected. Therefore, other things being equal, the success and seed setting of a single dominant species and following risk of weed problems in future crops is reduced when a sown species is present.

Management of crop margins in rotation by techniques which include sowing of a mixture of grasses, legumes and herbs, seems to be a good compromise to include for example in EU surplus regulation programmes. It is beneficial to most species of the arthropods, but also to wild plants without carrying too high a risk of weed problems in future crops.

ACKNOWLEDGMENT

The author thanks Gjorslev Estate for allowing field work on their farm and T. A. Watt for valuable comments on the manuscript.

REFERENCES

Hald, A.B.; Reddersen, J. (1990) Fugleføde i kornmarker - insekter og vilde planter (Bird food items in cereals - arthropods and wild plants). *Miljøprojekt* **125**, Copenhagen: Miljøstyrelsen, 108 pp.

McNeill, S. (1973) The dynamics of a population of *Leptopterna dolabrata* (Heteroptera: Miridae) in relation to its food resources. *Journal of Animal Ecology* **42**: 495-507.

Mead, R. (1988) The design of experiments: statistical principles for practical application, Cambridge: Cambridge University Press, 620 pp.

Prestidge, R.A.; McNeill, S. (1983) The role of nitrogen in the ecology of grassland Auchenorryncha. In: *Nitrogen as an ecological factor*, J. A. Lee, S. McNeill and I. H. Rorison (Eds) *British Ecological Society Symposium 1981*, **22**, pp. 257-281.

Rands, M.R.W. (1985) Pesticide use on cereals and the survival of grey partridge chicks: a field experiment. *Journal of Applied Ecology* **22**: 49-54.

SAS (1988) SAS Release 6.03 Ed. Cary, NC: SAS Institute Inc.

Siegel, S.; Castellan, N.J.Jr. (1988) Nonparametric statistics for the behavioral sciences, Singapore: McGraw-Hill. 2nd ed., 399 pp.

THE EFFECTS OF FLUAZIFOP-P-BUTYL AND CUTTING TREATMENTS ON THE ESTABLISHMENT OF SOWN FIELD MARGIN STRIPS

E. J. P. MARSHALL

Department of Agricultural Sciences, University of Bristol, Institute of Arable Crops Research, Long Ashton Research Station, Bristol BS18 9AF, UK

M. NOWAKOWSKI

Willmot Pertwee Conservation, West Yoke, Ash, Nr. Wrotham, Kent TN15 7HU, UK

ABSTRACT

A field experiment compared the effects of a graminicide, with and without mowing, on sown grass and wildflower strips at the edge of an arable field. Strips were sown in September 1991, treated with fluazifop-P-butyl ("Fusilade 5") in December and either left uncut or mown up to four times during the following spring and summer. The seed mixture contained a proportion of annuals which dominated the plots, especially in the first year after establishment. The graminicide did not adversely affect *Festuca* species, but reduced frequencies of other sown grasses. Two or more cuts in 1992 reduced the cover of annual species and as a result, plots were more diverse in 1992 and had the most sown dicotyledonous species in the following year. Where the annuals dominated and subsequently lodged, they were able to persist through the second year and plant diversity was reduced.

INTRODUCTION

Field margins constitute a network of semi-natural habitat in farm landscapes which interact with adjacent agriculture. Such semi-natural areas have a number of potential roles in more sustainable farming systems (Marshall, 1993). Expansion of the perennial ground flora at arable field edges has potential for increasing on-farm biodiversity, for controlling annual weed species of hedgerows that may colonise adjacent crops (Marshall, 1989) and for enhancing populations of beneficial insects (Thomas *et al.*, 1992). Previous work on the introduction of sown grass and wildflower strips has demonstrated that in fertile arable soils, suitable control of weed grasses in the first year is required. The herbicide fluazifop-P-butyl, will control many such grasses while leaving *Festuca* species unaffected (Marshall & Nowakowski, 1992). However, there were indications that sown grasses could also be adversely affected, while repeated cutting could reduce the frequencies of annual species. Confirmation of these results was sought in an experiment using the graminicide fluazifop-P-butyl and a variety of cutting treatments in the first year after sowing. In addition, natural regeneration (set-aside) and plots sown only with grasses were established.

METHODS

A replicated experiment with eleven plot treatments was established in September 1991 along a field margin bounded by a farm track. The experiment was located on Radcot Bridge

Farm, Oxfordshire (NGR: SU 275995). Each plot was 3 m wide and 8 m long; treatments were located at random along the margin within three replicate blocks. Treatment details are summarised in Table 1. Plots were either allowed to regenerate naturally from the soil seed bank or sown on 9 September 1991 with a grass-only mixture or a grass and wildflower mixture. The grass seed mixture contained 12 species and varieties and was sown at a rate equivalent to 37 kg/ha. The grass/herb mixture contained the same grasses, plus six annual and 23 perennial dicotyledonous species sown at the same rate. Herbicide was applied using an AZO sprayer fitted with a 3 m boom and flat fan jets delivering medium spray quality at 250 l/ha at 3 bars pressure. Mowing treatments were made with a pedestrian rotary mower cutting to between 3 and 5 cm, followed by raking off the cut vegetation. All plots were mown on 4 September 1992 and 21 April 1993.

In 1992, percentage cover of sown annual species was estimated for each species by two independent observers in early July, before the July mowing treatment. Between 17 and 22 July 1992, live vegetative presence/absence of species was assessed in six 0.1 m^2 quadrats in each of the 33 plots. Each species was given a score out of six dependent on the number of quadrats where it was recorded. Differences between treatments were tested using analysis of variance, with differences between means assessed using Least Significant Differences (LSD; P=0.05). Where necessary, data were transformed using a logarithm transformation ($\log_e$ N+0.5). Data for species groups (sown and unsown grasses, annual and perennial dicotyledons) and for all species were analysed as the sum of 0-6 scores divided by six to be expressed as numbers of species per quadrat. The presence/absence data were collected again on 8 July 1993, towards the end of the second season after sowing.

TABLE 1. Field edge plot treatments after sowing on 9 September 1991.

Code	Sowing	Herbicide and cutting treatment	
Setaside	Natural regeneration		
Grass	Grasses only	; isoxaben at 125 g AI/ha on 17 September 1991	
Untreated	Sown mixture	; Untreated	
X	"	; fluazifop-P-butyl at 94 g AI/ha on 2 December 1991;	Uncut
XA	"	; "	Cut in April
XAM	"	; "	Cut in April, May
XAMJ	"	; "	Cut in April, May, June
XAMJJ	"	; "	Cut in April, May, June, July
XAMJul	"	; "	Cut in April, May, July
XAJJ	"	; "	Cut in April, June, July

Mowing dates were: 18 April, 19 May, 22 June and 27 July 1992

RESULTS

Cover of sown annuals in July 1992

Sown annual species dominated ground cover in the first year in plots which were unmown or cut only once (Figure 1). Fluazifop-P-butyl alone had no significant (P=0.05) influences on the cover of sown annuals compared with the untreated control. However, mowing in June significantly reduced the cover of the dominant sown annuals *Papaver rhoeas, Centaurea cyanus, Agrostemma githago* and *Anthemis arvensis* in early July. A single cut in April, following fluazifop-P-butyl in December, also reduced cover of *A. githago* compared with controls. Mowing in April and May promoted cover of *P. rhoeas* and *S. alba* but reduced cover of *C. cyanus, A. githago* and *A. arvensis*.

Figure 1. Cumulative percentage ground cover of sown annual species (the sum of individual species cover) in field margin strips, some treated with fluazifop-P-butyl (X) and mown in April (A), May (M), June (J) or July (J;Jul).

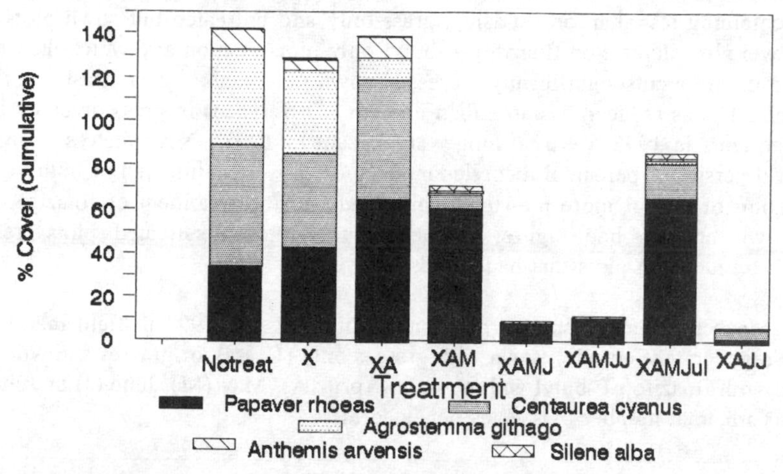

Species diversity in July 1992 and 1993

In 1992, total species diversity, measured as the mean number of species found per quadrat was least on set-aside, grass-only, fluazifop-P-butyl and fluazifop-P-butyl+April cut plots (Table 2).

TABLE 2. Mean number of plant species found per quadrat on plots at the edge of an arable field sown with different seed mixtures and treated in various ways (see Table 1 for treatment code details). SED = standard error of the difference between means (df=18).

Treatment code:	Set-aside	Grass	Un-treated	X	XA	XAM	XAMJ	XAMJJ	XA MJul	XAJJ	SED
1992	6.80	7.88	10.62	6.62	8.22	12.78	12.45	13.67	11.95	8.72	0.483
1993	5.95	7.17	9.45	9.33	7.05	8.22	8.33	8.05	9.28	7.67	1.112

Highest mean species number was recorded on plots treated with fluazifop-P-butyl and subsequently mown four times. However, these plots did not have significantly more species than plots mown only in April and May. Application of fluazifop-P-butyl reduced diversity compared with sown, untreated plots. Repeated mowing reversed this effect. The results from 1993 indicate an overall reduction in diversity with time, with the exception of fluazifop-P-butyl-only plots. Compared with the untreated control plots, there were no differences among cutting + fluazifop-P-butyl treatments, with the exception of the April only cutting

In 1993, these plots were just significantly less diverse than the control plots. A feature of the 1993 data was the increased variability, as shown by the SED value, despite the overall decline in diversity.

Examination of the six main groupings of the flora, sown and unsown grasses, perennial and annual (and biennial) dicotyledonous species, illustrated varying effects of the treatments (Figure 2). In 1992 and 1993, sown grasses were most frequent on grass-only and untreated plots and least diverse on fluazifop-P-butyl-only and set-aside plots. Mowing in April and May encouraged sown grasses. Weed grasses were well controlled by fluazifop-P-butyl, their diversity remaining less than on set-aside, grass-only and untreated but sown plots. Sown perennials were less diverse on fluazifop-P-butyl-only plots than on any which then received a cut. Two or more cuts significantly increased sown perennials. In contrast, the diversity of sown annuals was reduced by more than one cut. Differences in grass diversity between cutting treatments in 1992, were no longer significant in 1993. Nevertheless, a significant increase in diversity of perennial dicotyledonous species was maintained in both years with the application of two or more mowings, compared with no treatment or fluazifop-P-butyl alone. Sown annuals had highest frequencies on the latter-treated plots, remaining significantly reduced on plots that had at least two cuts.

Figure 2. Mean numbers of species per quadrat in 1992 and 1993 in field margins strips allowed to regenerate (Setaside), sown with grasses only (Grass), or grasses and wild flowers and treated with fluazifop-P-butyl and mown in April (A), May (M), June (J) or July (J,Jul). (SED bar is for total number of species per quadrat)

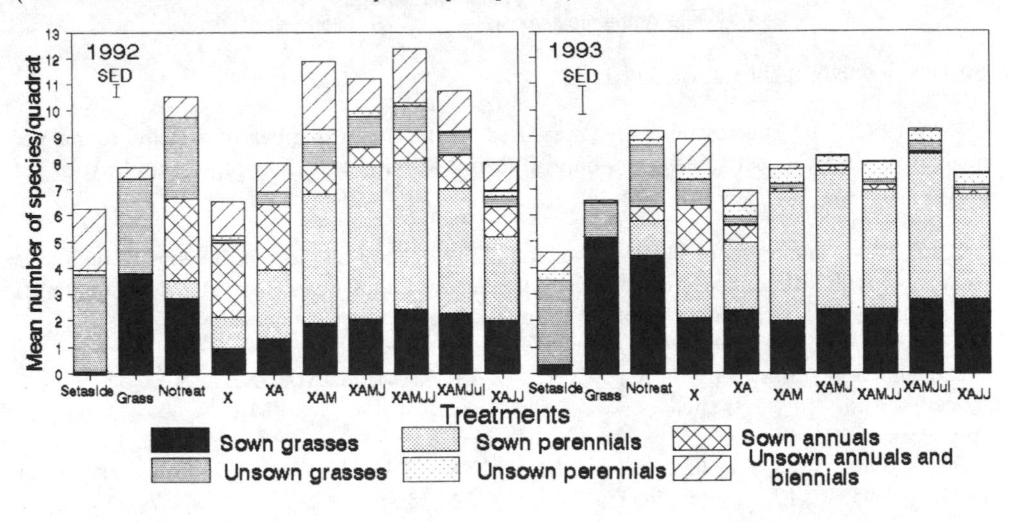

Individual species

The effects of treatments on selected species are shown in Table 3. The sown grass, *Festuca rubra*, was encouraged by the application of fluazifop-P-butyl. In contrast, *Trisetum flavescens* was reduced in frequency and *A. myosuroides* was largely controlled by the graminicide. The latter retained high frequencies on set-aside plots, but was reduced on sown grass-only and untreated plots in 1993. *Galium aparine*, a winter-germinating annual, was reduced in frequency in 1992 by two cuts. In 1993, the data were not significantly different between treatments, though the species had similar frequencies on fluazifop-P-butyl plots in both years. Sown *A. githago* was severely affected by cutting in 1992 and was absent in 1993 from those plots which had been mown at least twice. The sown perennial, *L. vulgare*, was encouraged in both years by mowing, with some increase in frequency on plots treated with the graminicide.

TABLE 3. Transformed mean frequency scores ($\log_e$+0.5) of selected plant species on field edge plots in the two seasons after sowing.

Treatment code:	Set-aside	Grass	Un-treated	X	XA	XAM	XAMJ	XAMJJ	XA MJul	XAJJ	SED
Festuca rubra - sown grass											
1992	-0.69	1.75	1.31	1.57	1.54	1.87	1.87	1.87	1.87	1.87	0.134
1993	-0.69	1.87	1.61	1.87	1.87	1.76	1.87	1.87	1.82	1.87	0.091
Trisetum flavescens - sown grass											
1992	-0.69	1.82	1.75	0.21	1.29	1.34	1.61	1.71	1.69	1.64	0.256
1993	-0.05	1.87	1.87	1.42	1.57	1.31	1.50	0.86	1.34	1.57	0.334
Alopecurus myosuroides - annual weed grass											
1992	1.87	1.87	1.55	-0.69	-0.69	-0.69	-0.33	-0.69	-0.69	-0.69	0.211
1993	1.82	0.49	1.12	0.58	0.21	-0.69	-0.69	-0.33	-0.69	-0.69	0.355
Leucanthemum vulgare - sown perennial dicotyledon											
1992	-0.69	-0.69	-0.33	-0.33	-0.16	1.71	1.71	1.71	1.54	1.21	0.374
1993	-0.69	-0.69	0.32	0.49	1.31	1.14	1.64	1.75	1.64	1.69	0.394
Agrostemma githago - sown annual											
1992	-0.69	-0.69	1.67	1.69	0.77	0.04	-0.33	-0.69	-0.69	0.69	0.339
1993	-0.69	-0.69	0.38	1.22	0.32	-0.69	-0.69	-0.69	-0.69	-0.69	0.365
Galium aparine - annual dicotyledonous weed											
1992	1.69	1.05	1.05	1.01	0.58	-0.69	-0.33	-0.33	-0.69	-0.69	0.418
1993	0.69	-0.33	0.04	1.01	0.53	-0.69	-0.69	-0.69	-0.69	-0.69	0.595

DISCUSSION

The sown annual species, *P. rhoeas, C. cyanus, A. githago* and *A. arvensis*, were highly competitive, especially in the year after sowing. Where only one cut in April or no cutting was applied, species diversity in 1992 was reduced, compared with two or more cutting treatments. This reflected poorer establishment of the sown grasses and perennial dicotyledonous species under cover of the annuals. In addition, the December application of fluazifop-P-butyl resulted in reduced frequencies of sown *T. flavescens, P. pratense* and *C. cristatus*. *Festuca* species were largely unaffected by fluazifop-P-butyl. Further work on the modification of timing and dose of the graminicide may allow control of weed grasses while minimising adverse effects on desirable perennial grass species. Overall diversity in 1992 and sown perennial species diversity in 1993 were significantly enhanced in plots mown at least twice, results also demonstrated by Smith & Macdonald (1992). This effect was almost certainly mediated by the suppression of sown annual species.

In previous work (Marshall & Nowakowski, 1992), sown perennial dicotyledonous species have responded positively to applications of fluazifop-P-butyl and the removal of weed grasses. In this experiment, there was greater competition from the annual dicotyledonous species and no obvious response to fluazifop-P-butyl alone. Mowing at leat twice after application of fluazifop-P-butyl produced greater diversity by reducing cover of these dicotyledonous annuals. It might therefore be argued that mowing could be used instead of a herbicide. This cannot be verified from the work described here, as comparisons between varied cutting treatments with and without a herbicide are needed. Under highly fertile conditions, as in this site, annuals should either not be sown or used at low rates. Nevertheless, the weed seed bank at the edges of fields is often large, capable of producing similarly competitive vegetation.

A feature of the data was the dominance of annual species through the second year on plots treated with fluazifop-P-butyl and not mown. Classical studies of secondary succession (e.g. Bard, 1952) predict the initial dominance by annuals, followed by rapid change to a biennial and perennial flora. This work shows that annual species may continue to dominate the flora under conditions of rapid growth, high canopy development, followed by lodging and shading-out of smaller plants. These conditions, which are most likely a reflection of high soil fertility common to much farmland, will not allow perennials dicotyledonous species to establish easily, delaying succession. This emphasises the importance of refining appropriate management and essential weed control to achieve the aim of rapid establishment of strips of perennial grass and wild flowers in field margins.

ACKNOWLEDGEMENTS

This work was supported by Willmot Pertwee Limited and by the UK Ministry of Agriculture, Fisheries and Food (Project BD0404). The site was kindly provided by Mr and Mrs Hitchin and the chemical by Zeneca plc.

REFERENCES

Bard, G.E. (1952) Secondary succession on the Piedmont of New Jersey. *Ecological Monographs*, **22**, 195-215.

Marshall, E.J.P. (1989) Distribution patterns of plants associated with arable field edges. *Journal of Applied Ecology*, **26**, 247-257.

Marshall, E.J.P. (1993) Exploiting semi-natural habitats as part of good agricultural practice. In: *Scientific basis for codes of good agricultural practice*. V.W.L. Jordan (Ed.) EUR 14957. Commission of the European Communities, Luxembourg. pp. 95-100

Marshall, E.J.P.; Nowakowski, M. (1992) Herbicide and cutting treatments for establishment and management of diverse field margin strips. *Aspects of Applied Biology* ***29***, *Vegetation management in forestry, amenity and conservation areas*. pp 425-430.

Smith, H. & Macdonald, D.W. (1992) The impacts of mowing and sowing on weed populations and species richness in field margin set-aside. *British Crop Protection Council Monograph No. 50, Set-Aside*, Thornton Heath: BCPC Publications, pp. 117-122.

Thomas, M.B.; Wratten, S.D.; Sotherton, N.W. (1992) Creation of 'island' habitats in farmland to manipulate populations of beneficial arthropods: predator densities and emigration. *Journal of Applied Ecology*, **28**, 906-917.

EFFECTS OF MANAGEMENT TREATMENTS ON CARABID COMMUNITIES OF CEREAL FIELD HEADLANDS

A. HAWTHORNE, M. HASSALL

School of Environmental Sciences, University of East Anglia, Norwich, NR4 7TJ.

ABSTRACT

The overall abundance and species composition of carabid beetles were compared in three different cereal field headland management regimes in the Breckland area of eastern England. Uncropped Wildlife Strips, which were cultivated but not sown, contained more species and a greater overall abundance of carabids than either sprayed headlands or 'Conservation Headlands', which were sown but which received reduced pesticide inputs. Both carabid abundance and species richness of the community were correlated with percentage cover of dicotyledonous plants and the abundance of other invertebrates. The species richness of the community was also strongly related to total vegetation cover and abundance of aphids and Collembola. Experimental reduction of vegetation in Uncropped Wildlife Strips lead to a decrease in abundance of most species of carabid except *Bembidion lampros*.

INTRODUCTION

Since the 1940's increased intensification of agricultural practices has lead to a general decline in the abundance of carabid beetles in arable farmland (Burn, 1988). However, during the last decade some reversal of this trend has occurred, with the increased interest in alternative farming practices. These include the use of Conservation Headlands, and Uncropped Wildlife Strips which have been included in the management prescriptions for the Breckland Environmentally Sensitive Area. Breckland is an arable area of land in eastern England which is historically associated with arable farming on a temporary basis. A unique community of flora and fauna has developed in association with areas of abandoned and extensive agriculture. The advent of modern agricultural techniques has led to the destruction and fragmentation of these habitats so that many species are now rare.

In this paper we examine the consequences of introducing Uncropped Wildlife Strips for the diversity and abundance of the carabid fauna in winter wheat fields and compare the carabid populations of an Uncropped Wildlife Strip with those of Conservation and fully sprayed headlands.

STUDY SITE

Two strips (120 m x 6 m each) of each of three headland treatments were arranged in a randomized block design along one side of a winter wheat field (18.6 ha). Sprayed headlands received pesticides and fertiliser applied at the same rates as the rest of the field. The Conservation Headland was treated as above but did not receive herbicides and insecticides, in accordance with specific guidelines for choice of compound and timing (Sotherton *et al.*, 1989), thus allowing weeds and associated invertebrates to flourish. The Uncropped Wildlife Strip was rotovated annually in September and left unsown, without pesticides or lime.

METHODS

Sampling of carabid community

Beetles were sampled using covered pitfall traps consisting of 60 mm diameter plastic cups containing ethylene glycol and water with a drop of detergent. An area immediately around the trap was cleared of vegetation to standardise traps in the different habitats. Whilst species varied in their trapability and activity, these parameters did not differ significantly between treatments (Hawthorne, in press), thus allowing comparisons of catches from different headland treatments to be made. Traps were placed 20 m apart in the middle of the headland treatments, 3 m from, and parallel to, the field boundary. Three or five traps were used for each headland treatment. Opposite each of these traps a further trap was established in the crop at 8 m from the field boundary. Catches were emptied every two weeks in 1990 from April until July, weekly from April to August in 1991 and weekly from February until July in 1992.

Measurements of environmental variables

The following environmental variables were monitored during 1991 and correlated with both overall carabid abundance and species richness:

The percentage cover of all vegetation, and bare ground were estimated on four dates in five 0.25 m^2 quadrats for each headland treatment, 3m from the field boundary. Vegetation height diversity, (using Simpson's Index) were calculated from the percentage cover in each 10 cm height intervals between ground level and 60 cm for each quadrat. Data were pooled where species were recorded at heights greater than 61 cm. Relative humidity was monitored at the soil surface using cobalt thiocyanate paper Measurements were made at two-weekly intervals from May until July. Cereal aphids were counted on 30 tillers selected at random in each headland treatment on the days of the vegetation survey. Ground-active invertebrates, including Collembola were recorded from pitfall traps.

Experimental manipulation of environmental variables

The influence of vegetation cover on species composition and abundance was examined in twelve 4 m x 4 m plots on the most heavily vegetated treatment, the Uncropped Wildlife Strip, in 1992 in a randomised block design. Four plots were sprayed, with "Gramoxone" at field strength (0.5 g paraquat per litre) on 14 May to create bare, undisturbed ground. Any subsequent vegetation was removed by hand. Four other plots were hoed to remove vegetation and create bare, disturbed ground. The remaining plots were left undisturbed as fully vegetated plots. After two weeks 2 x 2m areas at the centre of each plot were enclosed. This was done using lengths of 4 mm plywood buried 0.1 m deep to leave 0.15 m above ground. Ground-active carabids were sampled from both control and bare ground plots using covered pitfall traps emptied weekly during June and July.

RESULTS

Community structure

Rank abundance, pooled for the three years, were broadly similar for each headland treatments (Table 1). The biggest difference was that the highest ranking species in the Uncropped Wildlife Strip accounted for 28 % of the individuals, compared with 16.2 % and 18.6 % in the Conservation and sprayed Headlands respectively. In all

TABLE 1.

Catch totals per species and proportions of the overall carabid total in Conservation Headlands (CH), Uncropped Wildlife Strips (UH) and sprayed headlands (SH) for the 10 most highly ranked species, using pooled data. *A.dors = Agonum dorsale, P.mel = Pterostichus melanarius, B.lamp = Bembidion lampros, B. tetra = B. tetracolum, D.atri = Demetrias atricapillus, T. quad = Trechus quadristriatus, H.ruf = Harpalus rufipes, H.aff = H.affinus, A.bif = Amara bifrons, A.aen = A.aenea, A.sim = A.similata, A.fam = A. familiaris*

Rank	CH	%	Total	UH	%	Total	SH	%	Total
1	*A.dors*	16.15	238	*B.lamp*	28.03	697	*A.dors*	18.55	174
2	*P.mel*	14.18	209	*H.aff*	12.95	322	*B.lamp*	12.79	120
3	*A.bif*	13.77	203	A.bif	9.81	244	*H.ruf*	11.83	111
4	*T.quad*	8.68	128	*P.mel*	9.81	244	*P.mel*	10.34	97
5	*D.atri*	6.99	103	*H.ruf*	6.35	158	*D.atri*	9.28	87
6	*B.lamp*	6.38	94	*A.sim*	5.43	135	*T.quad*	7.57	71
7	*A.fam*	4.82	71	*A.den*	4.26	106	*H.aff*	3.20	30
8	*B.tetra*	4.75	70	*B.tetra*	3.70	92	*A.bif*	2.88	27
9	*H.ruf*	4.07	60	*T.quad*	3.38	84	*A.aen*	2.77	26
10	*A.sim*	2.85	42	*A.dors*	2.27	69	*A.sim*	1.92	18
	TOTAL		1474	TOTAL		2487	TOTAL		938

treatments six species contributed more than 5 % to the total catch, out of a total of 35 species for the normal sprayed Headland; 41 for the Conservation Headland and 43 species in the Uncropped Wildlife Strip (Table 1).

The most abundant species overall was *Bembidion lampros*, which was amongst the six most abundant species in each headland treatment. However it was more than twice as numerous in the Uncropped Wildlife Strip as any other species, and seven times more abundant there than in the sprayed headland. *Pterostichus melanarius* was the second most abundant species overall and amongst the top four species in all three treatments. *Agonum dorsale* was third most abundant overall but less common in the Uncropped Wildlife Strip than in the sprayed and Conservation Headlands, where it was the most abundant species. Some the less common species were restricted to one treatment. Those restricted to Uncropped Wildlife Strips included *Bembidion femoratum, Amara tibialis, Acupalpus meridianus, Bradycellus harpalinus, Bradycellus distinctus* and *Harpalus rubripes*. Whilst *Amara nitida, Amara spreta* and *Pterostichus angustatus* were only found in Conservation Headlands. Only one species, *Calathus piceus*, was restricted solely to sprayed headlands.

Relationships between environmental variables and community characterists.

Analyses of the relationships between carabid community parameters and environmental variables are summarized in Table 2. Carabid abundance was strongly correlated with porcentage cover of dicotyledonous plants and abundance of invertebrates. Species richness was most strongly correlated with total vegetation cover, % dicot cover and general invertebrate abundance, especially Collembola and aphids. Both carabid abundance and species richness were negatively correlated with relative humidity.

TABLE 2 Summary of significant correlations between carabid abundance and species richness with environmental measurements in 1991. Where ** = P < 0.01, *** = P < 0.001

	Carabid abundance	Species richness
No. dicot species	-	0.250**
% cover dicot species	0.302***	0.503***
No. monocot species	0.287***	0.253**
% cover bare ground	-0.272**	-0.386**
% cover all vegetation	0.247**	0.441***
Height diversity	-	0.254**
Collembola	0.282**	0.399***
Invertebrates	0.364***	0.458***
Aphids	0.226**	0.371***
Relative humidity	-0.399***	-0.420***

Experimental manipulation of environmental variables.

The overall number of carabids caught fell significantly when the vegetation cover was removed ($F_{(2,9)} = 16.55$; $P < 0.001$) (Table 3). One conspicuous exception to this trend was *B. lampros;* significantly fewer being caught in the fully vegetated plots than in the herbicide treated or hoed plots ($F_{(2,9)} = 79.71$, $P < 0.001$).

TABLE 3. Comparison of mean carabid density (m^{-2} ± one standard error) in fully vegetated, herbicide treated and hoed bare ground plots in 1992.

Species	Full vegetated control	Hoed bare ground	Herbicide treated bare ground
Harpalus affinis	18.31 ± 1.04	5.06 ± 0.60	4.38 ± 0.72
Amara similata	11.75 ± 3.56	0.75 ± 0.36	0.31 ± 0.21
Amara bifrons	9.19 ± 2.25	2.44 ± 0.53	0.88 ± 0.33
Amara aenea	6.44 ± 1.64	0.44 ± 0.26	0.88 ± 0.33
Harpalus rufipes	6.00 ± 1.91	1.94 ± 0.36	2.06 ± 0.56
Agonum dorsale	0.75 ± 0.29	0.31 ± 0.16	0.38 ± 0.19
Pterostichus melanrius	0.56 ± 0.42	0.13 ± 0.06	0
Bembidion lampros	7.56 ± 1.57	22.44 ± 1.61	37.06 ± 1.04
Bembidion tetracolum	0	0.19 ± 0.01	0.56 ± 0.29

DISCUSSION

Both the total carabid abundance and the number of species present were greatest in the Uncropped Wildlife Strips. This was probably because of their more complex structure and greater diversity of the vegetation which strongly influenced the abundance and species richness of carabid faunas (Speight & Lawton, 1976; Powell *et al.*, 1985). These parameters were strongly correlated with the percentage cover of all vegetation, but particularly with the cover by dicotyledonous species and with foliage height diversity. There are several probable reasons for this. Firstly plant tissues can be important directly as a source of food for carabids. Many *Amara* and *Harpalus* spp. are considered phytophagous (Hengeveld, 1980) while most species of carabids include some vegetation in their diet. A positive relationship exists between abundance and diversity for many insects and the species and height diversity of vegetation (Murdoch *et al.*, 1972). Thus the availability of potential invertebrate prey for carabids also increased with increasing complexity of the vegetation. The importance of this was confirmed by the strong positive correlations between carabid community characteristics and the density of Collembola and other invertebrates. High alternative prey densities could be important in encouraging polyphagous predators to persist in a field early in the season, when pest populations are low. A second indirect effect of increased vegetation complexity could be its influence in buffering fluctuations in microclimate (Speight & Lawton, 1976). This was considered to be critical in determining the distribution of species such as *P.melanarius* and *H.rufipes* (Skuhravy *et al.*, 1971) while a further benefit of structurally diverse vegetation was that it could provide shelter from predators and parasites (Lawton, 1978).

The experimental reduction of vegetation from the Uncropped Wildlife Strip provided confirmation that for most of the species the presence of complex vegetation was very important. However there were some species of carabid (*B. lampros*)which preferred and actively selected bare ground (Mitchel, 1963). The abundance of this species increased dramatically in the bare ground plots and it clearly preferred those with a smooth surface to the more disturbed hoed plots. This could account for why, early in the season, when numbers of this species normally peak, it was by far the commonest in the Uncropped Wildlife Strips, where little vegetation had developed.

Although one of the smallest carabid species caught in this survey, *B. lampros* is known to consume more aphids per unit of its body weight than any other species of carabids tested in the laboratory (Sopp & Wratten, 1988). Its very high abundance in the Uncropped Wildlife Strips could therefore be of considerable importance when evaluating their potential benefits in helping to control pest populations. The strong preference of *P. melanarius* for Uncropped Wildlife Strips later in the season may be of similar importance, because whilst less numerous than *B. lampros*, it is capable of consuming large numbers of aphids.

If these predators were to remain exclusively within the Uncropped Wildlife Strips they would be unlikely to have a very substantial influence on pests within the crop itself. This is not however the case as carabids sampled from 8 m out into the crop adjacent to the Uncropped Wildlife Strips, during 1992 were significantly ($F_{(2,180)} = 24.69$, $P < 0.0001$) more abundant than they were adjacent to either of the other two headlands. Corresponding counts of aphids densities showed that there were significantly less ($F_{(2,348)} = 3.32$, $P = 0.037$) in crops adjacent to Uncropped Wildlife Strips than there were adjacent to the other two headlands.

Such evidence of carabids dispersing out from the favourable conditions of the Uncropped Wildlife Strip is of wider significance in relation to the broader status of these beetles within the modern agricultural environment. Cultivation of crops is known to deplete the populations of some species (Blumberg & Crossley, 1983). If favourable conditions for larval survival and/or increased adult reproductive rates are provided by the Uncropped Wildlife Strip then these increased populations could provide a reservoir from which depleted populations can be replenished through dispersal.

Populations can survive in an area only if they have sufficient powers of dispersal to 'refound' subpopulations that have become extinct, or find new suitable habitats. In the modern agricultural landscape, suitable habitats are becoming increasingly fragmented (Mader, 1988). If appropriately managed headlands provide suitable conditions, albeit temporary, for a species to survive and reproduce, before establishing in a more permanent habitat, they may provide an important link between fragments and so help to ensure the survival of local populations.

REFERENCES

Blumberg, A.Y.; Crossley, D.A. (1983) Comparison of soil surface arthropod populations in conventional tillage, no tillage and old field systems. Agro-ecosystems, **8**, 247-253.

Burn, A.J. (1988) Assessment of pesticides on the impact on invertebrate predation in cereal crops. Aspects of Applied Biology, **17**, 279-288

Hawthorne, A.J. (in press). Validation of the use of pitfall traps to study carabid populations in cereal field headlands. Acta Jutlandica, (in press)

Hengeveld, R. (1980) Qualitative and quantitative aspects of the food of ground beetles (Coleoptera, Carabidae): A review. Netherlands Journal of Zoology, **30**, 555-563.

Lawton, J.H. (1978) Host-plant influences on insect diversity: the effects of space and time. In: Diversity of insect faunas. L. A. Mound and N. Waloff (Eds), Symposia of the Royal Entomological Society No. 9. Oxford.Blackwell. pp 105-125.

Mader, H-J. (1988) Effects of increased spatial heterogeneity on the biocenosis in rural landscapes. Ecological Bulletins, **39**, 169-179.

Mitchel, B. (1963) Ecology of two carabid beetles *Bembidion lampros* (Herbst) and *Trechus quadristriatus* (Schrank). I. Lifecycles and feeding behaviour. Journal of Animal Ecology, **32**; 289-299.

Murdoch, W.M.; Evans, F.C.; Peterson, C.H. (1972) Diversity and pattern in plants and insects. Ecology, **53**, 819-828.

Powell, W.; Dean, G.J.; Dewar A. (1985) The influence of weeds on polyphagous arthropod predators in wheat. Crop Protection, **4**, 298-312.

Skuhravy, V.; Louda, J.; Sykora, J.(1971) Zur verteilung der Laufkaufer in Feldmonokulturen. Beitrage zur entomolgie, **21**, 539-546.

Sopp, P.I.; Wratten, S.D. (1986) Rates of consumption of cereal aphids by some polyphagous predators in the laboratory. Entomologia Experimentalis et Applicata, **41**, 69-73.

Sotherton, N.W.; Boatman, N.D.; Rands, M.R.W. (1989) The 'Conservation Headland' experiment in cereal ecosystems. The Entomologist, 108, 135-143.

Speight, M.R.; Lawton, J.H.(1976) The influence of weed cover on the mortality imposed on artificial prey by predatory ground beetles in cereal fields. Oecologia, **23**, 211-223.

HABITAT CREATION IN LARGE FIELDS FOR NATURAL PEST REGULATION

B.J. HART, W.J. MANLEY, T.M. LIMB and W.P. DAVIES

Royal Agricultural College, Cirencester, Gloucestershire GL7 6JS

ABSTRACT

Field experiments at Harnhill Manor Farm, Cirencester, were initiated in Spring 1993 by the Royal Agricultural College, in conjunction with AFRC IACR Long Ashton Research Station, with the aim of extending the earlier Less Intensive Farming and Environment (LIFE) project funded by the European Community, to a farm scale. The studies use field margins to encourage beneficial arthropods which may enhance natural regulatory mechanisms for pest control. In the first year numbers of predatory spiders were significantly higher in sown plant margins than in sterile margins. Staphylinidae were found to be significantly less abundant in the sterile and enhanced fallow margins than in the other treatments.

INTRODUCTION

Increasing concerns for production costs and environmental impact is stimulating recent research on integrated farming systems. Small plot research at the AFRC Institute of Arable Crops Research, Long Ashton, into less-intensive farming and the environment in the UK (LIFE project) (Jordan *et al*, 1990) is being furthered by the Royal Agricultural College, by extending this research into commercial practice where production costs and financial returns have priority.

One of the main objectives of the lower input system is to investigate the feasibility of enhancing biodiversity, by reducing the size of large fields and using smaller cropping areas in an integrated rotation. This approach lends itself ideally to the use of flower or grass margins for division of large fields into smaller cropping areas, and to enhance beneficial arthropods. The work therefore provides an opportunity to monitor arthropod communities in a practical integrated arable farming situation. These studies should indicate whether the use of integrated systems is likely to enable exploitation of field margins to enhance natural regulatory mechanisms for pest control in practice.

METHODS

Field experiments were initiated at the Royal Agricultural College's Harnhill Manor Farm, Cirencester, in April 1993 for an initial period of three years. The entire low-input site consists of a 30 ha field (Figure 1). Since reduced field size is fundamental to the LIFE approach, the field was divided into cropping areas of approximately 5 ha using raised margins. Raised margins were preferred to flat margins due to enhanced properties of drainage and habitat value. The raised margins (0.3 m high) were divided into 5 treatment plots (4 m x 50 m) of plant mixtures (Table 1) sown in randomised blocks (Figure 1). Randomised blocks were used to overcome biasing of results due to variations in the field (e.g. soil, drainage). The randomised block design and the shortest division between cropping areas dictated the length of the margins. The maximum possible length (50 m) was used to limit effects on results caused by invertebrate transfer between strips. A margin width of 4 m has enabled the margins to be managed in accordance with standard practices using farm machinery.

Plant mixtures (Table 1) were chosen for their suitability to the soil type of the site, potential arthropod habitat value (both winter and summer habitats), ease of management, cost and low invasive nature. Treatment 1 is the standard grass mix recommended by the Game Conservancy Trust for beetle banks. Treatment 2, a wild flower and grass mix suitable to the soil type, was chosen for its low cost, ease of establishment and seed availability. Treatment 3, represents an "enhanced fallow" to investigate the qualities of a low density wild flower mixture, allowing for natural regeneration. The headland mix used in Treatment 4 compared standard farm practice routinely used on Harnhill Manor farm. Finally, the sterile margin (Treatment 5) was kept free of plant cover using herbicide (glyphosate), representing standard farm practice for field margins in other farming systems. The sterile and headland margins in effect provide control conditions.

Standard ecological methods (pitfall traps, soil samples, D-vac, mark and recapture) are being used to determine arthropod colonisation and holding capacity of each treatment throughout seasons. Arthropod dispersal into crops in the spring will be measured by sampling at regular distances into the cropping areas. Insect pests will also be monitored at regular distances into the crops. Pest numbers will be correlated with predator numbers and related to adjacent margin treatments.

A comprehensive soil and weed map is being made of the site. Details of agronomy, yield (using a global positioning crop yield monitoring system in addition to standard sampling techniques) and costs are being recorded. This information will be used when more detailed data on beneficial arthropod populations are available, to analyse the potential for natural pest regulation and the influence of this and other environmental factors on yield, as well as the costs involved.

Reported here are preliminary results from the first winter on the site. In October 1993, 3 x 0.04 m² soil samples were taken from each treatment plot. Arthropods were extracted from the samples using a modified version of the flotation technique of Sotherton (1984), identified and quantified. Results were subjected to analysis of variance after square root transformation (sqrt) to correct for non-normality in the data distribution.

Figure 1. Plan of integrated cropping system project area and wild plant margins at Harnhill Manor Farm.

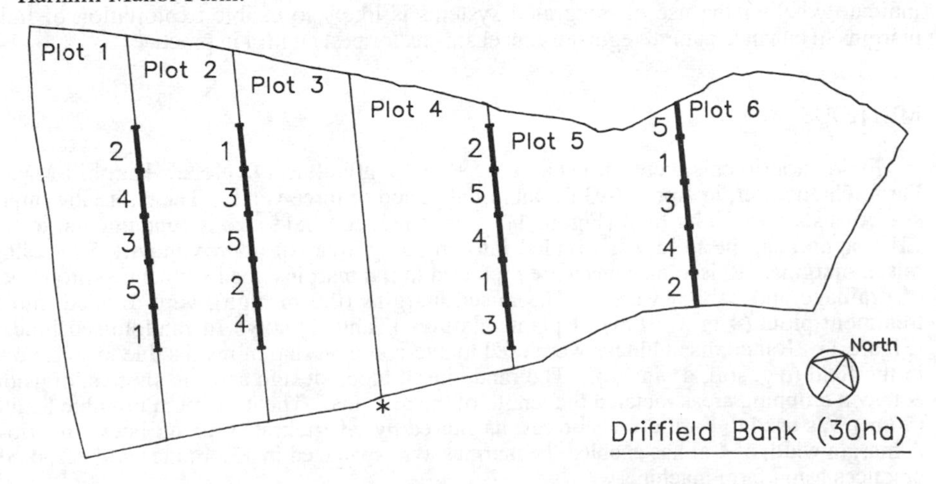

* = Tree strip to be planted. For details of treatments 1-5, see Table 1.

Table 1. Wild plant mixes and planting densities used in the integrated cropping systems study at Harnhill Manor Farm.

Treatment number	Planting Density (g/m^2)	Plant Species	% Total wt
1 (Grass)	4	*Dactylis glomerata*	50
		Holcus lanatus	50
2 (Wild flower)	4	*Plantago lanceolata*	5
		Galium verum	5
		Leucanthemum vulgare	5
		Daucus carota	5
		Festuca ovina	26.6
		Cynosorus crtistatus	26.6
		Agrostis castellana	26.6
3 (Enhanced fallow)	1.5	*Knautia arvensis*	20
		Centaurea nigra	20
		Geranium pratense	20
		Ranunculus acris	20
		Plantago lanceolata	20
4 (Clover headland)	4	*Lolium perenne*	91.5
		Trifolium repens	8.5
5 (Sterile)	-	No plants	-

RESULTS

Preliminary results on overwintering arthropods are shown in Table 2. Numbers of predatory spiders were significantly ($p = 0.02$) higher in sown plant margins than in sterile margins. Staphylinidae were found to be significantly ($p = 0.01$) less abundant in the sterile and enhanced fallow margins than in the grass and wild flower margins.

Table 2. Predatory beetles, spiders and mites found in soil samples analyses from field margin plots in October 1993.

Treatment	Carab. m^{-2} no.	Carab. m^{-2} sqrt	Staph. m^{-2} no.	Staph. m^{-2} sqrt	Spiders m^{-2} no.	Spiders m^{-2} sqrt	Mites m^{-2} no.	Mites m^{-2} sqrt	Total m^{-2} no.	Total m^{-2} sqrt
1	150	8.7	294	13.5	156	11.25	63	5.56	663	21.2
2	175	10.4	319	17.0	113	9.73	144	10.58	750	25.5
3	31	3.9	25	2.5	88	9.10	88	8.98	231	14.7
4	25	3.4	131	10.3	106	9.78	44	5.70	306	16.8
5	144	8.5	38	3.1	44	3.31	50	4.56	275	11.5
SED(12 df)		3.95		3.95*		2.10**		2.81		5.11

*, ** = significant at $p = 0.01$ and 0.02 respectively.

DISCUSSION

The numbers of predators found in this study are similar to those found by other workers in their first year (Wratten & Thomas, 1990).

Large variations in numbers were recorded for the same plant mix from different strips. This could perhaps be due to substantial differences in soils across the field (J. Conway, personal communication), or due to the early stages of arthropod colonisation of the strips. From previous studies it would appear that arthropod colonisation of newly sown wild plant strips takes up to 3 years (Thomas & Wratten, 1990).

The results suggest that the grass and wild flower margins may be more suitable for beneficial arthropods. These ongoing studies should clarify this, and establish the suitability of field margins for enhancing natural pest regulation in integrated cropping systems.

ACKNOWLEDGEMENTS

The authors wish to thank Cotswold Grain and Seeds and Emorsgate Seeds for supplying seed for the plant margins. We are also also grateful to Ms F. Collingborn for help with the winter sampling and to Dr M. Gooding for statistical advice.

REFERENCES

Jordan, V.W.L.; Hutcheon, J.A.; Perks, D.A. (1990) Approaches to the development of low input farming systems. In: *Organic and Low Input Agriculture*. R. Unwin (Ed.) *BCPC Monograph No.45*. Thornton Heath: BCPC Publications, pp. 9-18.

Sotherton, N.W. (1984) The distribution and abundance of predatory arthropods overwintering in farmland. *Annals of Applied Biology* **105**, 423-429.

Thomas, M.B.; Wratten, S.D. (1990) Ecosystem diversification to encourage natural enemies of cereal aphids. *Brighton Crop Protection Conference - Pests and Diseases 1990*, **2**, 691-696.

Wratten, S.D.; Thomas, M.B. (1990) Environmental manipulation for the encouragement of natural enemies of pests. In: *Organic and Low Input Agriculture*. R. Unwin (Ed.) *BCPC Monograph No.45*. Thornton Heath: BCPC Publications, pp. 87-92.

PLANT DISTRIBUTION ACROSS ARABLE FIELD ECOTONES IN THE NETHERLANDS

W. JOENJE, D. KLEIJN

Section Weed Science, Agricultural University, Bornsesteeg 69, 6708 PD - Wageningen, The Netherlands

ABSTRACT

Inventories made in crop edges and adjacent boundaries (no herbicide, no fertilizer; including fallow strips) on sandy soil in the eastern part of the Netherlands revealed the common plant species still occurring in today's field margins. The distribution of the species over crop edge and boundary is analysed and compared with historical data. Distribution classes and ecological status of the species are discussed, as well as the potential for restoration of species-richness in field margins.

INTRODUCTION

Current intensive farming practices in NW Europe are criticized for the widespread eutrophication and loss of biodiversity in the rural countryside Many species have disappeared, common species have become rare. In the Netherlands species such as *Cardamine pratensis, Papaver spp.* and *Centaurea cyanus*, very common until recently, are now considered valuable assets.

General concern has led to political consensus to implement measures that will restrict negative effects of agriculture and restore basic environmental qualities and nature conservation value (Joenje, 1991). In this context attention is increasingly focused on field margin management.

Why should we restore or maintain a certain level of diversity in the agricultural areas? Essentially three categories of arguments are being put forward: i) biodiversity for nature conservation purposes ii) diversity as indication of a safe environment related to human wellbeing, and iii) plant and animal species in communities along the crops *may* have beneficial effects as biocontrol agents (e.g. Welling, 1988) with economic significance in integrated pest management.

One means of restoring the diversity could be the establishment of enlarged field boundaries at the cost of arable surface (*viz.* Marshall *et al.* 1994). Such boundaries are expected to have a higher diversity potential which can be further increased by sensitive management. The actual diversity, however, is always the outcome of immigration and establishment. Species most probably stem from the local seedbank or from the direct vicinity of the (enlarged) field boundary. The potential diversity consists of all species available to colonise a particular spot. An impression of such a list can be obtained by making an inventory of species that grow in a larger area adjacent to the field boundary in roadsides, grasslands, hedgerows, etc.

In this paper we analyze species-lists of unsprayed and unfertilized margins, comparing the samples from both crop edge and boundary strip. The contemporary arable weed flora is compared with the flora of 1956 (Bannink *et al.*, 1974). Furthermore the short term, successional development on newly created boundary strips in the mid-east Netherlands is predicted on basis of presence, frequency and distribution of species in this inventory.

METHODS

Samples (with abundance classes 1-5, according to Tansley, 1935) were made in 19 arable field margins on sandy soils in the mid-east Netherlands, where a reasonable number of species is still present in a relatively small-scale landscape. As part of an experiment funded by the province of Gelderland, the crop edges did not receive artificial fertilizer or herbicides for the preceding one or two years. Weeds were controlled mechanically. Along a field margin two samples were made: one of the 3-4 m wide crop edge and one of the 3-5 m wide adjacent boundary. The arable fields, and thus the samples, were of different lengths depending on the size of the field. Four of these crop edges were kept fallow, the others were sown to various crops. The complete listing of the 38 samples is not presented here (copies from authors). References in the literature (Sissingh, 1950; Bannink *et al.*, 1974) with inventories of weed communities in the area were used for comparison. These data stem from arable fields in the same sandy region (plant cover and species-frequencies according to the Braun-Blanquet method); they give a relatively complete survey of the flora occurring in the 1956 crops.

RESULTS

A total of 221 species was recorded, comprising *ca.* 15 % of the Dutch flora, but belonging almost exclusively to the classes of 'common' and 'very common' species (Netherlands Central Bureau of Statistics, 1987). Twelve species were 'relatively common' and only three were 'less common', two of these being 'red list' species *(Stachys arvensis, Cuscuta europaea)* with limited and declining distribution. There were 28 species of trees and shrubs which are not considered further.

Arable weeds in 1956 and 1993

The species lists from the samples of 1993 and 1956 were compared. The most commonly found arable weed species are listed in Table 1. where the frequency ranking in the 1993-inventory is followed by the ranking in the 1956-inventory (Bannink *et al.* 1974). Taking the 10 most frequent species from the 1993-list, there are six species (+ in Table 1) in common with 1956 (*Chenopodium album, Bilderdykia convolvulus, Stellaria media, Viola arvensis, Galeopsis tetrahit, Elymus repens*). The remaining four most frequently occurring species of 1956 decreased strongly (- in Table 1). This resulted partly from specific control (*Apera spica-venti)*, but mostly from the known susceptability of 'old weeds' (*Anthoxanthum aristatum, Centaurea cyanus*) to intensified agricultural management.

The inventory of 1956 has another 27 species with low frequencies not found in the 1993 sample, among these are *Papaver rhoeas, Scleranthus annuus, Arenaria serpyllifolia* and *Veronica spp.* as well as *Avena fatua*. The total number of species found in crop edges in the 1993 inventory was as high as 174, apparently because the crop edges were included as opposed to the samples from the homogeneous crop in 1956. Furthermore the samples were larger (500 m^2 or more). The 1956 inventory was characterized by an even spread of species over the samples with frequencies well below 50%.

Species of crop edge and boundary

In the inventory of 1993, species were screened for their habitat preference. Ninety-eight species were found equally or more often in the

TABLE 1. Arable weed species in the 1993 inventory (no-input crop edges) in decreasing frequency of occurrence (A); ranking of species frequency in samples of 1956 (B), based on 119 plots of *ca*. 50 m^2, near Raalte.

	A	B	
Chenopodium album	1	10	+
Bilderdykia convolvulus	2	2	+
Stellaria media	3	4	+
Viola arvensis	4	1	+
Galeopsis tetrahit	5	9	+
Elymus repens	6	6	+
Polygonum hydropiper	7	12	
Spergula arvensis	8	18	
Myosotis arvensis	9	13	
Chamomilla recutita	10	14	
Vicia hirsuta	11	16	
Veronica arvensis	12	23	
Galium aparine	13	21	
Senecio vulgaris	14	15	
Apera spica-venti	15	5	-
Galinsoga parviflora	16	27	
Juncus bufonius	17	22	
Lamium purpureum	18	26	
Vicia sativa	19	8	-
Centaurea cyanus	20	3	-
Holcus mollis	21	24	
Polygonum lapathifolium	22	19	
Rumex acetosella	23	17	
Urtica urens	24	25	
Lamium amplexicaule	25	28	
Anthoxanthum aristatum	26	7	-
Agrostis stolonifera	27	11	
Aphanes microcarpa	28	20	

N.B.: The 1993 inventory also had eight frequent species (frequencies over 50%) which are not shown in the table, because they did not occur in the sample of 1956.

boundary than in the crop edge. This group of species, so-called boundary-species, is summarized in Table 2. It shows that 75% of the species was found in three or less boundaries; this of course is a common feature of many species distributions. About half of the species have a ruderal plant strategy (*sensu* Grime 1979; Grime *et al*. 1988) and are also characterized by high Nitrogen indicator-values (Anon. 1987).

The species of this inventory were then grouped in five classes according to their distribution over field boundary and crop edge:

1- Species limited to the boundary (13 spp.).
2- Species limited to the crop edge (20 arable weeds).
3- Species with the main distribution in the boundary, but also regularly found in the crop edge (22 spp.).
4- Species with the main distribution in the crop edge but also regularly found in the boundary (37 spp.).
5- Species equally occurring in the crop edge and the boundary (40 spp.). (another 66 species were found only once).

TABLE 2. Occurrence of the species with main distribution in the boundary in (n) of the 19 field margins and their plant strategies (according to Grime *et al.* 1988).

species	(n)	plant strategy
Urtica dioica	16	C
Holcus lanatus	13	C-S-R
Galeopsis tetrahit	12	R / C-R
& Holcus mollis		C
Agrostis capillaris	11	C-S-R
& Elymus repens		C / C-R
Galium aparine	10	C-R
& Dactylis glomerata		C / C-S-R
Arrhenatherum elatius	9	C / C-S-R
& Conyza canadensis		?
Ranunculus repens	8	C-R
& Lapsana communis		R / C-R
Festuca rubra	7	C-S-R
4 species	6	.
3 species	5	.
5 species	4	.
13 species	3	.
25 species	2	.
35 species	1	.

The five most frequent species of each distribution class are listed in Table 3. The first four distribution types are comparable with those described by Marshall (1985, 1989) analyzing normally cultivated crop edges. The fifth class, with species equally distributed over edge and boundary, may reflect the effect of unsprayed and unfertilized crop edge conditions. The high number of species in class 3 and 5 may indicate that a high proportion of boundary species has a ruderal plant strategy. Unfortunately there are no historical data of field boundary vegetation.

DISCUSSION

Since 1956 the arable weed flora of the sandy region studied has changed in a process of agricultural intensification and ruderalization whereby species associations from the syntaxonomical Class of the *Secalietea* shifted towards the *Chenopodietea*. This indicates nutrient enrichment of the field and the margin over the last four decades. The inventory of boundary communities is more dependent on differing local ecological conditions and cannot easily be compared with general data of an old species-inventory. It is expected, however, that boundary nutrient levels have also increased. This, in combination with disturbance from herbicide use, may well have led to communities dominated by a few species of grasses and herbs with characteristic ruderal/competitive strategy. This trend is by no means unique to the Netherlands. In an inventory of meadow boundaries in Wales (GB) on clay soils, six out of the 15 most frequent species were also found in the Netherlands, with the highest ranking *Urtica dioica* and *Holcus lanatus* in common (Marshall, E.J.P., pers. comm.).

We conclude that the present-day field-margin is characterized by (i) *Chenopodietea* species in the crop edge, (ii) a boundary with dominance of a small number of very common species with a high proportion of ruderals.

TABLE 3. The five most frequent species for each distribution class, with nB and nC: the number of Boundaries or Crop edges (in a total of 19 field margins) where a species was present; A: abundance (average of Tansley values 1-5).

species	nB	A	nC	A
class 1 (13 spp.)				
Arrhenatherum elatius	9	(3.0)	0	
Agrostis canina	3	(4.3)	0	
Deschampsia flexuosa	3	(2.7)	0	
Hieracium spec.	3	(2.7)	0	
Chamaemelum angustifolium	3	(1.7)	0	
class 2 (20 spp.)				
Echinochloa crus-galli	0		9	(2.2)
Galinsoga parviflora	0		6	(3.2)
Sonchus arvensis	0		5	(1.8)
Symphytum officinale	0		4	(2.0)
Polygonum lapathifolium	0		3	(2.7)
class 3 (22 spp.)				
Urtica dioica	16	(2.8)	8	(1.8)
Holcus lanatus	13	(2.8)	6	(1.7)
Holcus mollis	12	(3.4)	3	(1.7)
Agrostis capillaris	11	(3.5)	5	(1.4)
Galium aparine	10	(2.2)	6	(1.7)
class 4 (37 spp.)				
Chenopodium album	4	(3.0)	19	(2.7)
Polygonum persicaria	2	(1.5)	19	(3.0)
Polygonum convolvulus	5	(1.8)	18	(2.6)
Stellaria media	10	(2.5)	17	(2.9)
Viola arvensis	6	(2.2)	16	(2.5)
class 5 (40 spp.)				
Galeopsis tetrahit	12	(2.3)	13	(2.3)
Elymus repens	11	(2.9)	12	(2.7)
Ranunculus repens	8	(2.8)	7	(1.6)
Lapsana communis	8	(2.1)	6	(2.2)
Plantago major	4	(2.3)	6	(1.5)

What does this mean for the species diversity to be expected in a newly created strip / enlarged field boundary? Here a so-called old field succession is likely to take place, starting with a rapid colonization and dominance of arable weeds, including perennials. In subsequent years the vegetation will probably be increasingly dominated by perennial ruderal species, notably grasses, especially from distribution-classes 3 and 5. Their dominance is likely to reduce establishment of other species from the seedbank, or from nearby or more distant provenances. Thus, the diversity to be expected on a new boundary strip will generally be moderate. However, in a few decades the increase in species diversity could be substantial, depending on the presence of seed sources in a larger area (Schmidt, 1993). However, Borstel (1974) reported decreasing diversity.

If current input levels of agrochemicals and management practices were reduced, dispersal barriers could be bypassed by the introduction of desired species in the **first** year of the succession. This would greatly enhance the establishment of these species (even when sown with grasses). Such improved boundary communities could then resume their importance as indicators of a safe environment for sustainable agricultural production.

ACKNOWLEDGEMENTS

We thank the Province of Gelderland and the farmers participating in the crop edge management programme for help and use of their fields. The cooperation with W. Meijer and A. Pancras (Province Gelderland) and with A.J.W. Rotteveel (Plant Protection Service) is gratefully acknowledged.

REFERENCES

Anon. (1987) *Botanical database*. Voorburg: Centraal Bureau voor de Statistiek, 121 pp.

Bannink, J.F.; Leys, H.N.; Zonneveld, I.S. (1974) *Akkeronkruidvegetatie als indicator van het milieu, in het bijzonder de bodemgesteldheid.* Bodemkundige studies 11. Mededelingen Stiboka, Wageningen, 88 pp.

Borstel, U. (1974) Untersuchungen zur Vegetationsentwicklung auf ökologisch verschiedenen Grünland- und Ackerbrachen hessischer Mittelgebirge (Westerwald, Rhön, Vogelsberg). *Dissertation Universität Giessen.*

Grime, J.P. (1979) *Plant strategies and vegetation processes.* Chichester: John Wiley & Sons, 222 pp.

Grime, J.P.; Hodgson, J.G.; Hunt, R. (1988) *Comparative plant ecology.* London: Unwin Hyman Ltd.

Joenje, W. (1991) Perspectives for nature in dutch agricultural landscapes. *Proceedings British Crop Protection Conference Weeds - 1991*, 365-376.

Marshall, E.J.P. (1985) Weed distributions associated with cereal field edges - some preliminary observations. *Aspects of Applied Biology*, **9**, 49-58.

Marshall, E.J.P. (1989) Distribution patterns of plants associated with arable field edges. *Journal of Applied Ecology*, **26**, 247-257.

Marshall, E.J.P.; Thomas, C.F.G.; Joenje, W.; Kleijn, D; Burel,F.; Le Coeur, D. (1994) Establishing vegetation strips in contrasted European farm situations. *Field Margins - Integrating Agriculture and Conservation*, N. D. Boatman (Ed.), B.C.P.C. Monograph *No. 58*, Farnham: BCPC Publications.

Schmidt, W. (1993) Sukzession und Sukzessionslenkung auf Bracheäckern - Neue Ergebnisse aus einem Dauerflachenversuch. *Scripta Geobotanica*, **20**, 65-104.

Sissingh, G. (1950) Onkruidassociaties in Nederland. *Verslagen Landbouwkundig Onderzoek*, Wageningen, 56.15.

Tansley, A.G. (1935) The use and abuse of vegetational concepts and terms. *Ecology*, **16**, 284-307.

Welling, M. (1988) Auswirkungen von Ackerschonstreifen. *Mitteilungen der Biologische Bundesanstalt für Land- und Forstwirtschaft*, Berlin-Dahlem, Heft 247.

PRELIMINARY FINDINGS FROM A STUDY OF SOWN AND UNSOWN MANAGEMENT OPTIONS FOR THE RESTORATION OF PERENNIAL HEDGE-BOTTOM VEGETATION

F.A. DUNKLEY

Department of Biology, University of Southampton, Southampton, Hampshire, SO9 3TU.

N.D. BOATMAN

The Game Conservancy Trust/Allerton Research and Educational Trust, Loddington House, Loddington, East Norton, Leicestershire, LE7 9XE.

ABSTRACT

The methodology and preliminary findings of a three year experiment designed to evaluate the effectiveness of various management options for restoring a botanically degraded hedge bottom are described. Two types of management option are being investigated. The first makes the assumption that by stopping damaging practices, the natural process of succession will restore the perennial vegetation. The second approach involves the removal of existing vegetation using a broad-spectrum herbicide, and its replacement by sown perennial grasses. There are three unsown treatments using the first approach and six sown treatments. The early findings show some significant differences between the relative frequencies of benign perennial grasses, barren brome (*Bromus sterilis*) and annual forbs between treatments. False oat-grass (*Arrhenatherum elatius*) was the most abundant perennial grass in both the unsown plots and the treatments where it was sown in a mixture with other grasses. Secondary succession in the unsown plots has resulted in a change in distribution of some species.

INTRODUCTION

Traditionally hedges and their associated flora have provided a diverse wildlife habitat in an increasingly uniform farm environment. They are an important nesting site for both songbirds and game birds (O'Connor, 1987); an over-wintering refuge for polyphagous predatory arthropods (Sotherton, 1984) and they provide pollen and nectar for other beneficial insects (van Emden, 1965; Fussell & Corbet, 1992). However, hedge-bottom vegetation is now commonly impoverished in terms of species richness and is also often a potential source of pernicious weeds, particularly cleavers (*Galium aparine*) and barren brome (*Bromus sterilis*) (Boatman & Wilson, 1988). There is evidence that damage has been caused by a number of farming practices such as inaccurate fertiliser application, drift of herbicides and cultivation close to the field edges. To make matters worse farmers have attempted to deal with the problem by spraying out the hedge-bottom vegetation with broad-spectrum herbicides (Marshall & Smith, 1987).

This project differs from other studies of management techniques for the restoration of a benign hedge-bottom vegetation (e.g. Smith & Macdonald, 1992), in that it is primarily concerned with the evaluation of different perennial grass species for their effectiveness in excluding annual weeds such as *B. sterilis* and *G. aparine* and in providing over-wintering habitat for polyphagous predators. Yorkshire fog (*Holcus lanatus*) and cocksfoot (*Dactylis glomerata*) were chosen because they have been shown to support high numbers of over-wintering predacious arthropods (Thomas *et al*, 1991); red fescue (*Festuca rubra*) and false oat-grass (*Arrhenatherum elatius*) were chosen because they are natural colonisers of hedgerows and form a dense sward sward which is likely to resist invasion by annual weeds and may also provide a suitable habitat for over-wintering predators. Our aim is to provide farmers with a workable strategy for dealing with hedgerow weeds and encourage natural enemies. Here we report our preliminary findings.

METHODS

Establishment of the experimental site

In October 1991 an experiment was set up at Down Farm, Headbourne Worthy, Winchester (SU 470 336), using a stretch of damaged hedge-bottom vegetation, dominated by *B. sterilis* and *G. aparine*. The hedge itself was reasonably well-maintained. Nine different treatments were prescribed widening the hedge bottom to 2m. These were repeated three times as 2m x 10m plots in a randomised block design along the length of the hedge. All farming practices were excluded from the extended hedge bottom after the field was ploughed in the autumn of 1991, thus regeneration of the vegetation in the unsown treatments was from this time. Sowing was carried out in September 1992 following removal of the existing vegetation with glyphosate. All the treatments were mown biennially in the autumn.

TABLE 1. Treatments

A. Control: No treatment
B. No sowing or herbicide treatment. Mown three times a year (June, July and September) during establishment.
C. One application in December 1992 of mecoprop-P at 1380g a.i. per hectare and quizalofop-ethyl at 0.075g a.i. per hectare.
D. Sown with cocksfoot (*Dactylis glomerata*).
E. Sown with red fescue (*Festuca rubra* ssp. *rubra*).
F. Sown with false oat-grass (*Arrhenatherum elatius*).
G. Sown with Yorkshire fog (*Holcus lanatus*)
H. Sown with a mixture of all four grasses in D - G above.
I. Sown with a mixture of all four grasses and cut three times a year as in treatment B.

Mecoprop-P was applied for selective control of *G. aparine*, and quizalofop-ethyl for selective control of *B. sterilis* (Boatman, 1992 and unpublished data). Seed rates were determined according to the weight of the seed so that roughly an equal number of seeds was sown for a given area. Thus *F. rubra* and *D. glomerata* were sown at 10g (2g) m^{-2}, *H. lanatus* at 5g (1g) m^{-2} and *A. elatius* at 20g (5g) m^{-2}. Amounts in brackets are the quantities sown in the mixtures. By March 1993 it was apparent that, with the exception of *A. elatius*, there was poor establishment of all the sown treatments in the first block.

It was necessary to have reasonable establishment of all the treatments to ensure sufficient replication for assessing over-wintering arthropods, therefore it was decided that all the sown treatments in that block should be raked over and re-sown at half the original seed rate. This was carried out at the end of March 1993.

Botanical assessments

Botanical assessments were made of the unsown treatments in June 1992 and of of all the treatments in June 1993. A comb-shaped rectangular quadrat, 2m x 0.5m, was placed in a transect across each plot, perpendicular to the hedge, at 2m intervals, thus sampling five transects per plot. The "comb" was divided along its length by prongs into 40 sections of 5cm x 10cm. Presence or absence of each species was recorded, giving a value of 0 or 1 per section, and a total of 0 - 40 per transect.

Mean percentage frequencies of perennial grasses of the species sown, *B. sterilis* and annual forbs were compared using a two-way ANOVA of arcsine transformed data followed by the Tukey test. Confidence limits for the percentage frequencies of different grass species found in the sown mixtures were derived from the binomial distribution.

RESULTS

Mean numbers of species found in the unsown treatments are shown in Table 2. There was a small, non-significant change in species number between the two years. Perennial and annual species were found in roughly equal proportions in the two years. In the unsown treatments *A. elatius* was by far the most abundant perennial grass in both years. In 1992 it was mainly limited to within one metre from the hedge base, however, in 1993, it had spread across the plots. The opposite was true of *B. sterilis* which had an extensive distribution in 1992, but by 1993 was confined to the outer half of the plots and had been almost eliminated from the herbicide treatment. The distribution of perennial grasses and *B. sterilis* was different where the vegetation had been removed and re-sown, represented by the sown mixture (Treatment I) in Figure 1.

TABLE 2. Mean number of species per plot found in unsown treatments

(Standard errors of the means are given in parentheses)

Treatment	1992	1993
Control	19.3 (2.89)	15.7 (0.58)
Frequent cutting	18.0 (0.58)	15.3 (1.20)
Selective herbicide	20.3 (3.18)	18.0 (1.16)

Significant differences were found between treatments in the frequency of desirable grasses, *B. sterilis* and annual forbs (Figure 2), however in the case of *B. sterilis* and annual forbs there was also a highly signicant block effect ($P<0.01$).

In both treatments sown with a mixture of grasses the percentage frequency of *A. elatius* was significantly higher than that of *F. rubra* ($P<0.05$), which was significantly higher than that of the other two species ($P<0.01$). In the uncut mixture (Treatment I) percentage frequency of *D. glomerata* was significantly higher than *H. lanatus* ($P<0.05$).

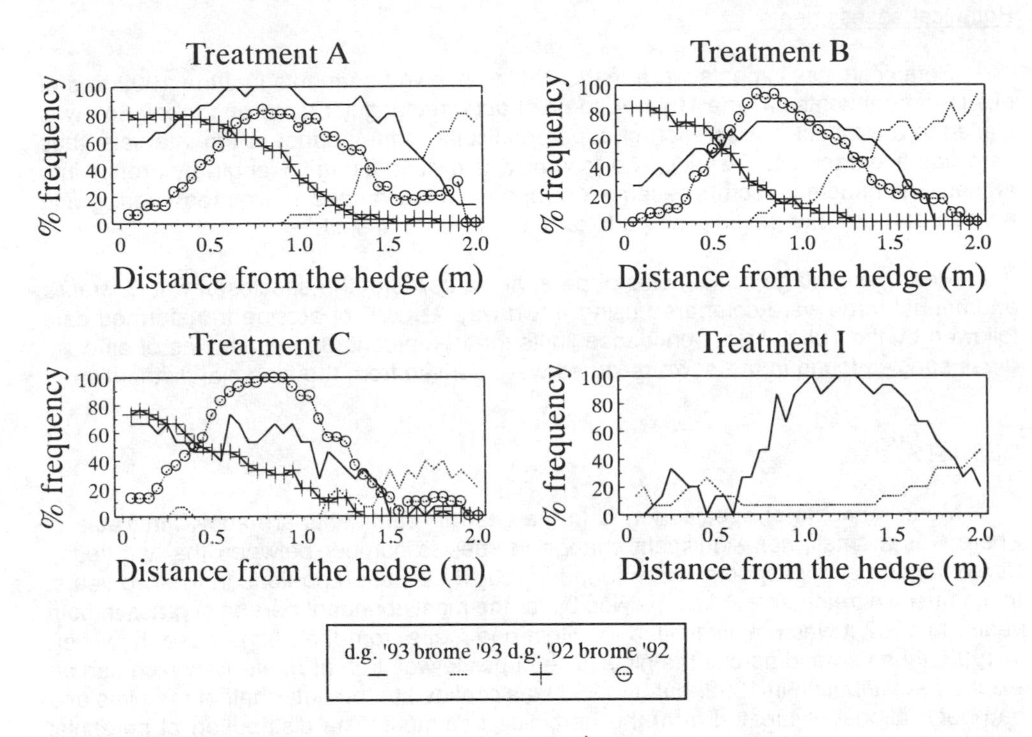

FIGURE 1. Changes in the distribution of selected* perennial grasses (p.g.) and *B. sterilis* (brome) in Treatment A (control), Treatment B (frequent cutting) and Treatment C (selective herbicide) between 1992 and 1993, and in Treatment I (sown grass mixture) in 1993. * *A. elatius*, *F.rubra*, *D.glomerata* and *H.lanatus*.

DISCUSSION

Successional change in the control and unsown cutting treatment resulted in an increase in both frequency and distribution of selected perennial grasses. Dominance of these plots and the treatments where a mixture of grasses was sown by *A. elatius* is in agreement with findings in other work (Marshall, 1990; Grubb, 1982; Mahmoud & Grime, 1976). This grass species possesses many characteristics of dominance (Grime, 1977; 1987): greater height, a root storage system ensuring that resources are available for rapid reallocation at the start of the growing season, the development of dense tussocks that exclude other species and an ability to spread and invade. In consequence *B. sterilis* was reduced in both frequency and distribution in these unsown plots with a trend towards a reduction in species number, although with the annual/perennial ratio unchanged. The hedge grew considerably between 1992 and 1993, and this was the most likely reason for the reduced grass frequency directly beside the hedge.

It would be unwise to draw any conclusions regarding the herbicide treatment at this stage, although in other experiments selective herbicide use has been shown to be effective in controlling *B. sterilis* and *G. aparine* in field margins (Boatman, 1992).

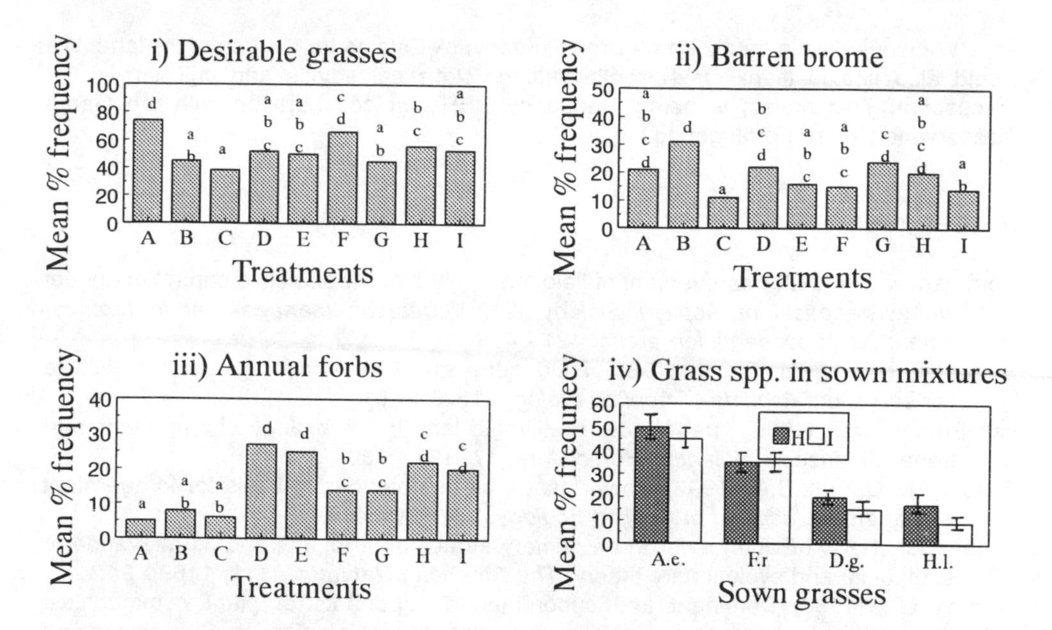

FIGURE 2. i),ii) and iii) Mean percentage frequencies of plant types found in the different treatments (see Table 1). Those with the same letter showed no significant difference (P>0.05) by two way ANOVA of transformed data followed by Tukey test. iv) Mean frequencies of different grasses found in treatment H (sown mixture with frequent cutting) and Treatment I (sown mixture with annual/biennial cut). Error bars indicate 95% C.I.

The best perennial grass establishmant was by *A. elatius* in the control and in the treatment where it was sown as a pure stand. The competitive ability of *A. elatius* is greatest in an undisturbed environment (Grime, 1987), hence its poor performance where there was frequent cutting in treatment B. Despite our view that we had chosen a hedge with a reasonably uniform vegetation and structure, there were highly significant block effects in terms of weed control. Re-sowing of the first block, slope of the field and differences in the original vegetation were probably all contributory factors. As we expected, there was a higher frequency of annual forbs in the sown than unsown treatments in this first year of establishment, as these were able to establish at the same time as the grass seedlings.

Although only tentative conclusions can be drawn from these preliminary data, we have demonstrated the competiveness of *A. elatius* in a hedgerow environment. We have also shown that non-intervention can be an effective, if slow, option for restoration of perennial hedge-bottom vegetation. However, there may be some trade-off between species diversity and weed control (Smith & Macdonald, 1992); the prevention of weed invasion by competitive grasses is likely to mean that other benign and beneficial perennial plants will be unable to establish and compete.

ACKNOWLEDGEMENTS

We would like to thank Simon Brown of Meyrick Estates for the use of his land. We would also like to thank Dr Tom Sherratt for statistical advice and criticism of the manuscript. This project is being funded by SERC in collaboration with the Game Conservancy Trust, Fordingbridge.

REFERENCES

Boatman, N.D. (1992) Improvement of field margin habitat by selective control of annual weeds. *Aspects of Applied Biology*, **29**, *Vegetation management in forestry, amenity and conservation areas*, 431 - 436.

Boatman, N.D.; Wilson, P.J. (1988) Field edge management for game and wildlife conservation. *Aspects of Applied Biology*, **16**, 53 - 61.

van Emden, H.F. (1965) The role of uncultivated land in the biology of crop pests and beneficial insects. *Scientific Horticulture*, **17**, 121 - 136.

Fussell M.; Corbet, S.A. (1992) Flower usage by bumblebees: a basis for forage plant management. *Journal of Applied Ecology*, **28**, 451 - 465.

Grime, J.P. (1977) Evidence for three primary strategies in plants and its relevande to ecological and evolutionary theory. *The American Naturalist*, **111**, 1169 - 94.

Grime, J.P. (1987) Dominant and subordinate components of plant communities: implications for succession , stability and diversity. In: *Colonisation, Succession and Stability*, P.J. Gray; M.J. Crawley; P.J. Edwards (Eds), pp. 413 - 428.

Grubb, P.J. (1982) Control of relative abundance in roadside *Arrhenatherum*: results of a long-term garden experiment. *Journal of Ecology*, **70** (3), 845 - 861.

Mahmoud, A.; Grime, J.P. (1976) An analysis of competitive ability in three perennial, grasses. *New Phytologist*, **77**, 431 - 435.

Marshall, E.J.P. (1990) Interference between sown grasses and the growth of *Elymus repens* (couch grass). *Agriculture, Ecosystems and Environment*, **33**, 11 - 22.

Marshall E.J.P.; Smith B.D. (1987) Field margin flora and fauna; interaction with agriculture. In: *Field Margins: British Crop Protection Council Monograph* No. 35, J.M. Way; P.W. Greig-Smith (Eds), pp. 23 - 33.

O'Connor, R.J. (1987) Environmental interests of field margins for birds. In: *Field Margins: British Crop Protection Council Monograph* No. 35, J.M. Way; P.W. Greig-Smith (Eds), pp. 35 - 48.

Smith, H.; Macdonald, D.W. (1992) The impacts of mowing and sowing on weed population and species richness in field margin set-aside. *Set-Aside: 1992 British Crop Protection Council Monograph No. 50*, pp.117 - 122.

Sotherton, N.W. (1984) The distribution and abundance of predatory arthropods overwintering on farmland. *Annals of Applied Biology*, **105**, 423 - 9.

Thomas, M.B.; Wratten, S.D.; Sotherton, N.W. (1991) Creation of 'island' habitats in farmland to manipulate populations of beneficial arthropods: predator densities and emigration. *Journal of Applied Ecology*, **28**, 906 - 917.

ESTABLISHING VEGETATION STRIPS IN CONTRASTED EUROPEAN FARM SITUATIONS

E. J. P. MARSHALL, C.F.G. THOMAS
Department of Agricultural Sciences, University of Bristol, Institute of Arable Crops Research, Long Ashton Research Station, Bristol BS18 9AF, UK

W. JOENJE, D. KLEIJN
Section of Weed Science, Agricultural University, Bornsesteeg 69, 6708 PD Wageningen, The Netherlands

F. BUREL, D. LE COEUR
Laboratoire d'Évolution des Systèmes Naturels et Modifiés, Université de Rennes, Campus de Beaulieu I, 35042 Rennes Cedex, France

ABSTRACT

Replicated field experiments are being made in the UK, The Netherlands and France to assess the impact of introduced field margin strips and to investigate the dynamics of field edge ecotones. Comparisons are being made between cropping to the boundary and sown strips of *Lolium perenne*, unsown natural regeneration and sown grass and wildflower mixtures. This paper describes the results of initial studies on fauna and flora and the sampling protocols employed. Initial studies on the development of the vegetation show differences between crop and other plots, while the influence of the field weed seed bank is apparent in the first year.

INTRODUCTION

Field margin habitats form a network of semi-natural ecosystems in farm landscapes which interact with adjacent agriculture (Marshall, 1988; Joenje & Kleijn, 1994). Such semi-natural areas have important functions within landscapes (Burel & Baudry, 1990), including acting as corridors for certain species (Burel, 1989). There are also a number of roles for field boundaries for crop and environmental protection within more sustainable farming systems (Marshall, 1993). Maintenance or expansion of a diverse perennial ground flora at arable field edges has been suggested as an ecological approach (Marshall & Smith, 1987) for increasing on-farm biodiversity, for controlling annual weed species of hedgerows that may colonise adjacent crops (Marshall, 1989) and for enhancing populations of beneficial insects (Thomas *et al.*, 1992). Previous work on the introduction of sown grass and wildflower strips has demonstrated that in fertile arable soils, control of weed species in the first year may be required (Marshall & Nowakowski, 1991). However, with the exceptions of work by Smith & Macdonald (1989) and Marshall & Nowakowski (1991;1992), there has been little work published on the impact of introduced vegetation strips. Classical work on secondary succession (Bard, 1952; Falinska, 1991) would indicate that a perennial flora should develop on regenerating plots, borne out by studies by Smith *et al.* (1994). However, the effects of soil fertility, weed seed burden and management may influence the direction of the new communities. The approach of creating vegetation strips forms part of a European Commission-funded programme on the ecology of field boundary habitats (reference: AIR3-

CT 920476). This paper describes the establishment of replicate experiments on contrasted margin strips in the UK, The Netherlands and France.

METHODS

A protocol for designed experiments on margin strips was agreed, setting the minimum for valid comparisons between sites, but allowing each group to expand their assessments and treatments. Either two fields with three replicate blocks each or a single field with six replicates have been established (Table 1). The sites have been selected with the same exposition, roughly perpendicular to the prevailing winds (South West). The side with the North East aspect has the plots, each 3 or 4m wide and a minimum of 8m long, arranged along it. The three basic treatments are: 1) a control with the arable crop planted up to the existing boundary, 2) a planted strip of *Lolium perenne* and 3) a plot for natural regeneration. The adjacent crop is typical of the local crop rotation. After crop harvest, non-arable crop plots are mown. The arable plots are treated in the same way as the rest of crop, receiving fertiliser and pesticide applications. Records of farm operations, including machinery and fertiliser used, are kept.

TABLE 1. Field edge plot treatments established in spring 1993 in three European areas.

Site	Aspect	Boundary Type	Soil	Replication	Crop	Treatments
UK						
A	NE	Tall hedge	Clay	3	Spring barley	Crop; Ryegrass; Regeneration; Grass+Wildflowers
B	SW	Tall hedge	Clay	3	Spring barley	Crop; Ryegrass; Regeneration; Grass+Wildflowers
C	NE	Hedge	Sandy loam	3	Winter wheat	Crop; Ryegrass; Regeneration; Grass+Wildflowers
The Netherlands						
D	NE	Ditch	Sand	3	Spring wheat	Crop; Ryegrass; Regeneration; Wildflowers; Crop with no inputs
E	NE	Ditch	Sand	3	Winter wheat	Crop; Ryegrass; Regeneration; Wildflowers; Crop with no inputs
F	S	Unpaved road	Sand	3	Triticale	Crop; Ryegrass; Regeneration; Wildflowers; Crop with no inputs
G	N	Unpaved road	Sand	3	Spring barley	Crop; Ryegrass; Regeneration; Wildflowers; Crop with no inputs
France						
H/I	NE	Tall hedge	Clay + silt	3+3	Buckwheat	Crop; Ryegrass; Regeneration

The occurrence of the flora is recorded in 0.5m by 2.0m permanent quadrats, using a modified Braun-Blanquet score (0-9) for each species present in early and mid-season. The permanent quadrats are arranged in five positions across the boundary, plots and adjacent crop

(Figure 1). Position 1 is located within the pre-existing boundary; Position 2 and 3 are within the sown plot, 0.5m from the inner and outer edges. Position 4 is 0.5m into the crop from the plot, while Position 5 is 12m from the plot. Maximum standing crop is assessed in July/August depending on site, by cutting above-ground vegetation in 0.5m x 0.5m quadrats in the boundary, the plot and the crop. Grasses, dicotyledons, crop and grain are separated for fresh and dry weight (24h at 80°C) determination. Soil seed banks are being estimated in the arable and regeneration plots from 20 core samples (diameter 2.5cm) per plot taken from two depth profiles, 0-5cm and 5-20cm. Samples are bulked, mixed and 0.5kg soil subsamples washed over 0.25mm sieves before flotation in saturated calcium chloride and seed identification under the binocular microscope. Soil samples are taken at the same time, in order to characterise soil chemistry at the different sites. Fauna, particularly Carabidae and Staphylinidae, are being examined in pitfall trap catches from Position 1 in the boundary, mid-plot between 2 and 3, and Positions 4 and 5 in the crop. Traps are open for three days every three weeks during the summer.

Figure 1. Layout for a single plot with adjacent boundary and crop, showing the position of permanent quadrats and pitfall traps.

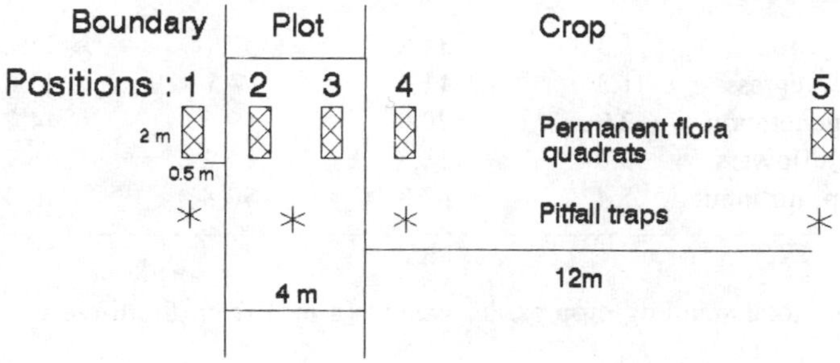

Boundary
Plot
Crop
Positions : 1 2 3 4 5
2 m
0.5 m
Permanent flora quadrats
Pitfall traps
4 m
12m

The data are analysed using analysis of variance and differences between means assessed using LSDs. Data are transformed using squareroot or logarithms, where the data are not normally distributed. Simple classification of species data from quadrats into similar groups was also made using the TWINSPAN program (Hill, 1979).

RESULTS

Flora

The botanical composition of the three UK sites was assessed in June 1993 and the data subjected to TWINSPAN analysis. The resulting classification gave five end-groups, the first of which indicated the pre-existing boundaries of the three UK sites were similar, with *Galium aparine* as the indicator species. Sites A and B were located within the same field, while site C had a different crop and weed flora. Thus the crop areas and crop plots were classified into the same group in sites A and B, but different to site C. The sown and natural regeneration plots were dominated by weed species in sites A and B, notably *Sinapis arvensis*. Other indicator species were *Anagallis arvensis, Polygonum aviculare, Ranunculus repens* and *Sonchus asper*. In site C, the indicator species for sown and natural regeneration plots were *Poa annua* and *Plantago major*, weeds of pastures, where there was also little differentiation between sown and natural regeneration plots.

A summary of percentage bare ground on UK and Dutch sites is given in Table 2. In the UK, after logarithm transformation of the data and analysis of variance, there were significant differences (P=0.05) between the three sites, with poorest establishment and most bare ground in site A. Within the UK sown and unsown plots, natural regeneration plots had the greatest area of bare ground, significantly so at position 3. In the Netherlands, the crop plots had the greatest area of bare ground, with ryegrass providing greatest ground cover and regeneration plots having intermediate values.

TABLE 2. Percentage bare ground in July on field edge strips established in spring 1993.

Country	Position: Sowing	1 Hedge	2 Plot	3 Plot	4 Crop edge	5 Crop
UK	Crop	4.3	20.9	5.9	4.4	11.6
	Ryegrass	4.6	15.4	1.7	23.2	4.7
	Regeneration	9.3	27.8	41.7	22.4	3.1
	Grass+flowers	11.1	8.1	8.6	19.9	4.3
Netherlands	Crop	1.3	35.8	37.1	34.2	29.2
	Ryegrass	1.3	11.3	7.5	43.8	30.8
	Regeneration	2.9	20.0	19.6	44.2	30.4
	Wildflowers	0.8	18.8	14.2	46.7	32.5
	Crop, no input	5.4	43.8	50.8	40.4	31.3

Estimates of total standing crop as dry weight (g m^{-2}) from the three centres are shown in Table 3.

TABLE 3. Mean total dry weights (g m^{-2}) of above-ground harvests from samples taken at positions 1, 2, 3 and 5.

Country	Position: Sowing	1 Hedge	2 Plot	3 Plot	5 Crop
UK	Crop	505	798	925	987
	Ryegrass	387	472	611	1030
	Regeneration	510	367	507	931
	Grass+flowers	396	461	429	888
Netherlands	Crop	388	822	886	851
	Ryegrass	438	336	397	821
	Regeneration	416	393	484	884
	Wild flowers	386	406	483	880
	Crop, no inputs	356	633	603	877
France	Crop	198	342	548	792
	Ryegrass	295	228	226	810
	Regeneration	362	259	214	793

Examination of the total dry weights of vegetation in the UK after squareroot transformation, showed that mean values at position 1 in the boundary and position 5 in the crop, were not significantly different between sowing treatments. At positions 2 and 3 (plot), the crop had significantly greater biomass than any other sowing. Ryegrass plots tended to have greater biomass than the remaining treatments. There was evidence of differences between fields in the UK, with lowest biomass in site A, where soils were particularly heavy and establishment poor. Similar effects were found in the Dutch sites (Table 3), with crop plots having higher biomass than the sown grass, wildflowers or natural regeneration. The crop plots with no inputs had lower biomass than those receiving agrochemicals. In France (Table 3), where the crop was buckwheat, the buckwheat plots adjacent to the hedgerow had considerably lower dry weights than 15 m into the field, possibly because of shading.

Fauna

Initial examination of pitfall catches has been limited to grouping animals into Coleoptera, Araneae and others. Data on the catches of Coleoptera over five trapping periods in the UK are given in Table 4. There was evidence that there was some differentiation between arable crop plots and other sown plots as the summer progressed, though the patterns were not clear.

TABLE 4. Mean total individual beetles, expressed as the logarithm of beetle number [log(n+1)] from two traps per trap site, on four different vegetation strips in the UK from five trapping periods. SE = standard error of treatment means.

Treatment	10/5/93	27/5/93	17/6/93	8/7/93	15/8/93
Crop	0.88	1.20	1.59	0.69	0.70
Ryegrass	0.93	0.82	1.23	0.57	0.74
Regeneration	0.80	1.30	1.55	0.63	0.71
Grass+flowers	0.94	1.15	1.40	0.92	0.93
S.E.	0.083	0.095	0.087	0.123	0.098

DISCUSSION

Early establishment varied between sites and within countries. These effects were probably a reflection of soil conditions at sowing, soil type, climate and seed banks. The developing flora in regeneration plots were often a dominant component of sown plots within the same field. In the Netherlands, the crop areas had least ground cover, probably because of the sandy soils. The data showed that the arable crops were able to achieve greater standing biomass, compared with the spring sowing of ryegrass and grass + wildflower mixtures in the first year. Competition from weed species can affect the establishment of sown field margin strips (Marshall & Nowakowski, 1991). The effects of dicotyledonous weeds, notably *Sinapis arvensis*, on early establishment of sown swards in the UK have yet to be analysed. However, there was little evidence of lasting effects, as evidenced by the floral composition or visual examination of the plots. Plots sown with *Lolium perenne*, in particular, developed high levels of ground cover over the first year. Natural regeneration plots were more open, and ground beetle numbers in traps were highest on these plots in May and June, reflecting greater activity. Further studies are in progress to follow successional development of the flora, vegetation structures and their associated fauna.

ACKNOWLEDGEMENTS

This research is funded by the Commission for the European Union within the Third Framework AIR Programme under contract AIR3-CT 920476/477. EJPM is also supported by the UK Agricultural & Food Research Council.

REFERENCES

Bard, G.E. (1952) Secondary succession on the Piedmont of New Jersey. *Ecological Monographs*, **22**, 195-215.

Burel, F. (1989) Landscape structure effects on carabid beetles spatial patterns in western France. *Landscape Ecology*, **2**, 215-226.

Burel, F.; & Baudry, J. (1990) Structural dynamics of hedgerow network landscapes in Brittany, France. *Landscape Ecology*, **4**, 197-210.

Falinska, K. (1991) *Plant demography in vegetation succession.* Tasks in Vegetation Science 26. Kluwer Academic Publishers, Dordrecht. 210 pp.

Hill, M.O. (1979) TWINSPAN - a FORTRAN program for arranging multivariate data in an ordered two-way table by classification of the individuals and attributes. New York: Section of Ecology and Systematics, Cornell University, Ithaca.

Joenje, W. & Kleijn, D. (1994) Plant distribution across arable field ecotones in the Netherlands. In: *Field Margins - Integrating Agriculture and Conservation. BCPC Monograph No. 58,* Thornton Heath: BCPC Publications, this volume.

Marshall, E.J.P. (1988) The ecology and management of field margin floras in England. *Outlook on Agriculture*, **17**, 178-182.

Marshall, E.J.P (1989) Distribution patterns of plants associated with arable field edges. *Journal of Applied Ecology*, **26**, 247-257.

Marshall, E.J.P. (1993) Exploiting semi-natural habitats as part of good agricultural practice. In: *Scientific basis for codes of good agricultural practice.* (ed. Jordan V.W.L.) *EUR 14957.* Commission of the European Communities, Luxembourg. pp. 95-100

Marshall, E.J.P. ; Nowakowski, M. (1991) The use of herbicides in the creation of a herb-rich field margin. *Brighton Crop Protection Conference - Weeds 1991*, **2**, 655-660.

Marshall, E.J.P.; Nowakowski, M. (1992) Herbicide and cutting treatments for establishment and management of diverse field margin strips. *Aspects of Applied Biology* **29**, *Vegetation management in forestry, amenity and conservation areas.* pp. 425-430.

Marshall, E.J.P.; Smith, B.D. (1987) Field margin fauna and flora; interaction with agriculture. In: *Field Margins* (eds) Way, J.M. & Greig-Smith, P. W. *BCPC Monograph No.35* , Thornton Heath: BCPC Publications, pp. 23-33.

Smith, H.; Macdonald, D.W. (1992) The impacts of mowing and sowing on weed populations and species richness in field margin set-aside. In: *Set-Aside, BCPC Monograph No. 50,* Thornton Heath: BCPC Publications, pp. 117-122.

Smith, H.; Feber, R.E.; Macdonald, D.W. (1994) The role of seed mixtures in field margin restoration. In: *Field Margins - Integrating Agriculture and Conservation. BCPC Monograph No. 58,* Thornton Heath: BCPC Publications, this volume.

Thomas, M.B.; Wratten, S.D.; Sotherton, N.W. (1992) Creation of 'island' habitats in farmland to manipulate populations of beneficial arthropods: predator densities and emigration. *Journal of Applied Ecology*, **28**, 906-917.

THE EFFECTS OF RESTORATION STRATEGIES ON THE FLORA AND FAUNA OF OVERGROWN HAWTHORN HEDGES AND METHODS OF REPAIRING GAPS IN OVERMANAGED HAWTHORN HEDGES

T. HENRY

Greenmount College of Agriculture and Horticulture, Antrim, BT41 4PU

A. C. BELL and J. H. McADAM

Department of Agriculture, Newforge Lane, Belfast, BT9 5PX

ABSTRACT

The effects of different restoration strategies on the wildlife value of overgrown hawthorn hedges revealed that plant species richness was higher for all treatments compared to the control, though only the coppiced treatment was significantly higher. Similarly, all treatments increased the number of invertebrate groups trapped, though only the lay treatment was significantly higher. Comparison of methods of repairing gaps in overmanaged hawthorn hedges indicated no requirement to replace the soil with fresh topsoil, though watering, if required, and good weed control were very important for successful plant establishment.

INTRODUCTION

Family farms and field sizes in N. Ireland are small (mean areas 25.4 ha and 1.8 ha respectively), livestock production being the predominant agricultural enterprise. This has resulted in a landscape characterised by 150,000 km of hedgerows (Department of Agriculture for Northern Ireland (DANI), 1992a). However, many hedges have been poorly managed over the years and have become gappy (Hegarty, 1992). Such hedges fail to provide stockproof barriers and are of limited conservation value.

Poorly managed hedges can be categorised as those suffering from neglect and which have grown out of control, and those which have been overmanaged by frequent cutting and are becoming exhausted (Osborn, 1987).

Provided the gaps are not too large, overgrown hedges can be restored by laying if the stems are not thicker than 50-100 mm at the base (Hellewell, 1991). Where the stem bases are thicker than 100 mm, coppicing is recommended. Cutting an overgrown gappy hedge to a standard 1.0 - 1.5 m height is not encouraged since the base will remain thin and not stockproof (Hellewell, 1991).

Routine annual cutting of hedges reduces their vigour and also causes gappiness. DANI (1992b) suggests that the gaps in such hedges may be difficult to repair because of 'thorn sickness' in the soil. It is suggested that, if planting with hawthorn, the soil in the gaps should be replaced with fresh topsoil or, alternatively, different species such as blackthorn or holly should be planted (Brooks, 1988). To date, no research has been conducted into methods of establishing shrubs in existing hedges (Osborn, 1987).

The objectives of this study were (1) to assess the effects of different restoration techniques on the flora and fauna of overgrown hawthorn hedges, and (2) to determine the best method of planting up gaps in overmanaged hawthorn hedges. This paper presents early results from a long-term experiment.

METHODS

The effects of restoration strategies on the flora and fauna of overgrown hawthorn hedges

In 1990 and 1991 a randomised block experiment was set up in 14 overgrown, gappy, predominantly hawthorn hedges at 10 sites throughout N. Ireland (Figure 1). Both sides of all hedges bordered permanent pasture. In each hedge (= block), 25 m lengths of the following treatments were imposed:

1. Unchanged control - **Control**
2. Laid - **Lay**
3. Cut to 1.5 m high with stem bases nicked to encourage sprouting - **Pollard**
4. Coppiced with gaps interplanted with the same species (hawthorn) - **Coppice-ISS**
5. Coppiced with gaps interplanted with different species (blackthorn, hawthorn, hazel, beech and holly) - **Coppice-IDS**

Hedges were fenced on both sides to exclude stock and mechanical trimming planned for every third year. Those sites established in 1990 were trimmed in February 1993.

A list of all plant species occurring within each treatment plot (extending 1 m out on both sides of the hedge) was made during the summer of 1991 and 1992. Plots represented 20 m lengths of hedge, i.e. terminal 2.5 m sections were omitted. Fauna were monitored using five shelter traps (20 cm long x 5 cm diameter open-ended plastic cylinders filled with rolled corrugated cardboard) in each treatment plot, placed in the hedgerow canopy for 28 days in May 1992 and May 1993. Sample catches were identified to the main invertebrate groups, namely Arachnida, Crustacea, Insecta (Coleoptera, Collembola, Dermaptera, Diptera, Hemiptera, Hymenoptera, and Lepidoptera) and Myriapoda.

Repairing gaps in an overmanaged hawthorn hedge

An overmanaged hawthorn hedge was removed by coppicing and this revealed that many of the stumps were rotten and would not have resprouted. A randomised block experiment involving four treatments x two replicates (8 plots each 8 m in length) was imposed on the hedge in February/March 1992. Where feasible, two apparently healthy coppiced stumps were left untouched within each plot.

Hawthorn quicks (3 yr old nursery plants) were planted in all treatments which were:

1. Quicks planted in the existing soil - **Control**
2. Quicks planted in the existing soil and watered during the 1992 growing season - **Watered**
3. 500 g of organic composted farmyard manure (Dungstead, Abbey Organics, Portglenone) incorporated into the existing soil around the roots of each quick - **Fertilised**
4. Existing soil removed and quicks planted into fresh topsoil obtained from an adjacent field (trench 600 mm wide and 450 mm deep) - **Replace soil**.

A double staggered row method of planting was used, with 250 mm between plants and 300 mm between rows. Weed control was achieved using glyphosate (Roundup), propyzamide (Kerb granules), and hand weeding. Chemicals were used in accordance with the manufacturers' recommendations.

Stem diameter (mm) at 250 mm above ground and height (cm) were recorded for all quicks in May 1992 and July 1993. Stem diameter was measured using digital calipers and calculated as the mean of two measurements taken at right angles to each other. Survival of quicks was assessed in July 1993.

Data analysis

Data from both experiments were analysed using randomised blocks ANOVA. Where differences among means were significant, these were compared using Least Significant Range (LSR = $Q_{0.05}$ x SE mean) (Parker, 1979). Plant species composition data for the hedges in the restoration experiment were analysed using TWINSPAN (Hill, 1979a) and DECORANA (Hill, 1979b).

RESULTS

The effects of restoration strategies on the flora and fauna of overgrown hawthorn hedges

Ordination of floristic data by DECORANA with TWINSPAN groups superimposed, generally discriminated among sites, but not among treatments (Figure 1). However, in both 1991 and 1992, all restoration treatments resulted in an increase in the mean number of plant species, though only the coppiced/interplanted with hawthorn treatment had significantly ($P<0.01$, $F_{4,52} = 3.82$) more species than the control (Figure 2, 1992 data). There were significantly ($P<0.05$, $F_{4,52} = 2.97$) more invertebrate groups associated with the lay treatment than with pollard in 1993 (Figure 3). The lay and two coppice treatments had greater numbers of invertebrate groups than the control, though the differences were not significant.

Figure 1. TWINSPAN groupings superimposed on a DECORANA plot of hedgerow flora

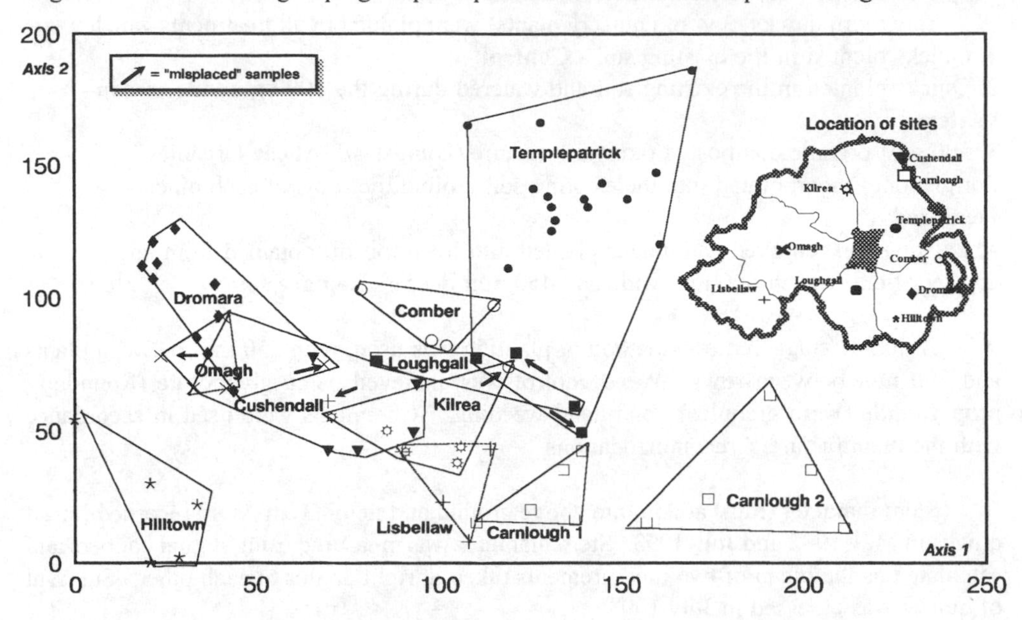

Figure 2. Plant species richness, 1992 *(Least significant ranges shown, P<0.05)*

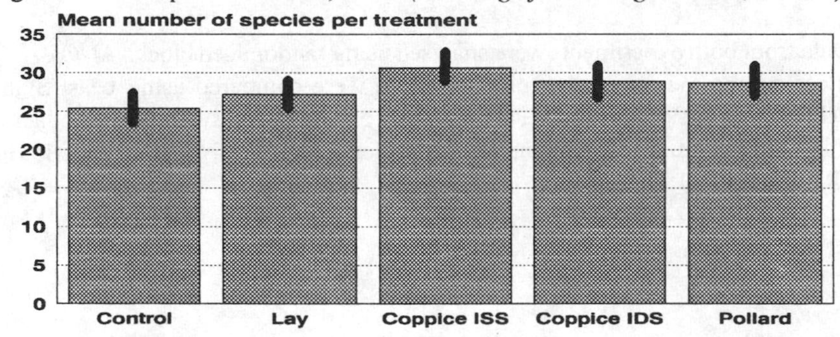

Figure 3. Invertebrate groups, 1993 *(Least significant ranges shown, P<0.05)*

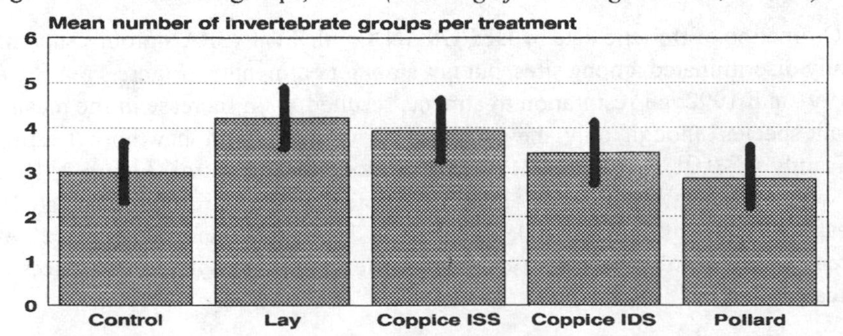

Repairing an overmanaged hawthorn hedge

There were no significant differences in the increases in either height or stem diameter at 250 mm among the four gapping up treatments (Table 1).

TABLE 1. Mean increase in height (cm) and stem diameter at 250 mm (mm) between May 1992 and July 1993 for four treatments (40 - 60 hawthorn quicks per block, N=2)

Treatment	Control	Watered	Fertilised	Replace soil	SE mean
Height (cm)	24.05	23.60	29.50	29.15	3.02
Stem diam. (mm)	3.34	3.38	3.68	3.87	0.36

However, the watered treatment displayed significantly higher ($P<0.05$) quick survival than the control. The three months immediately post planting were much drier than the previous ten year mean rainfall. Fertilised and replace soil treatments showed higher quick survival than the control, but the difference was not significant.

DISCUSSION

Positive management is urgently required to restore the quality of Northern Ireland's hedges, most of which were planted between 1700 and 1850. DANI offers 80% grants to farmers for hedge restoration in the Environmentally Sensitive Areas (ESAs). Grant aid under the 'Hedgerow Incentive Scheme' (Countryside Commission, 1992) is available to English farmers for laying and coppicing hedges. However, to date, no information has been available on the effect of different restoration techniques on the wildlife value of hedges or on methods of repairing gaps.

This study has shown that the highest plant species richness occurred in the coppiced treatments, probably as a result of increased light penetration encouraging new species establishment. At some lowland sites, this has resulted in excessive weed growth which may be detrimental to hedge development and maintenance of forbs in the long term. All restoration treatments increased numbers of plant species compared to the control. The highest number of invertebrate groups was found in the lay treatment, presumably due to the greater density of the hedgerow canopy. One limitation of a sampling method such as shelter trapping is that it will inevitably be selective for particular groups. There may be occasional occurrences of groups not readily taken by this method, but the degree of replication in this experiment would mitigate against this source of error.

These differences among treatments were interesting at this early stage of a long term trial. It is likely that as the canopy develops in the coppiced treatments, these will become a better habitat for invertebrates.

Osborn (1987) requested a reliable method for establishing shrubs in gaps in overmanaged hedges. Despite advice to replace the soil with fresh topsoil (Brooks, 1988; DANI, 1992b), our results reveal no benefit in terms of hawthorn growth from replacing or fertilising the soil. The most important factor affecting survival was soil moisture. Watered quicks displayed significantly higher survival compared to the control. However, these results should be interpreted with caution as they were recorded over a 16 month period from a newly-planted hedge.

The results of this study indicate the dynamic nature of hedgerow ecology and suggest that a combination of coppicing and laying, combined with interplanting where appropriate, are the most beneficial strategies in achieving conservation objectives.

ACKNOWLEDGEMENTS

The authors thank Jenny Thompson, Jill Forsythe, John Anderson, and Roger Martin for assistance with sampling and recording, and Eugene McBride and staff of the Experimental Farm at Greenmount College for carrying out the restoration work on hedges.

REFERENCES

Brooks, A. (1988) *Hedging: a practical conservation handbook.* Wallingford: British Trust for Conservation Volunteers, p. 52.

Countryside Commission (1992) *Handbook for the Hedgerow Incentive Scheme.* Cheltenham: Countryside Commission, 8 pp.

DANI (1992a) *Field boundaries in the landscape*. Belfast: HMSO, 11 pp.

DANI (1992b) *Managing gappy and overgrown hedges*. Belfast: HMSO, 10 pp.

Hegarty, C.A. (1992) *The ecology and management of hedges in Northern Ireland.* Unpublished PhD thesis, University of Ulster, Coleraine, 260 pp.

Hellewell, M. (1991) Managing farm hedges. In: *Farm Woodland Management*, J. Blyth, J. Evans, W.E.S. Mutch and C. Sidwell, (Eds) Ipswich: Farming Press Books, pp. 45-47.

Hill, M.O. (1979a) *TWINSPAN: a FORTRAN programme for arranging multivariate data in an ordered two-way table by classification of the individuals and attributes*. New York: Cornell University, 80 pp.

Hill, M.O. (1979b) *DECORANA: a FORTRAN programme for detrended correspondence analysis and reciprocal averaging*. New York: Cornell University, 52 pp.

Osborn, A. (1987) Management for conservation of wildlife - FWAG experiences. In *Field Margins*, J.M. Way and P.W.Greig-Smith (Eds), BCPC Monograph No. **35**, Thornton Heath: BCPC Publications, pp. 105-108.

Parker, R. E. (1979) *Introductory statistics for biology.* Studies in Biology, No 43, 2nd edition, London: Edward Arnold, p. 72.

PROBLEMS WITH THE RE-ESTABLISHMENT OF THORN HEDGES IN THE CAMBRIAN MOUNTAINS ENVIRONMENTALLY SENSITIVE AREA

JOHN WILDIG, BERNARD GRIFFITHS

ADAS Pwllpeiran, Cwmystwyth, Aberystwyth, Dyfed SY23 4AB

TIM MILSOM

Central Science Laboratory, Ministry of Agriculture, Fisheries & Food,
Tangley Place, Worplesdon, Surrey, GU3 3LQ

ABSTRACT

Hedges are an important feature of middle hill land within the Cambrian Mountains Environmentally Sensitive Area but many are in poor condition through lack of management. Re-establishment of Hawthorn (*Crataegus monogyna*) hedges by planting is hindered by grazing animals. Double fencing of hedges will relieve grazing pressure but this often results in competition between ground vegetation and young hawthorns, to the detriment of the latter. Weed control should alleviate the competition but methods need to be effective under conditions of high exposure and high rainfall. A field experiment was carried out at ADAS Pwllpeiran to evaluate the effects of three types of weed control (glyphosate, propyzamide, mulching) on survival and growth of hawthorn. The treatments plus a control were laid out along a newly planted hawthorn hedge, at 290 metres altitude, in a randomised block design. Mulching slightly enhanced the growth of young hawthorn plants whereas propyzamide had no discernible effect. Glyphosate had an adverse effect on growth and survival of hawthorn.

INTRODUCTION

Hedgerows are an important feature of the upland landscape in the Cambrian Mountains of central Wales and have historical interest. Many date back to the period of enclosure of middle hill-land or Ffridd in the 19th century (Davies and Davis, 1973), when labour was readily available to carry out planting and management. However, the condition of these hedges deteriorated during the 20th century through lack of management which, in turn, was the result of the vicissitudes of the hill farming industry and changes in land use from sheep walk to forestry (Parry and Sinclair, 1985). During the 1960s and 70s, hill farming underwent an unprecedented expansion but this led to further deterioration of hedges because stocking densities on middle hill-land were increased to very high levels with the result that the remnants of hedges were browsed out. Moreover, the decline in farm labour and the economics of hill farming made fencing a more attractive option to hedge management.

In recent years, there has been a renewal of interest in hedges on farms in the Cambrian Mountains Environmentally Sensitive Area (ESA) partly because of their shelter value for stock but also because of their importance as landscape features. These concerns are reflected in the management prescriptions within the ESA (Anon, 1989). However, the condition of many hedges is now so poor that replanting is required to restore them.

Re-establishment of hawthorn (*Crataegus monogyna*) hedges by planting is hindered by grazing animals. Though this problem can be remedied by double fencing, the reduction of grazing pressure within fenced enclosures may often result in intense competition between existing ground vegetation and hawthorns, to the detriment of the latter. The objective of the experiment was, therefore, to determine whether the survival and growth of young hawthorn plants could be enhanced by weed control on an exposed site with high rainfall. Three weed control strategies, two chemical and one physical, were compared.

METHODS

The experiment was carried out at ADAS Pwllpeiran, which is situated within the Cambrian Mountains ESA.

A post and wire fence was erected parallel to, and 2m apart from, an existing fence along the margin of a pasture, at about 290m altitude. In January 1991, a double row of hawthorns with 25cm spacings between plants and rows was planted down the middle of the enclosure between the two fences. The hawthorn plants were obtained from a local supplier but were not a native variety. Each hawthorn was cut back to 5cm in height to mitigate the effects of die-back due to exposure to wind.

The hedge enclosure was split into 12 plots, each of which was 10 metres long and contained 80 hawthorn plants. Four treatments were applied to the plots in a randomised block design with three replicates. The treatments were as follows: (i) a contact broad-spectrum herbicide (glyphosate) applied in May 1991 in relatively calm conditions at 1.4kg/ha AI, using a knapsack sprayer fitted with a shielded spray-head; (ii) a residual broad-spectrum herbicide (propyzamide) applied in December 1991, using a granular formulation at a rate of 1.52 kg/ha AI; (iii) a mulch mat of 400 gauge black polythene, laid down at the time of planting; (iv) a control, where no vegetation management was carried out.

Twenty hawthorn plants from each plot were marked at the start of the experiment for repeated growth measurements. Growth was assessed in June 1992 and again in November 1993 from measurements of the height of the plant and diameter of the main stem. Mean values were taken for each plot and used as dependent variables when assessing treatment effects by analysis of variance. The variance of the means was checked for homogeneity across treatments. Growth increments for each plot were calculated by subtracting the height (or stem diameter) for each marked plant on a given sampling date from that on the next sampling date and then taking the mean of the differences. Ground vegetation cover in each plot was estimated visually to the nearest 5%.

RESULTS

Nearly half of the young hawthorns (47%) planted in January 1991 had died by June 1992, when the first growth assessments were made. However, the percentage surviving in each plot varied from 11% to 81% and differed significantly between treatments (Table 1).

The hawthorns grew from a uniform 5cm in January 1991 to 41 cm, on average, by June 1992, although the height increments varied from 12cm to 54cm between plots. Their height in June 1992 was not correlated with ground vegetation cover (Kendall rank correlation coefficient = 0.0177, n=12, P=0.47) but it did differ significantly between treatments (Table 2).

TABLE 1 Survival of hawthorn plants between January 1991 and June 1992, expressed as percentage surviving per plot, in relation to weed control treatment

Treatment	Mean of plot means	Standard Error	n
Propyzamide	69.6	2.3	3
Glyphosate	16.3	3.8	3
Polythene mulch	75.0	3.6	3
Control	52.1	9.1	3

ANOVA $F_{3,8} = 24.15$, $P<0.001$

TABLE 2 Height of hawthornplants in June 1992, expressed as mean height per plot (cm), in relation to weed control treatment

Treatment	Mean of plot means	Standard Error	n
Propyzamide	44.4	3.2	3
Glyphosate	21.7	2.6	3
Polythene mulch	53.7	2.5	3
Control	45.3	0.9	3

ANOVA $F_{3,8} = 30.83$, $P<0.001$

By November 1993, the average height of the hawthorns was 63cm but the growth increments between June 1992 and November 1993 varied from 2.5cm to 34cm between plots and differed significantly between treatments (Table 3). The rate of increase in stem diameter over the same period also differed significantly between treatments (Table 3).

TABLE 3 Increases in height[1] and stem diameter[2] of hawthorn plants between June 1992 and November 1993, expressed as mean difference per plot, in relation to weed control treatment

Treatment	Height (cm) Mean & standard error of plot means		Stem diameter (cm) Mean & standard error of plot means		n
Propyzamide	22.9	2.5	0.43	0.041	3
Glyphosate	8.5	3.0	0.13	0.055	3
Polythene mulch	29.8	2.2	0.60	0.003	3
Control	24.5	0.4	0.34	0.053	3

1.ANOVA $F_{3,8} = 16.51$, $P<0.001$. 2. ANOVA $F_{3,8} = 19.96$, $P<0.001$.

The treatment effects were explored further by making *a posteriori* comparisons of the treatment means using Tukey's test. On average, hawthorn survival in the glyphosate plots was lower than that in the others (Table 1). Growth between 1991 and 1992 in the glyphosate plots was also slower (Table 2) and this effect carried over into 1993 (Table 3). There were no differences between the other treatments and the control except in 1993 when stem diameters of plants in the mulched plots had increased by a greater amount than those in the control.

DISCUSSION

These findings imply that, at the level of replication used in the experiment, mulching at the time of planting produced a slightly greater increase in growth of young hawthorn plants but that propyzamide, applied 10 months after planting, did not have a detectible effect. Glyphosate appeared to have an adverse effect on growth and survival and it is known that broad-leaved trees and shrubs can be particularly susceptible to this herbicide in summer (Robinson 1985). Even though a shield had been fitted to the spray head, and spraying was done on a day with little wind, it is thought that droplets of glyphosate may have been deflected by the luxuriant ground vegetation onto the young hawthorn plants with the result that some were killed outright, while the growth of others was retarded. However, recent experiments with tree seedlings suggest that alternative contact herbicides, particularly haloxyfop, may be less damaging (Clay *et al.,* 1992). Fluazifop-P-butyl, which has been used for the selective control of grasses in field margins (Marshall and Nowakowski, 1992), may also be appropriate.

The data suggested that mulching enhanced growth between 1992-93 and this method may be a viable alternative to herbicides, particularly in exposed sites where the combination of high rainfall and high winds make spraying impracticable. In this experiment, polythene silage pit sheets were cut up for mulch mats mainly because they were cheap, available on the farm and appeared to be suitable for the task. However, other materials which are readily available within the ESA, such as sheeps' wool, might be equally effective and should be evaluated.

ACKNOWLEDGEMENTS

The work was funded by the Land Use Conservation and Countryside policy group of the Ministry of Agriculture, Fisheries & Food. Steve Langton (CSL) gave statistical advice. Chris Feare, Roger Trout (CSL) and Carey Coombs (ADAS) co-ordinated the project. We thank them all.

REFERENCES

Anon. (1989). *Environmentally Sensitive Areas: Wales.* First report on monitoring the effects of the designation of environmentally sensitive areas. Cardiff: Welsh Office Agriculture Department,

Clay, D.V.; Goodall, J.S.; Williamson, D.R. (1992). Evaluation of post-emergence herbicides for forestry seedbeds. *Aspects of Applied Biology*, **29**, 139-148.

Davies, P.W.; Davis, P.E. (1973). The ecology and conservation of the Red Kite in Wales. *British Birds*, **66**, 183-224.

Marshall, E.J.P.; Nowakowski, M. (1992). Herbicide and cutting treatments for establishment and management of diverse Field margin strips. *Aspects of Applied Biology*, **29**, 425-430.

Parry, M.L.; Sinclair, G. (1985). *Mid Wales Uplands study.* Cheltenham: Countryside Commission.

Robinson, D. (1985). Efficacy of glyphosate in nursery stock and amenity horticulture. In: *The Herbicide Glyphosate* E. Grossbard, and D.J. Atkinson, (Eds)., pp. 339-354. London: Butterworths.

THE RE-INTRODUCTION OF NATIVE AQUATIC PLANTS INTO FENLAND DITCHES

S. J. TOWN, S. R. RUNHAM

ADAS Arthur Rickwood Research Centre, Mepal, Ely, Cambridgeshire, CB6 2BA

ABSTRACT

Drainage ditches are a necessary part of the fenland farming scene, forming the commonest type of field margin. They can provide a valuable habitat for flora and fauna. When maintained routinely by dredging (slubbing) or complete clearance, it can take many years for a ditch to regain a diverse range of flora. A trial was set up at ADAS Arthur Rickwood Research Centre, in Mepal Fen, to investigate the suitability of artificially introducing a range of native, aquatic plants to a depleted ditch. The small-scale replicated trial was implemented from 1989 to 1993. Six plant species (*Alysma plantago-aquatica, Butomus umbellatus, Caltha palustris, Iris pseudacorus, Nymphoides peltata* and *Sagittaria sagittifolia*), which were either already occurring locally or in the nearby Ouse washes, were chosen and planted into a typical 'slubbed' fenland ditch. Species were assessed twice yearly (in early summer and autumn) for survival, proliferation and colonisation and also for their ability to compete with other vigorous species which colonised naturally. Results suggested that the *Iris pseudacorus* and *Sagittaria sagittifolia* (the two most prolific), together with the *Alysma plantago-aquatica* and *Butomus umbellatus,* could be successfully introduced into a cleared or depleted ditch; whilst the *Nymphoides peltata* and majority of *Caltha palustris* plants gradually disappeared.

INTRODUCTION

Drainage ditches are vital to the well-being of fenland farming and have to be dredged or cleared routinely to ensure the free drainage of the land. They also provide an important semi-natural 'corridor' for wildlife in an area of intensive arable farming, often with few other suitable habitats (Milsom *et al.*, in press). Following a maintenance operation, it can take a considerable time for a ditch to regain a diversity of plant species; meanwhile, the fauna associated with a particular species can disappear.

The ditches can be more rapidly restored and ecologically enhanced by introducing a wide variety of suitable native plants. Planting can be used to increase species diversity in comparison with natural colonisation. It can also be used to favour dominance of particular species, limiting opportunities for undesirable species to colonise the whole ditch.

Drainage operations in recent years, have been so extreme that wetland habitats are now much reduced. Consequently, many aquatic species have declined drastically (Evans, 1991).

Conservation of some plants and associated animal species could be encouraged by transplanting. Taller species intercept blown leaves before they reach the water's surface and also catch and store seeds to provide winter food resources, together with nesting habitat, for birds, including wildfowl (Brooks, 1981). Planting can also protect the ditch banks from erosion. Landscape architects are becoming more aware of the need to include aquatic plants as an integral component of river improvement schemes (Spencer-Jones & Wade, 1986).

METHODS

The ditch selected was a typical fenland drainage ditch, 6 m wide, with a mean water depth of 0.8 m, situated at ADAS Arthur Rickwood, Mepal, Cambridgeshire. The adjacent fields consisted of a loamy peat up to 0.9 m deep with a 30% organic matter content over fen clay of the Adventurer's Shallow series. There was a willow hedge on the north-west side which was removed in spring 1992 and replaced by a 2 m margin of uncropped land. A margin of at least 2 m was maintained, by regular rotavation, on the cropped south-east side of the ditch throughout the experiment. The bank tops were trimmed in spring 1992 but the remainder of the ditch was left uncut throughout the experimental period. The water was maintained at a fairly constant depth in spite of drought affecting nearby ditches during the summer of 1991.

The ditch was mechanically cleared of plant debris and silt in 1988 and a 22 m section divided into 1 m wide plots. The experimental design consisted of a randomised complete block with three replicates. There was a 1 m wide guard at each end and also between each replicate. The six species selected for re-introduction (Table 1) occurred naturally either in nearby ditches or in the nearby Ouse Washes. They were obtained as pot-grown transplants and two plants were planted into the ditch margin, one on opposite sides of each plot, at their preferred depth (see Table 1) on 7 November 1989.

The establishment of each species was recorded, followed by twice yearly assessments to determine growth rate. This was done using a scoring system of 0-3 (where 0 = dead or disappeared, 1 = alive, 2 = growing well and spreading within the plot, 3 = growing very well and spreading outside the plot). The final growth rate score data, assessed on 22 September 1992, was subjected to analyses of variance. The pH of the water in the ditch was recorded once each year. Weather records were taken at a meteorological site 1 km from the experimental ditch site.

RESULTS AND DISCUSSION

All plants, except two of the original six fringed water lilies, survived transplantation and the winter of 1989/90 (temperatures down to -10.0°C) and were growing vigorously on 10 July 1990. *I. pseudacorus* was the most prolific, having begun to spread outside some of the plots after two years and outside all of the plots by the third year (Figure 1). *S. sagittifolia* also spread outside the plots by June 1992. *A. plantago-aquatica* colonised well within their plots and occasionally spread outside, but one disappeared, probably due to a substantial colonisation of the same plot by *T. latifolia*. *B. umbellatus* also colonised their plots well and occasionally spread to adjacent plots.

TABLE 1. Aquatic plant species chosen for re-introduction.

Species	Common English name	Growth habit	Habitat requirement *(Nutrient requirement)	Value to wildlife
Alysma plantago-aquatica	Water-plantain	Submerged emergent (floating)	Shallow margins up to 0.75 m deep. (Eut)	Shelter and perches for invertebrates e.g. dragonflies.
Butomus umbellatus	Flowering rush	Emergent (submerged) (floating)	Shallow margins on clay 0.5-1.5 m. (Eut to mes)	Nesting by wildfowl.
Caltha palustris	Marsh marigold	Emergent	Margins, away from fast flow. (Eut to oli)	Shelter and perches for insects.
Iris pseudacorus	Yellow flag iris	Emergent	Shallow margins at or above water level. (Variable nutrient)	Shelter and perches for invertebrates. Limited for nesting.
Nymphoides peltata	Fringed water lily	Floating submerged	Still to slow water 0.5-3.0 m deep, neutral or alkaline. (Eut)	Shelter and perches for fish and invertebrates, attractive to wildfowl.
Sagittaria sagittifolia	Arrowhead	Submerged floating emergent	Shallow to moderately deep 0.1-1.0 m. Tolerant of low pollution level only. (Eut-mes)	Harbours invertebrates, vegetation eaten by wildfowl.

* Key to Nutrient Requirement
eut - eutrophic, requires an ample supply of nutrients, especially nitrate and/or phosphate.
mes - mesotrophic, requires a moderate supply of nutrients.
oli - oligotrophic, requires a low supply of nutrients. (Haslam *et al.*, 1982; Newbold *et al.*, 1989).

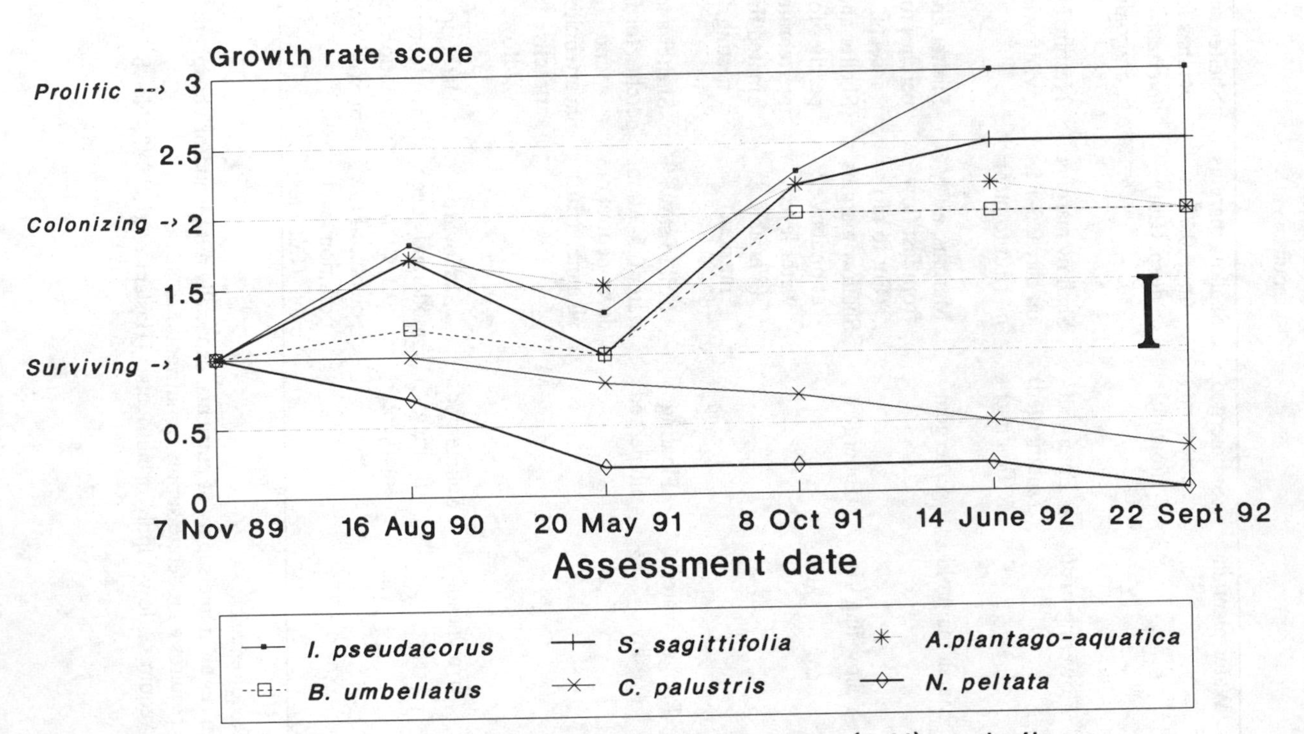

N.B. *Vertical bar represents SE mean (6df) excluding I. pseudacorus and N. peltata.*

Figure 1. Mean score of growth rate of species of aquatic plants introduced into a fenland ditch over a period of three years.

C. palustris and *N. peltata* did not perform well after transplantation. *C. palustris* gradually declined over the three year period until only two plants remained in September 1992. *C. palustris* appeared to have suffered from competition with vigorous species which naturally occurred in the ditch section. These were primarily *J. effusus, T. latifolia* and *A. plantago-aquatica*, all of which had colonised the *C. palustris* plots by September 1992. *C. palustris* may also have been adversely affected by the acidity of the water, the pH of which ranged between 4.0 and 5.0 during the course of the experiment, as they tend to occur more frequently in slightly alkaline water.

N. peltata lacked vigour following transplantation and had disappeared by September 1992 (Figure 1). This species prefers neutral or alkaline water and would have suffered from the levels of acidity recorded. The acidity in this ditch can be related to the fen clay in adjacent soils which itself can have very low pH levels (below 3.0). Some fenland ditches border peat soils over sand and gravel, and have higher pH levels.

Vigorous growth by some dominant competitive species may have been encouraged by nutrients contained in the water. These were recorded as between zero and 17 mg/l for nitrate-N and zero and 7 mg/l for phosphate-P, in a related experiment on a nearby site. It is not known to what extent the growth of each species was influenced by these, or other available nutrients.

The plants would possibly have been more prolific if planted in blocks or clumps so that species which might not survive mixed planting can establish themselves before coming into competition with more vigorous species. This may apply particularly to rhizomatous species, which naturally interlock to form a tough blanket which outcompetes other species (Brooks, 1981). This was possibly one of the contributing factors towards the success of *I. pseudacorus.*

At the end of the experiment, the floristic and amenity value of the ditch had been enhanced by the survival of introduced species which apart from *A. plantago-aquatica,* did not occur naturally in other parts of the ditch following slubbing in 1988.

It was not possible to analyse the data from all six species due to extremes of *I. pseudacorus* and *N. peltata*, which data were not normally distributed, even following transformation. Therefore, the remaining four species only were subjected to analyses of variance. There were no significant differences between *A. plantago-aquatica, B. umbellatus* and *S. sagittifolia*, which all proliferated equally well. There was a significant difference between each of the above three species and *C. palustris*, which colonised relatively poorly.

CONCLUSIONS

The results suggest that *I. pseudacorus, S. sagittifolia, A. plantago-aquatica* and *B. umbellatus* could all be successfully introduced into a newly cleared or species poor ditch of similar condition. The successful establishment of these species provided both a useful habitat for invertebrates such as dragonflies, and shelter and nesting sites for wildfowl and other birds. It would be advantageous to expand the experiment using a range of fenland ditches with a diversity of available nutrients and levels of pH, nutrients and pollution.

ACKNOWLEDGEMENTS

The plants were kindly donated by London Aquatics Ltd. We are grateful to Kevin Hand and John Crowe for their preliminary work on this experiment.

REFERENCES

Brooks, A. (1981) *Waterways And Wetlands Practical Conservation Handbook*, Wallingford, Oxford: BTCV Publishers, pp. 142-144.

Evans, C. (1991) The conservation and management of the ditch flora on RSPB reserves. *RSPB Conservation Review*, 5, pp. 65-71.

Haslam, S.; Sinker, C.; Wolseley, P. (1982) *British Water Plants*, J H Crothers (Ed), Taunton, Somerset: FSC Publications, pp. 295-316.

Milsom, T. P.; Runham, S. R.; Sherwood, A. J. M., Rose, S. C.; Town, S. J. (in press) Management of fenland ditches for farming and wildlife: an experimental approach. *Icole Conference, Nature Conservation and the Management of Drainage System Habitat*, 23-25 September, 1993

Newbold, C.; Honnor, J.; Buckley, K. (1989) Nature conservation and the management of drainage channels. Peterborough: Nature Conservancy Council in conjunction with the Association of Drainage Authorities, pp. 13-18.

Spencer-Jones, D.; Wade, M. (1986) Aquatic plants - a guide to recognition, Romsey, Hants: Borcombe Printer, pp. 6-7.

Session 5
Political Aspects

Session Organiser
& Chairman DR S WEBSTER

AN OVERVIEW OF UK POLICY AND GRANT SCHEMES

DR JANET DWYER

Countryside Commission, John Dower House, Cheltenham, Glos. GL50 3RA

ABSTRACT
There has been a recent coming-together of agriculture and countryside policy, and a search for new policies to encourage countryside conservation alongside viable farming, using regulations, incentives and cross-compliance. Environmental payment schemes have been the most popular of these. Beginning with SSSI management agreements, we have developed many schemes. ESAs now cover 15% of UK farm land, and embrace landscape, wildlife and historic objectives. Newer schemes include nationally-available ones such as Countryside Stewardship in England; and Tir Cymen and the Hedgerow Restoration Scheme in Wales. Originally, schemes often targeted marginal areas, or land which was not farmed. Today, schemes increasingly work hand in hand with productive farming. Government, on behalf of the public, can use schemes to buy environmental goods from farmers in the same way that others buy crops and livestock. In future, the UK could benefit from better integration of schemes, and a further increase in expenditure. But schemes alone cannot integrate agriculture and conservation. A more radical reform of the complexities of the CAP, including set-aside and livestock regimes, is also vital. Farmers, conservationists and politicians must work together to build a balanced policy framework for the future.

INTRODUCTION

Before I start, I should like to thank the conference organisers for inviting me to give this paper. As an officer in the Countryside Commission, whose remit covers the English countryside, and not Scotland, Wales or Northern Ireland, the title at first gave me some concern. However, my thanks to contacts in partner organisations for helping me to broaden my paper to deal with the UK as a whole. I should also say that it gives my own personal view, and not necessarily the Countryside Commission's official policy.

CHANGING PRIORITIES

UK policy for agriculture and the countryside has reached a new stage of development. From a post-war period when agriculture was seen as a benign influence upon the countryside, and during which time farming underwent major changes in pursuit of increased 'food from our own resources', the 1970s and 80s saw a growing appreciation that in many ways, farming and conservation policies had been in conflict with one another.

The policy drive to produce more food from the land had led many farmers to change their farming patterns and practices in ways which had reduced the diversity of our landscapes and wildlife. By encouraging greater food production, alongside reductions in farm labour, these changes had made it difficult for farmers to devote time and effort to take care of the many beautiful and historic features of the countryside which had ceased to have economic value. At the same time, there was a growing awareness of the need to conserve the countryside, and to encourage people in our burgeoning towns, cities and suburbs to appreciate it and use it responsibly. We have all learned much from this legacy of change and conflict.

But in the past ten to fifteen years there has been a general recognition of the importance of helping farmers to re-invest time and effort in their role as 'stewards of the countryside'. This has, I believe, led to a gradual coming-together of agriculture and conservation policies, increasingly building links between the production of food and fibre, and the generation of a beautiful, diverse and enjoyable countryside at the same time.

The government has been seeking new policies which can encourage both the conservation and enhancement of the countryside **alongside** productive farming: if you like, putting together some building blocks for a new, more multi-purpose strategy for rural areas (- although I don't think you could claim that we have anything like a coherent strategy at the moment!).

In preparing the Countryside Commission's recent policy statement, 'Paying for a Beautiful Countryside', I spent a lot of time reviewing the various new policies. I feel that these policies now have real potential to help transform the relationship between modern farming and the countryside, for the better. They should help to create a future policy framework which encourages farmers to farm responsibly and for the benefit of society as a whole, through a mixture of mechanisms. These are:

> **regulation**, which should set those basic environmental and countryside standards that we all expect from responsible and safe farming;
>
> **incentive**, providing appropriate support for the creation of new or better countryside benefits, for example through payment schemes; and
>
> **cross-compliance** (that rather clumsy word) or more accurately, **adjusting other policies**, to make sure that other support for farming or rural development does not compromise the aims of these first two mechanisms.

PAYMENT SCHEMES

Regulations have certainly been a major preoccupation for farmers over the last year or so. But among the new 'agriculture and countryside' policies, I would suggest that countryside payment schemes have so far been the most popular mechanism: schemes which pay farmers for making positive contributions to the countryside, managing important areas and features under threat, or upgrading the quality of damaged or neglected areas.

The first countryside payments, made by the Nature Conservancy Council on Sites of Special Scientific Interest (SSSIs), offered certain farmers management agreements to secure sensitive management of land which was of special value for wildlife. Some agreements have been in existence for many years, but the system expanded after a new payment formula was introduced in the 1981 Wildlife and Countryside Act. SSSI agreements have been limited only to 'special sites', and each agreement has been tailor-made to the needs of the site. Although these features make it a relatively costly system, SSSI management agreements now cover over 180,000 ha of 'special' land in the UK.

Through the late '80s and into the 1990s, there have been a growing number of schemes and grants which encourage farmers to produce a wider variety of 'countryside products'. The Ministry of Agriculture and its counterparts in Wales, Scotland and Northern Ireland have developed an impressive fleet of Environmentally Sensitive Areas, schemes which will soon cover 15% of the UK's farmland. ESAs are designed with the help of farming and conservation interests, to preserve the characteristic landscapes and habitats of particular designated parts of the country, mainly in the hills and uplands, and certain 'fragile' lowland landscapes. Although they began with relatively limited aims and small resources, ESAs now take a very broad view of integrating agriculture and conservation within their boundaries, helping farmers to care for landscape, wildlife and history.

The Countryside Commission set up its first payment scheme, at the request of the Department of the Environment, in 1989. This was the Countryside Premium scheme, offered for five-year set-aside land in Eastern England. Farmers were offered a menu of standard payments to adopt management which would create benefits on set-aside land, such as areas for quiet recreation, and grassland rich in wildflowers. In 1991, again at DoE's behest, we launched the Countryside Stewardship scheme, a more bold, experimental scheme which targets landscapes all over England, looking to encourage new forms of countryside management. This scheme has been able to attract land throughout 'the wider countryside', not just the 'special areas' where other schemes are on offer. Also, in 1992 we launched the Hedgerow Incentive Scheme, another national scheme which adds to Ministry of Agriculture funds to offer special payments to farmers to restore neglected hedges. In Wales, our sister organisation the Countryside Council for Wales (CCW) has recently launched two similarly ambitious, 'wider countryside' schemes - Tir Cymen and the Hedgerow Restoration Scheme - although for the time being Tir Cymen is only available in three pilot areas.

To add to its tailor-made SSSI agreements, English Nature, successor in England to the Nature Conservancy Council, launched a Wildlife Enhancement Scheme in 1992, which offers standard agreements to farmers in larger SSSIs, similar to the agreements offered in ESAs and Countryside Stewardship. And in National Parks and some Regions, Counties and Districts, local authorities have launched farm and conservation schemes of their own.

The motive behind all these schemes is really to show that it is possible for government, on behalf of the public, to buy environmental goods from a farmer, in the same way that others buy crops and livestock. Schemes encourage farmers to produce valuable things which conventional markets for food and fibre do not, nowadays, give them any real incentive to produce. The activities they fund, and the management they promote, have become much more varied and sophisticated over the years, as experience has shown us how to improve scheme performance and value for money.

INTEGRATING AGRICULTURE AND CONSERVATION

In the early days, conservation grants and schemes were often confined to marginal areas or to land which in some cases had been hardly farmed at all. For example, those awkward corners which farmers had, until recently, found unprofitable to drain or plough, where wildlife had flourished undisturbed; or those marginal farms on poor quality soils or in harsh climates, where farming practices were still dictated largely by natural constraints. Early schemes tried to identify and protect these 'special areas' from damage, because of recent changes which had increased the profitability of bringing them into production, or of greatly increasing their production.

Today, particularly with the more recent schemes and the expanded range of ESAs, we can attract the profitable farmer who wishes to do something positive for the environment. It is my impression that we are moving into an era where schemes must increasingly work hand in hand with productive farming, in the wider countryside, beyond the 'special' areas. Increasingly, all kinds of farmer can see the attraction of mixing certain kinds of countryside scheme within their overall business management plan. With help from schemes and with the good technical advice provided by many local and national groups, farmers can identify opportunities to improve the countryside for themselves and their neighbours. It may also bring them added financial benefits - in the form of new business ventures, a better public image, and a more attractive, and therefore valuable, asset to pass on to future generations.

On the whole, farmers seem to like countryside schemes. Spending on schemes has grown from four or five million pounds a year in the early 1980s to over 60 million pounds in 1992. The Government proposes that this figure should reach nearly 100 million pounds a year by 1996, and a range of new schemes have been launched this year under the UK Agri-Environment programmes.

The Countryside Commission, and many others, are keen to see the budget increased significantly beyond this, by the end of the decade. I am sure that there are a lot more countryside benefits which these schemes could generate, which would still represent good value for taxpayers' money. However, we must also try to simplify the picture by scheme integration; because there are now so many schemes, each with different rules and targets, that it can be very difficult for farmers to keep abreast of what is on offer. I would like to see Government moving towards a single 'menu' or shopping-list of countryside benefits during the next few years, in each of the four UK countries. Such a menu could be available throughout each country, but on a discretionary basis, so it would be up to farmers to offer to supply the various items on that menu for government agencies, as buyers, to choose between.

So much for payment schemes. But we mustn't be led, by all this, to think that they are the be-all and end-all of what is needed to integrate agriculture and conservation, in the future. Other, major issues will need to be tackled. Whilst we have been pouring efforts into schemes, greater resources have been poured into other policies which have, in many cases, constrained or undermined the performance of our schemes. In particular, Government needs to press for changes to the 'big spender' elements of the CAP, to make sure that these, which still dwarf EC expenditure on countryside schemes, are not undermining countryside quality.

For example, the 1992 CAP reform introduced set-aside to control arable production. I am pleased that the Countryside Commission has now 'gone public' in saying that this offers few real environmental benefits, and soaks up resources that might be better directed to other schemes. For example, schemes which involve new land uses, like tree-planting and meadow restoration, would bring many environmental benefits, whilst also taking land out of arable.

The CAP also involves many direct payments to farmers, such as beef and sheep premia; which because they are paid on **headage**, have encouraged some farmers to stock or crop at intensities which, as we increasingly recognise, cannot be sustained without damaging the land and the countryside. I believe that many farmers, as well as conservation bodies, would be happier with a livestock regime which would make it possible and profitable to keep fewer stock on their land, and thus to be able to manage both stock and land more carefully. Achieving these sorts of change will require imagination and co-operation from all interested groups.

FIELD MARGINS AS AN EXAMPLE

Just for the record, I thought you might find it interesting to look at how various policies currently offer some help for field margin management in the countryside. I hope you will excuse me if these details relate only to England.

It is possible to have a kind of 'field margin' option within Set-aside (which is not a countryside scheme) - but only if you count a 20 metre wide strip as a field margin. Under rotational or non-rotational set-aside, the payments for these wide bands are the same as for all set-aside land. The environmental benefit of these strips depends very much upon the interest of the farmers who create them, but with help from organisations such as FWAG, RSPB and the Game Conservancy, you may achieve some benefits.

Among countryside schemes, if you have non-rotational set-aside, or are in an ESA, you could also look to open up your field margins to public access, in which case you might qualify for payments for this. The payments in ESAs (which incidentally, could include field edges on pasture land as well as arable) are for a 10 metre wide accessible margin around the field. For the Countryside Access scheme on NRSA, the planned payment will be slightly lower for a 10 metre wide accessible margin, but it will be on top of the basic set-aside payment. These schemes will offer you specific cash, for a specific new countryside benefit.

There are field margin and conservation headland options in some of the ESAs where arable farming is common - the Breckland ESA has the most comprehensive options. Here, you can choose a 6 metre or 12 metre margin or headland. Similarly, in any part of the country, if you wish to enter Countryside Stewardship, you'd be looking at a 6 metre margin, available if your land is in one of the priority landscape types for Stewardship - for instance along a waterside, on the coast, or on chalk - or, wherever you have hedges which you are going to restore or protect through the scheme, you can opt for a 6 metre strip, or a 2 metre strip on which a lower payment is offered.

In both ESAs and Stewardship, the specific field margin payments would be in addition to other payments you might get under the scheme, such as payments for hedge laying, tree planting or other capital work. In these schemes, you are paid directly in return for providing specific countryside benefits.

Beyond this, I know that a growing number of farmers who are not in schemes are experimenting with field margins and conservation headlands, mainly as a management tool and a means to more efficient use of the land. As you know, on some soils, crop yields drop dramatically as you get to the field edge, and in the current economic climate you may well ask yourself whether it is worth cultivating, sowing and spraying right up to that edge when the return it can give you may be very small. A 1 or 2 metre unsprayed, uncultivated or grass-sown strip around the field might make both good business sense, and could do a lot for the wildlife and landscape of your farm - and maybe also the sporting value - without any special grants. As someone who works to improve the grants system, it is important for me to remember that there is life beyond grants!

Focusing here on field margins illustrates the rather piecemeal way in which different policies can affect one aspect of the farm. It is a long way from the simple, integrated countryside strategy that I believe we need to work for, but on the other hand, at least it offers quite a bit more for the countryside than we had twenty years ago!

CONCLUSIONS

In conclusion, the various countryside agencies share the Government's concern, that more thorough environmental reform of the complexity and bureacracy of the CAP must be a central aim of future policy, for both farmers and conservationists alike.

In the meantime, we must work together to find practical and profitable ways to combine farming and conservation, in everyday land management. The countryside agencies and farmers together must continue to help government to set appropriate environmental standards through regulations and codes of practice, and to consider ways in which we could perhaps develop new markets for countryside benefits. We should recognise the valuable work done by farming and voluntary organisations, such as experiments in integrated crop management which may benefit the countryside, and work to develop new techniques and skills for habitat restoration.

Because we seek a thriving, diverse and beautiful countryside for everyone to enjoy, I believe the Countryside Commission and other countryside agencies are willing to work in partnership with farmers and other land managers, towards this goal.

POLICY OR SCHEME AFFECTING FIELD MARGIN	MINIMUM WIDTH OF STRIP	MANAGEMENT AND OTHER CONDITIONS	PAYMENT RATE PER 100M PER YEAR
Set Aside - rotational or non-rotational	20m	former arable land only: natural regeneration or sown strip, several cuts allowed	£22.50
Countryside Access scheme on non-rotational Set Aside only	20m strip of set aside land, with 10m wide access strip along it	as for set-aside withat least one cut annually	£145/mile which is around £9
ESA field margins: Breckland ESA	6m or 12m strip	cultivate annually or bi-annually to create a seedbed (Aug-Feb)	£350/ha, which is £21 or £42
ESA conservation headlands: Brecklands ESA	6m or 12m strip	limited herbicide & pesticide use, headland is not rolled until after harvest	£110/ha, which is £6.75 or £13.50
South Wessex Downs ESA	6m strip		£60/ha which is £3.75
Access in ESAs	10m access strip along field edge or elsewhere	arable or pasture land eligible cut at least once annually	£274/mile, which is around £17
Countryside Stewardship (now merged with Hedgerow Incentive Scheme)	6m strip along field margin with or without hedge	must be either on a landscape type eligible for Stewardship, or alongside hedges entered into the incentive scheme grazed or mown, follow Game Conservancy guidelines if on unhedged arable	£35 basic payment. With linear access, £45-55, depending on type of access plus lump sum per new path or bridleway
Countryside Stewardship - new for 1994:	min. 2m strip alongside hedge	as for the 6m strip	£15

FIELD MARGINS AS A NATURE CONSERVATION OBJECTIVE IN THE NETHERLANDS AND GERMANY FOR NATURE CONSERVATION; POLICY, PRACTICE AND INNOVATIVE RESEARCH

TH.C.P.MELMAN

Governmental service for Land Management of the Ministery of Agriculture, Nature Conservation and Fisheries. Postbox 20022, 3502 LA Utrecht, The Netherlands.

ABSTRACT

To be an attractive form of agricultural nature management, field margin management has to fulfil at least two qualities. Firstly, it has to be effective ecologically. Secondly, the possibility to combine it with current use of the adjacent field is a prerequisite for farmers to participate in great numbers.

From many investigations it appears that field margins have a higher species richness than the rest of the field. Furthermore, is that there are several indications that specific field margin management, of which the absence of fertiliser or herbicide use are important elements, is indeed ecologically effective. This holds both for grassland (Melman & Van Strien, 1993) and arable land (De Snoo, 1993; Wolff-Straub, 1985; Schumacher, 1980).

Amongst others based on these findings, both in the Netherlands and most Bundesländer of Germany field margins are recognised as an objective for nature conservation. For several years margin management has been part of the national programmes concerning nature conservation issues within agriculture.

This paper deals with the Dutch policy of field margin management, the actual practice and with the innovative experiments to introduce this type of management on a larger scale. In addition some information about the German situation is presented.

THE NETHERLANDS

Nature conservation on agricultural land in the Netherlands has a relatively short history. As an element of policy it had its birth in 1975 when the so called Relation Paper was published. This note focuses on the nature conservation value of agricultural land. An important part of the programme, based on the Relation Paper is that farmers (in selected areas) are invited to adapt their farming practice to a more nature-friendly one (on an voluntary basis). They have the opportunity to conclude a management agreement, for which they are financially compensated (extensification of management reduces their income). For a more general paper about the Dutch programme based on the Relation Paper see Baldock (1993).

Position margin management

In the first period, management agreements were directed at field-wide agreements with no specific attention for margins; their conservation qualities had not yet been recognised. Once this was the case, there was a reluctance to implement margin management within the regulation. Reasons for this were (1) the difficulty of inspecting compliance of the farmers not using fertiliser and/or herbicides; (2) doubt about the ecological effects in the long term.

The aspect of compliance inspection is hard to solve, especially for grassland. The chance to see the farmer while applying fertiliser is very small and inspection of the margin after application is very laborious and is possible only for a few days. After this the pellets disappear. Another method of inspection is monitoring the vegetation a number of years. If one doesn't recognise any difference in colour, growth or composition between the vegetation of the margin and the rest of the field one may go and speak to the farmer and talk about the way he manages the margins. However, it may be clear that this is no base for judicial consequences, i.e. to withdraw the compensation paid in the foregoing years in case he has not been able to save the margins from fertiliser effectively. For arable land inspection is less problematic. From the first year on cessation of herbicide-use is easy visible. However, until recently the attention of nature conservation was mainly aimed at grassland; arable land was treated stepmotherly.

On the second point the results of the research mentioned above gave sufficient indication that specific margin management is effective; model studies for grassland show that the influence of the management on the adjacent parcel is limited, even in the long term (Melman & Van Strien, 1993). Also for arable margins several indications for robust results are available (Dover *et al.*, 1990).

Notwithstanding the difficulty of compliance inspection for margin management in grassland effectively, it is incorporated in the programme because of its potential ecological meaning and its modest costs to the farm business. Concerning the use of fertiliser trust is an important element of the agreement between the farmer and the public authority.

Increasing recognition of the significance of field margin management

In the course of time field margin management is assigned an increasingly independent position in agricultural nature conservation. This can be illustrated by several alterations which are carried through in the regulation:

- the possibility to conclude margin management exclusively instead of inclusively; at first one was obliged to combine margin management (of which no use of fertiliser is the most important element) with field wide management (of which postponement of mowing and grazing is the most important element). This obligation has now been removed.
- greater choice of widths; at first one had a choice between a fixed width of 3, 5 or 10m. This range is widened from 2 to 12m with steps of 1m. Recently, for a specific situation a flexible width has been introduced, varying within a margin so that there is a possibility that the curving character of a landscape element is bordered by a straight line of arable land. To calculate the rate of compensation the average width is calculated which mest be a maximumed of 12m. These elements of increasing flexibility reflect not only the growing importance of margin management, but perhaps especially the Dutch bureaucratic style of working. Nevertheless, for the Dutch situation

it is a way of finding solutions. Each square metre is important, whether for agriculture or for nature conservation.

Table 1. Some features of field margin management agreements in The Netherlands. Additions/alterations from mid-1994 between sq.brackets.

width of margin	scope of prescriptions	payment (dfl/ha.yr)
2 - 12 m	grassland: fertiliser herbicides sowing deposition of mud (from adjacent ditches)	1300 [1600]
	arable land: fertiliser herbicides/pesticides [crops] [mech. weeders]	1500-2200 [2200-2800]

Innovative research

All the elements of the more flexible approach of margin management helped to lower the threshold to enter a management agreement. The participation is growing steadily (table 2) . Yet, despite all our efforts we felt that the flexible, inviting character of field margin management still was not optimal. So there was a challenge to look for further improvement. Therefore an experiment was designed in which some new elements of application of margin management were to be tried out. To make this understandable I have to tell you some specific details of the Dutch situation in which agricultural nature conservation is embedded.

Table 2. Parcticipation in field margin agreements in course of time.

year	length [km] (area. [ha])	average width [m]
1989	311 (141)	4,5
1990	704 (321)	4,6
1991	953 (443)	4,7
1992	1239 (585)	4,7
1993	1595 (810)	5,1

In the first place the area in which management agreements can be concluded is limited up to 100,000 ha (i.e. exclusive of the area in which reserves, to be withdrawn from current agriculture, are planned). In several regions the ecological ambitions are for the greater part fulfilled when only margin management is practised; nature conservation-oriented management on the rest of the field is not necessary.
Secondly, it is felt by the farmers that their fields being designated as an area wherein management agreements can be concluded induces a (prescriptive) planological decision and thus is a threat for the freedom of conduct of business. They feel to be doomed to be second class farmers,

because they fear that works of land-improvement will not be carried out any longer. This is in spite of the procedure which ensures that application of the Relation Paper programme follows planological decisions and does not induce these.
Thirdly, farmers feel the current system of agreements sometimes acts like a straightjacket. The management agreement contains mainly statements of what one is not to do (notwithstanding the voluntary character of entering into an agreement) and thus is not very challenging in positive sense.

The starting point for the experiment is that margin management is ecologically sufficiently effective and as such does not need special attention. The main problem to be solved is how the barrier to farmers' participation can be lowered. The design of the margin experiment tries to join in the three above-mentioned points.

Modest claim on area

To meet first and second points, in the experimental areas the margins are the only elements for which an agreement can be concluded. This reduces the amount of hectares needed in an area by 75-90%. This way of using the available hectares enables to exploit a far larger area for agricultural nature conservation. Furthermore, farmers, hopefully, no longer feel so heavily threatened, because nature conservation has no interest in the field itself. Only the margins are targeted, which because of their qualities have less agricultural importance than the rest of the fields.

Limited quantity personnel available

Another point of attention at the start of the experiment is the quantity of personnel which is needed. As a rule the traditional management agreements require personal visits to the farmers. Without this approach farmers would not recognise agreements as a serious option. To minimize time needed for recruiting farmers we tried to design a promotion campaign which would rouse enough enthousiasm to conclude an agreement without visiting the farmers. The campaign included information via local press and attractive brochures.

Nature result payment

The third point has been taken into the experiment by introducing another concept of paying the farmers. In the current system one is paid for creating nature-friendly conditions (e.g. no fertiliser). Payment is assured whether or not nature reacts to these favourable conditions. In the experiment the concept of nature result payment is introduced. Farmers are paid for positive results only, whether or not as a consequence of their efforts. This basis for payment has been discussed for several years and has fierce supporters and opponents (Van Strien *et al.*, 1988; Clausman & Melman, 1991). We considered the time was ripe to try this concept in practice, at least for the botanical aspect on which the experiment is focused. The basic idea is that paying according to this concept farmers are approached in their quality as entrepeneurs. They are free to choose the way they achieve nature benefits on their fields. There are no prescriptions as to how to reach them. This might be a more challenging way for farmers to contribute to nature conservation. Notwithstanding the possible advantages complications should also be mentioned. These concerns questions like:

- how is the nature result to be defined?
- who is going to assess the results? (skill and time)

- how and when are the results to be announced to the paying authorities?
- how are the given results to be inspected? (e.g. in what time-scale after the announcement); this is especially important given the limited quantity of public authority personnel;
- what is to be paid per unit of nature result?
- what to do in case announcer and controller disagree?
- how to react when farmers haven't sufficient results in relation with their efforts? (frustration!)

Some questions had to be answered, at least tentatively, before the start of the experiment (see also the poster of Kruk *et al.* presented during this symposium).

We defined the nature result in terms of the presence of a number of selected plant species. The selection was based on: (a) the indication for the nature value of the whole vegetation (if you find one of the species, you may expect a diverse vegetation), (b) recognisability (conspicuous flowers), (c) attractiveness (no problematic weeds), (c) distribution (preferably throughout the country), (d) rareness (not too rare, to prevent farmer-frustration) and (e) flowering time (more or less simultaneously). This resulted in a list of 19 species for grassland and 15 for arable land. We composed a booklet in which the entire selection is presented. This is a handbook for the farmers to be able to assess the results. As a consequence of payment concept we considered the farmer responsible for announcing the results. Concerning the control/verification of the announced results, this has to be done as soon as possible after the results are reported. This minimizes the risk of not being able to find the species. We try to do it within two weeks after the announcement is received (announcement once a year, before a fixed date).

Quality classes of nature

The payment for the results depends on the number of species found, that are on the list. Three quality-classes are distinguished (table 3). Scoring below the lowest class (less than two species) gives no money at all. Payment (grassland) for the first class (2 or 3 species) is ƒ0,15/m margin of two m width, for the second quality class (4 or 5 species) it is equal to the traditional compensation payment (ƒ0,25/m margin) and for the third and highest class ƒ0,35/m margin. The species are to be found on a piece of 100m length, appointed by the paying party (government). Such a sample has to be taken on each 1000m margin for which an agreement is concluded. It is essential that the payment is entirely determined by the results: there are no prescriptions about the management. It is up to the farmer how to reach the results. The only condition is that the results arise from natural regeneration (not planting or sowing).

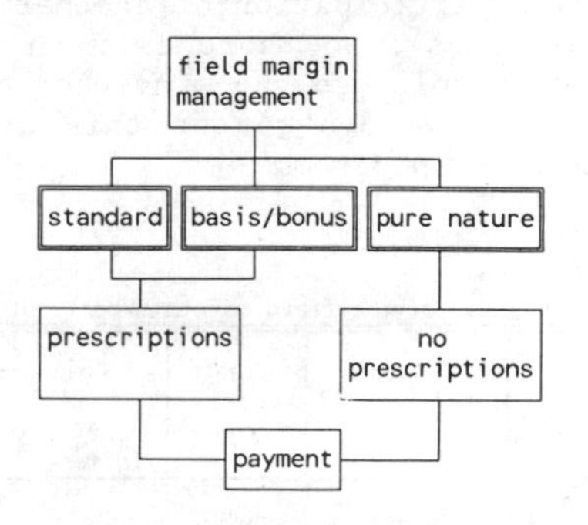

Fig. 1. Three payment concepts of management agreements

Beside this completely result dependent payment (called 'pure nature' payment) we offered the possibility of traditional compensation payment (called 'standard' payment) and also a hybrid between these two, the so-called 'base-bonus' payment (fig. 1; table 3). For these two ways of participation the farmers have to fulfil the normal management prescriptions (e.g. no fertiliser). The farmers are fully free to choose between these three variants. Moreover, they may change from year to year. For instance, after a few years of 'standard' payment they might find it more profitable to join pure nature-result payment, because they know they have enough species for the more profitable high-quality class.

Table 3. Payment for field margin agreements in relation to quality class (number of species on the list) according to or three rewarding concepts. Mind the distinction between grass and arable.

quality class / payment concept	lowest (0-1 species)		low (2-3 species)		middle (4-5 species)		high (≥6 species)	
	grass (2m)	arable (6m)	grass (2 m)	arable (6 m)	grass (2 m)	arable (6 m)	grass (2 m)	arable (6 m)
pure nature	0,00	0,00	0,15	0,50	0,25	0,90	0,35	1,30
base-bonus	0,15	0,50	0,15	0,50	0,25	0,90	0,35	1,30
standard payment	payment as middle class							

First results of the experiment

The experiment was started by us actively in two provinces; one location (Zuid-Holland) concerned grassland and the other (Gelderland) arable land. The other provinces also had the opportunity to join. This resulted in participation of another six provinces (we have 12 provinces in the Netherlands). So there is much interest in this phenomenon. It feels very comfortable to be invited by farmers for the sake of nature conservation. We don't meet this attitude in the normal practice of the Relation Paper programme.

Table 4. Progress of the field margin experiment in the different provinces of The Netherlands.

Province	culture	start date	standard		base-bonus		pure nature		total	
			nr of farmers	km	nr of farmers	km	nr of farmers	km	nr of farmers	km
Zuid-Holland	grass land	nov"92	2	5	-	-	8	107	10	112
Gelderland	arable	dec"92	5	2	-	-	-	-	5	2
Drenthe	arable	mar"93	3	8	-	-	-	-	3	8
Friesland	grass land	sep"93	29	97	1	3	7	24	37	124
Groningen	arable	sep"93	3	4	-	-	-	-	3	4
Brabant	grass land	nov"93	no data available							
Overijssel	grass land	nov"93	2	4	-	-	-	-	2	4
Total			44	121	1	3	15	131	60	254

Notwithstanding their interest in the experiment, concrete participation of the farmers has so-far been moderate on average but varies strongly from province to province (table 4). (Per province the experimental area involves about 30-50 farmers). The main limiting factor is probably fear for planological consequences (designation), due to the rising nature value of the field margins by the nature aimed management. What starts on voluntary basis may lead to obligations which one cannot escape, is what farmers think. Even letters of the authorities stating that there will be no relation between the experiment and the planning decisions in their region hardly persuades doubting farmers. Concerning uptake statistics it has to be mentioned that so far the farmers are approached by written material and public meetings only; they are not visited at home yet as is the case with normal management agreements. So, the participation cannot be compared with that of the regular programme. Later in the experiment such visits will take place and then there will be an indication as to how far participation depends on the way farmers are approached. Anyhow, the participation is already sufficient enough to present some interesting features.

In the first place it is striking that participation is higher on grassland than on arable land. Probably, stopping the use of fertiliser and herbicides has more severe consequences on arable crops than on grassland yield. A second feature is that the majority of the participants chooses the traditional standard payment. Notwithstanding the attractiveness of pure nature payment they prefer the certainty of payment. However, the intitial choice might be a strategic one: first financial certainty via standard payment and later higher income via pure nature payment. So, in the next years the pattern may change; farmers are free to change from year to year.

A remarkable phenomenon linked to the nature result payment was the growing interest of the farmers in nature itself. We organised some field excursions to teach the farmers how to recognise the species for which payment was to be made. The floristic richness of the margins was enteriley new for them and learning the species promoted their interest and involvement. It might be that coupling the paying to the presence of species delivers an extra stimulus for performing nature-oriented management. However, substantial conclusions on this can only be drawn after a number of years.

Significance for policy

So far the experiment appears to be a stimulus for margin management. Its popularity is rising both among farmers and policy-makers. For this reason the Dutch minister of nature conservation (the Secretary of State, to be precise) asked for special attention to be given to field margin management. Therefore, in Spring 1994 it will be examined which elements of the experiment can be put into real, operational practice. Possibly, the selective use of nature-conservation hectares for margin management is the first to be introduced. Concerning the incentive concept it will be too early to decide on introducing nature result payment; some extra years experience will be necessary. The same holds for the more extensive method of promoting agreements. There is a continuous drive to optimize the cost-effectiveness relation of the personnel manpower. Besides the margin experiment there are some other activities on this item. The whole of these will contribute to the future system of acquisition.

GERMANY

Because Germany consists of sixteen Länder which all organise their own programme of nature conservation, it is not too easy to gather information on this subject together. Notwithstanding the quality journals about nature conservation, publications on national scale do not always cover all the Länder; programme parts dealing with field margins are not always treated into the same level of detail; schemes are continuously developed, so information quickly becomes out-of-date; an finally, not all programmes are completely embedded in EC-policy and not easy to trace. So, it is not the ambition of this paper to present a complete overview. (For a more general paper about the German programmes see Baldock (1993)).

Table 5. Some features of the field margin programmes in the Bundesländer of Germany (based on Anonymous (1991)).

land use	Grassland margins	Arable land margins					
scheme-info ▸ ▾ Länder	programme	programme	width	scope of prescriptions	payment (DM per ha per year)	area under agreement (ha)	start date
Baden-Württemberg[1]	No	Yes	≥5m? 3-6m?	crop-margin: crop(density) herbicides fertiliser bound. strip: peren. char. mowing freq.	up to 1400	0,7	1987
Bayern	Yes	Yes	3-5m	herbicides fertiliser	1000	2829	1985
Berlin	No	Yes	3-5m	herbicides fertiliser	700-1150	13	1987
Bremen	No	Yes	3-6m	herbicides fertiliser mech. weeders	700-1500	1,7	1987
Hamburg	No	No?	? only parts	herbicides fertiliser	450-1400	50	1987
Hessen	No	Yes	3-5m	herbicides fertiliser	900-1300	295	1986
Niedersachsen[1]	No[2]	Yes	3-10m	crops herbicides fertiliser mech. weeders	1000-1500	174	1987
Nordrhein-Westfalen[1]	No	Yes	2-10m	herbicides fertiliser	750-1200	ca 850	1985
Rheinland-Pfalz	No	Yes	2-5m	herbicides fertiliser	1250	76	1986
Saarland	No	Yes	?	crops herbicides fertiliser	1100	4,6	1987
Schleswig-Holstein	Yes	Yes	3-6m 5-24 (fallow)	herbicides fertiliser mech. weeders 1-2 yr fallow no sowing	300-800 700+10 for each soil qual. point	1,6	1986

[1] more recent information used (1992/1993) for some of the items

[2] There are also policies for margins along water courses, but here, ecological conservation is on an obligatory basis and therefore falls outside the scope of this paper.

In Germany the attention given to field margin management dates from the late seventies. It was especially the research of Schumacher which drew attention to the ecological potency of the margins of arable land. Gradually in most Bundesländer a field margin programme has been implemented as a part of the nature conservation activities (table 5). About the field margin programmes the following remarks can be made.

- The basis of the margin programmes is, just as in the Netherlands, the ecological effectiveness and the modest demands margin management makes upon agriculture.
- For the most part the programmes are focused on arable land. Only two Länder show also (implicit) interest for grassland margins.
- Concerning arable land there are also some differences with the programme of the Netherlands. Besides the crop margin in some Länder also a boundary strip is targeted. This strip is intended to provide both an element with an ecological value on its own and an element of connectivity within the agricultural landscape.
  This boundary strip, which is thought to have a perennial character, does not always exist already but might need to be created. For this purpose special seed mixtures are developed, containing the ecologically appropriate species, originating from the local neighbourhood.
- The incentive concept is the 'traditional' one: a management agreement is based on prescriptions about the use of the margin and the participants get a fixed amount of money per unit area per year, irrespective of the concrete nature results.
- In most Länder farmers can enter into an agreement if their fields are entered into the programme. In some cases, however, an additional ecological check is performed; for example margins need to have a minimum number of (a selection of) species. Also the connectivity properties (especially concerning the boundary strip) can be taken into account before a margin can be entered into an agreement.
- Many leaflets, brochures etc. have been made to promote margin management. At least this is the case for Baden-Württemberg, Nordrhein-Westfalen and Niedersachsen.
- The impression is that during the execution of the programmes improvements have been made, aimed at the (1) enhancing participation; (2) possibility to inspect the agreements; (3) ecological effectiveness.
- Some aspects of the programmes have been criticized by a national working group (Anonymous, 1991). Some of the points probably also apply to margin management. These are: (1) programmes are designed for short term; (2) shortage of (quality) personnel; (3) insufficient inspection of the agreements; (4) programmes too much directed at margins only (there is more to do for nature conservation). (For a review per state see also Hepburn & Weins (1993)).

CONCLUDING REMARKS

Both in the Netherlands and in Germany management of field margins is given close attention in programmes for nature conservation. Important aspects of margin management are the ecological effectiveness and the possibity to fit with current agricultural practice. In Germany the field margin programmes concern mainly arable land, whereas in the Netherlands so far attention has been directed to grassland margins.

In both countries there are several activities taking place to enhance the attractiveness for farmers. The changes deal with the management prescriptions, the width of the margins and the payment-level.

In the Netherlands an experiment has recently begun in which amongst others a new paying-concept has been introduced: nature result payment. In this concept payment is based on the actual nature results. The farmer is free to choose how to manage the margins. This might be more challenging for farmers than the 'traditional' condition payment.

A problem, especially in the Netherlands, is the fear of farmers for designation of their fields as a result of their nature aimed management. This appears to be a serious barrier to farmers concluding a management agreement, irrespective of the basis of payment.

REFERENCES

Anonymous (1991) Extensieverungsförderung; Bilanz und Folgerungen. *Natur und Landschaft*, **66**, *no 2, 91-92, (einschließlich tabellarische Übersichten im Beilage)*.

Baldock, D. (1993) Incentives for environmentally sound farming in the European community. *Interim report of the Countryside Commission, Cheltenham, England*.

Clausman, P.H.M.A.; Melman, Th.C.P. (1991) Instruments for combining intensive dairy farming and nature conservation in the Netherlands. *Landscape and Urban Planning*, **20**, *205-210*.

De Snoo, G.R. (1993) Onbespoten akkerranden in de Haarlemmermeer. *Landinrichting*, **33**, *no 4, 31-36*.

Dover, J.W.; Sotherton, N.W.; Gobbett, K. (1990) Reduced pesticide inputs on cereal field margins: the effects on butterfly abundance. *Ecological Entomology* **15**, *17-24*.

Hepburn, I.; Weins, Ch. (1993) Implementation of the agri-environmental regulation (EC 2078/92) in Germany. *Studies in European agriculture and environment policy. Interim report*. BirdLife International/RSPB.

Kruk, M.; Twisk, W.; De Graaf, H.J.; Ter Keurs, W.J. (1994) An experiment with paying for conservation results concerning ditch bank vegetation in the western peat district of the Netherlands. *Symposium 'Field margins-integrating agriculture and conservation'*, Warwick, UK.

Melman Th.C.P.; Van Strien A.J. (1993) Ditch banks as a conservation focus in intensively exploited peat farmland. In: *Landscape ecology of a stressed environment*, Vos, C.C. and P. Opdam (Eds), IALE studies in landscape ecology 1, London: Chapman & Hall, pp. 122-141.

Schumacher, W. (1980) Schutz und Erhaltung gefährdeter Ackerwildkräuter durch Integration von Landwirtschaftlicher Nutzung und Naturschutz. *Natur und Landschaft* **55**, *447-453*.

Van Strien, A.J.; Dorenbosch, M.M.; Kruk, M.; Ter Keurs, W.J. (1988) Natuurproductiebetaling. *Landschap* **5**, *no 2, 131-138*.

Wolff-Straub, R. (1985) Schutzprogramm für Ackerwildkräuter. *Schriftenreihe des Ministers für Umwelt, Raumordning und Landwirtschaft des Landes Nordrhein-Westfalen*.

HEDGES - A FARMER'S VIEW

O.P.DOUBLEDAY

Bax Farm, Tonge, Sittingbourne, Kent, ME9 9BU

A.CLARK, B.McLAUGHLIN

National Farmers' Union, 22 Longacre, London, WC2E 9LY

ABSTRACT

A variety of hedge functions and their different management systems are discussed. The diversity of management systems appropriate to these different hedges highlights the difficulty of achieving the active preservation (as opposed to passive retention, with inevitable deterioration) of hedges through legislative regulation. While some farmers consider hedges as potential sources of problems, such as rabbits and an increased proportion of low yielding headlands, most apple growers value their windbreak hedges, which provide valuable shelter and are an important part of their integrated pest management strategies. This suggests that when farmers are well informed of the benefits that may be derived from a hedge they appreciate them more. However, the considerable costs associated with hedgerow management pose problems for farmers which must be addressed by society if more hedgerows are to be positively managed and thereby saved from deterioration..

Over time farmland hedges have acquired an extensive range of values, often different to those for which they were originally created. In this paper we will consider the variety of functions that hedges can fulfil and use these functions or purposes to define appropriate management of these hedges. Possibly the most common agricultural function of a hedge has been as a stock-proof barrier, to enclose sheep and cattle. Indeed, the density of hedges in the UK still reflects this use, with a strong bias of hedges found in dairy farming regions (Bunce *et al.*, 1993). Such hedges normally consists of a high proportion of species with thorns, such as hawthorn (*Crataegus monogyna*) blackthorn (*Prunus spinosa*) dog rose (*Rosa canina*) sweet briar (*Rosa rubiginosa*) and holly (*Ilex aquifolium*). These hedges also provide valuable shelter to livestock. In the past, when labour costs were, in real terms, cheaper than they are now, the planting and maintenance of these hedges was an efficient use of a farm's resources.

Similarly, the shelterbelts around apple and pear orchards also have a definite value to a fruit growers. They can provide shelter and thereby warm orchards, improving pollination and reducing water losses from the fruit trees. Shelter from hedges gives useful protection against the damaging effects of wind, which would otherwise cause marking of fruit by rubbing. Indeed, high winds can cause significant losses of mature fruit due to wind drop. In addition, pesticide application is easier in sheltered conditions. Ideally an orchard windbreak should have early leaf cover, in order best to provide protection during the blossom period of the fruit tree, when insect pollination is critical.

Hedges may act as possible sources of disease inoculum, pests or beneficial predators. Understanding these relationships is likely to significantly influence a farmer's appreciation of his hedges. For example, hawthorn (*Crataegus monogyna*) is a bad orchard shelter species since it is often subject to infection by *Erwinia amylovora* (Billing, 1981), the bacterium which causes Fireblight, an important disease (Northover, 1987) to which pears are particularly susceptible. The entomology of fruit orchards appears to be more clearly understood than that of many arable crops. This may be because stable perennial orchards, with largely resident populations of insects, are easier to study than annually cropped land which is often subject to dramatic changes, such as ploughing and rotations. It may also reflect the fact that many insect pests are of critical importance to fruit growers, since the damage they cause can destroy the value of a crop, which must be of the highest quality if it is to satisfy the demands of the market (Doubleday, 1992; Doubleday and Wise, 1993). For these reasons, the characteristics of a hedge as a potential refuge for either pests or helpful predators are of considerable interest to fruit growers. A brief review of some of these characteristics for orchard hedges will be given here, since we believe that they illustrate some general principles, which may have a significance for general agriculture.

Oak (*Quercus robur)* might be considered an inappropriate orchard shelter species, since it acts as a host for *Blastobasis decolorella* (Easterbrook, 1985) and winter moth (*Operophtera brumata*) (Solomon, 1987). In contrast, various hedge species can act as a refuge for certain beneficial insects, for example alders (*Alnus* spp.) often harbour significant populations of anthocorids, such as *Anthrocoris nemoralis* and *A. nemorum* and mirids such as *Blepharidopterus angulatus* (Solomon, 1981). These are significant predators of aphids (*Dysaphis plantaginea, Rhopalosiphum insertum*), fruit tree red spider mite (*Panonychus ulmi*) and other apple and pear pests. Sallow (*Salix caprea*) is also a useful component of an orchard windbreak, since it attracts large numbers of anthocorids (*Anthrocoris nemoralis* and *A. nemorum*) during its flowering period, which is very early. When the anthocorids leave the sallow they may be useful in controlling pear psylla (*Psylla pyricola*) (Solomon, 1981; Solomon *et al.*, 1989). The husbandry of these sorts of predators, and of the predatory phytoseiid mite *Typhlodromus pyri*, has becoming an important part of the integrated pest management of fruit orchards (Solomon, 1987; Doubleday, 1992). In response to problems associated with pesticide resistance (Solomon, 1987; Doubleday, 1992), and problems of registration of certain pesticides (Doubleday and Wise 1993), the majority of fruit growers in the UK have adopted an integrated pest management approach. Alder has emerged as the favoured windbreak tree, since it combines a useful early leaf habit with excellent properties as a refuge for beneficial insects but not for pests. It also suffers relatively little from rabbit damage. *Alnus cordata* and *A. incana* are somewhat preferred to the native English alder (*A. glutinosa*), in view of their increased resistance to drought (Baxter, 1979; Solomon, 1981). Good orchard management also extends to control of non tree species. Eliminating plantains (*Plantago* spp.), docks (*Rumex* spp.) and knotgrass (*polygonum* spp.) from orchards and their vicinity denies food to rosy apple aphid (*Dysaphis plantaginae*) and dock sawfly (*Ametostegia glabrata*). The removal of docks and knotgrass is likely to have the greatest effect, since the dock sawfly largely attacks apples close to where it has been feeding (Solomon, 1981).

We believe that an understanding of the relationship between field margin plants and arable crops will be of increasing interest and importance to farmers. It is known, for example, that black bean aphid (*Aphis fabae*) overwinters in the spindle tree (*Euonymus europaeus*) and lettuce root aphid (*Phemphigus bursarius*) overwinter in poplars (*Populus* spp.). However, if farmers knew of the specific refugia properties of their field margins and hedges it is likely that they would appreciate and manage them accordingly. The Game Conservancy Trust and University of Southampton (Harwood *et al.*, 1994) have done pioneering work in this area, but more needs to be done to demonstrate the links between hedge, margin and crop in contemporary crop management.

Hedges may also have a recreational or sporting function. For example, a hedge may be valued as a means of showing and driving game species, such as grey partridges (*Perdix perdix*) and red-legged partridges (*Alectoris rufa*). Height would clearly be an advantage for such a hedge. In addition, it may be managed to optimise its value for the breeding of partridges and pheasants (*Phasianus colchicus*). Ideal nesting habitat consists of grassy field boundaries, either side of a hedgerow, fence or boundary structure, preferably at least one metre wide, in order to reduce the risk of predation by foxes (*Vulpes vulpes*) patrolling the crop edge. It is helpful if the nesting site is higher than the surrounding field, as this improves drainage and reduces the risk of incubating hens becoming waterlogged and consequently deserting the nest. Grassy banks are therefore ideal for this. The best grasses for this nesting habitat have a tussocky habit, such as cock's-foot (*Dactylis glomerata*), Timothy (*Phleum pratense*), Yorkshire fog (*Holcus lanatus*). The residual material from these grasses provides both cover and nest construction material in the early spring. These perennial grassy swards also inhibit growth of some important arable weeds, such as cleavers (*Galium aparine*) and barren brome (*Bromus sterilis*). One metre wide strips of ground kept free of vegetation between the grass bank and the field are useful cordons sanitaires inhibiting emigration of weeds into the cropped area. They also offer useful possibilities for the control of predators such as foxes and useful areas where chicks can dust or, after wet weather, dry. Actually, the presence of a hedge next to this grassy bank is not of great significance for nesting purposes. However, if there is a hedge it should be kept reasonably low (about 2 metres) in order to avoid impoverishing the ground flora by shading out. The presence of isolated large trees is not very helpful, since these may act as predator perches. The appropriate management of field margins and field headlands for optimising their value as game habitat has been described elsewhere (e.g. Sotherton and Rands, 1987; Aebischer and Blake, this volume). The sporting interest in hedges may reside in their value as jumps for horses. Again, such hedges require specific management - it would be highly inappropriate to allow them to grow to the height that might be desired for showing partridges!

Some hedges have historic value. Often they may mark a Parish boundary, or a farm boundary. In many cases understanding of a hedge's historic character increases a landowner's or farmer's appreciation of the hedge, and causes him to value it more highly than he might otherwise have done. Hedgerows and field boundaries provide a valuable link with our past and the centuries of agricultural development that has shaped our countryside. Landscape historians are able to 'read' the landscape and determine not only the previous use of farmed land but also identify the phases of this use. The countryside has been referred to as a 'palimpsest' (a manuscript on which two or more texts have been written, the former one being erased to make way for the next). To the trained eye the countryside may be distinguished as 'planned' (post 1750) or 'ancient' (pre 1750) (Rackham, 1986). The geometric patterns of regular, species-poor hedgerows clearly differentiates the Georgian enclosures from the winding, blowzy, species-rich hedgerows of Saxon or Medieval origin. For the farmer these differences may have little significance, yet to society such differences may be of greater importance.

Hedges may also be planted, or retained, because of their wildlife interest. In general the older a hedge is the greater degree of biodiversity it contains (Pollard *et al.*, 1974). Indeed, it is gratifying to note that this test is now being used by 'The Archers', of Radio fame. A comprehensive recent survey (Bunce *et al.*, 1993) has shown that boundary habitats - hedges, streamsides, road verges etc - are valuable refugia for plant and animal biodiversity. In some landscapes up to 85% of total biodiversity is found in these linear features. A considerable body of research (e.g. Pollard *et al.*, 1979; Carr and Bell, 1991) has shown that the management of this wildlife is not a simple process. For example, when planting hedges care must be taken to match species to soil and regional conditions, to locate the new boundary features with reference to both agronomic and ecological considerations, and to ensure that the farmer has the appropriate technical advice to ensure the survival of the hedge. Similarly, hedgerow management research has shown that hedges which a farmer might consider to be overgrown, with a broad scrubby base, are often better bird habitats than neat, annually trimmed hedges. The retention and management of hedges has attracted much attention, and bodies such as the Farming and Wildlife Advisory Group (Anon, 1991), the Game Conservancy Trust (Anon, 1994) and MAFF (Manning, 1986) have issued guidance on the appropriate management systems, such as biennial trimming, coppicing, laying and selective herbicide use. In general, this advice stresses that the key to successful management for wildlife conservation is a diversity of management styles, to ensure a wide range of different habitat reflecting the varying breeding, feeding and shelter requirements of different wildlife species.

Some hedges are planted, or retained, for their amenity value. For new development, planning permission is often granted subject to the provision of an earth bund, hedge, spinney or other similar sight and or sound screen. Where the intention is to screen development a hedge is likely to be designed with a significant quantity of evergreen species, or species such as beech (*Fagus sylvatica*) which retains its leaf in the winter. It is also likely that such a hedge will be allowed to grow to a significant height.

Given these interesting and useful functions of hedges, why are they not more esteemed by farmers? This is an important question, since the answer may provide clues as to what may encourage farmers to value their hedges more, and give them appropriate management. Firstly, it should be acknowledged that, with the exception of stockproof hedges and orchard windbreaks, none of the functions that we have discussed are strictly agricultural. Moreover, regrettably hedges can act as breeding habitat for rabbits (*Oryctolagus cuniculus*) which are agricultural pests. Rabbit meat is now worth very little. In contrast, between the World Wars rabbits were worth about 1 shilling, which was more than an hour's wage for an agricultural labourer (G.L.Doubleday, pers comm). This means that although controlling rabbits was economically attractive it no longer is. Although *Myxomatosis* dramatically reduced rabbit populations when it first emerged, it no longer appears to be as effective in population control as it was. The actual amount of damage caused by rabbits depends both upon the level of infestation and the specific location. Rabbit damage to rough grazing and cereal fields is less economically damaging than their damage to intensive vegetable or fruit production. Rabbits (and hares, *Lepus capensis*) are particularly damaging in recently planted woodland and in fruit orchards, when they bark unprotected trees, often causing their death. Thus apple growers are forced to resort to expensive rabbit proof perimeter fencing, or protecting trees with individual guards or painting them with repellant paints. However, as we have seen, in spite of this complication most growers plant and manage shelterbelt hedges around their orchards.

Hedges, particularly badly managed hedges, can act as a reservoir of weeds which can cause problems to arable crops. This often happens as a result of accidental or intentional spraying of hedge bottoms with herbicides. This disrupts the equilibrium of the flora, as does accidental application of fertilizers, particularly nitrogenous fertilize.. The recovery of such 'disturbed' habitats is a costly and skilled exercise (see other papers in this volume). Another potential disadvantage from hedges arises from the fact that, by making fields smaller, they increase the proportion of headland within a field. Headlands normally have lower yields than the rest of the field, possibly as the result of increased soil compaction, resulting from the inevitable turning that occurs on headlands. Indeed, the reduced work rate caused by the increased turning in smaller fields was a common justification for the removal of hedges in the 1970s and even 1980s. However, analysis in 1980 revealed that this argument was not very strong for fields greater 10-20 ha in size (Sturrock and Cathie, 1980).

However, the most serious obstacle to the better management of hedges is expense. For example, in order to maintain a hedge's stockproof qualities it needs rotational management over 10-20 years. Trimming alone cannot maintain such hedges. Gaps appear in the base of these hedges, a process aggravated by livestock browsing. Traditionally hedges are then 'sided up' and allowed to grow 2.5-3.0 metres before laying to re-establish a thick stockproof base. This process can take up to 5 years to regenerate the hedge, and in some circumstances the hedge has to be double fenced during this period to avoid further livestock damage! Laying a hedge, at £8-12 per metre, is prohibitively expensive for many farmers when compared to the alternative cost of £2-3 for replacement fencing. Some of the costs of hedgerow planting and management are given in Table 1.

Table 1. Costs of Hedgerow Management (per metre)

Operation	Cost	Frequency
Trimming	£0.10	biennial
Laying *	£8-12	every 15 years
Coppicing *	£2.50	every 10 years
Planting *	£3-5	
Fencing	£2-4	every 15 years

* these may all require additional double fencing to prevent livestock browsing.
Source: Nix and Hill (1993)

The consequences of these costs are shown in the recent survey of UK land (Barr *et al.* 1993), which demonstrated that the majority of hedgerow losses were in fact due to change of hedgerows to different boundary types, such as "relict hedge". These "losses" of hedges, particularly stockproof hedges, have caused considerable public anguish, since the general public consider hedges as the central feature of the "English Countryside". The resulting pressure to retain hedges fails to address, or even understand, their different functional values, the dynamics of their management or the cost of their management. A response to this has been pressure for a regulatory framework to prevent the removal of hedges.

Clearly the loss of hedges over the period 1984-1990 cannot be explained entirely by their reclassification as relict hedges. Hedges have been removed, but it is important to recognise that not all of these losses have resulted from agricultural change. Development, especially road construction and improvement, is a common cause of hedge removal. In the agricultural context, however, it is clear that more hedges are being "lost" as a result of their reclassification than as the outcome of a concerted programme of removal associated with agricultural improvement. We would also suggest that in the vast majority of cases, reclassification of hedges reflects the increasing costs of their management rather than any deliberate strategy of neglect by farmers. In that context therefore we seriously question the value of legislative control as a means of ensuring the appropriate management of hedges. It is clear that the management of an orchard windbreak hedge consisting of alders (*Alnus* spp) and no hawthorn (*Crataegus monogyna*) will be very different from the management of a stockproof hedge in which hawthorn may well predominate. Similarly, the management of a tall hedge used to show partridges will differ from that of a hedge managed in order to maximise its nesting potential for partridges, or managed with a view to fox hunting. There are also substantial regional differences in the methods of hedge management - for example, hedges are not usually laid in Kent, but are subject to regular trimming and coppicing. Just as legislation cannot deliver appropriate management of hedges, legislation cannot prevent natural processes, such as growth and eventual decline.

Farmers will be less inclined to remove hedges, or other boundary features, if they appreciate that they are performing a useful function, such as giving shelter or acting as refugia for beneficial insects. Good research and vigorous work by extension services could help to promote such benefits. However, the passive retention of hedges is not enough to protect them from deterioration, since hedges need positive management, which we have shown is expensive. What is needed is recognition of the value of hedges for agricultural reasons, as well as for their cultural and wildlife reasons. How are these non-market values of hedges to be taken into account? The most efficient way would appear to be through attractive incentives for appropriate management. Regrettably the schemes that might offer appropriate incentives, the MAFF funded Environmentally Sensitive Areas (ESA) management agreements and the DoE funded Countryside Commission's Countryside Stewardship Schemes, are only available to a very small proportion of UK farmers. The Countryside Commission administers a Hedgerow Incentive Scheme, but unfortunately this has limited funding and has attracted less than 500 Km of hedgerow. The recent halving of MAFF grant aid for hedgerow management is therefore very unfortunate. If the publicly expressed desire to retain hedgerows is to be achieved there is a need for more research to demonstrate the potential benefits of hedges, promulgation of appropriate management advice through extension services and other agencies, and financial incentives.

ACKNOWLEDGEMENTS

The authors thank Dr M Solomon of Horticulture Research International, East Malling, Kent for valuable discussions on aspects of orchard pest management and Dr N Sotherton of The Game Conservancy Trust, Fordingbridge for useful discussions on game habitat.

REFERENCES

Aebischer N.J.; Blake K.A. (1994) Field margins as habitats for game. In: *Field margins - integrating agriculture and conservation*, BCPC monograph (this volume)

Anon (1991) *A hedgerow code of practice*. Published by the Farming and Wildlife Advisory Group, National Agricultural Centre, Stoneleigh.

Anon (1994) *Guidelines for the management of field margins (conservation headlands and field boundaries)*. Publ. Game Conservancy Trust, Fordingbridge.

Barr C.J.; Bunce R.J.H.; Clarke R.T.; Fuller R.M.; Furse M.J.; Gillespie M.K.; Groom G.B. Hallam C.J.; Hornung M.; Howard D.C.; Ness M.J. (1993) *The Countryside survey 1990 - main report*. Department of the Environment, London.

Baxter S.M. (Ed) (1979) *Windbreaks*. Pinner, Ministry of Agriculture, Fisheries and Food.

Billing E (1981) Hawthorn as a source of fireblight bacterium for pears, apples and ornamental hosts. In: *Pests, Pathogens and vegetation*, J.M.Thresh (ed), Pitman

Bunce R.G.H.; Howard D.C.; Hallam C.J.; Barr C.J.; Benefield, C.B.; (1993) The Ecological Consequences of Land Use Change. Institute of Terrestrial Ecology, Merlewood. Publ. Department of the Environment, London.

Carr S.; Bell M. (1991) *Practical Conservation - Boundary Habitats*. Open University.

Doubleday O.P. (1992) Role of crop protection agents in farming systems: protecting the apple. In: *Food Quality and Crop Protection Agents*, L.G.Copping and B.T.Grayson (eds), BCPC monograph **49** pages 69-76

Doubleday O.P.; Wise C.J.C. (1993) Achieving quality - a grower's viewpoint. In: *Crop Protection: crisis for UK Horticulture*? D.Tyson (ed), BCPC pages 71-78

Easterbrook M.A. (1985) The biology of *Blastobasis decolorella* (Wollaston) (Lepidoptera: Blastobasidae), a potentially serious pest of apple. *Entomologist's Gazette*, **36**, 167-174

Harwood R; Hickman J.; MacLeod A.; Sherratt T.; Wratten S. (1994) *Managing field margins for hoverflies*. (this volume)

Manning P. (1986) *Hedgerows*. Agricultural Development and Advisory Service, publ MAFF.

Nix, J.; Hill P. (1993) Farm management pocket-book (twenty third edition). Wye College, Kent

Northover, C.J. (1987) *Fireblight*, an open letter from Plant Health Division, MAFF, accompanying a MAFF advisory leaflet "*Fireblight*".

Pollard E.; Hooper, M.D.; Moore, N.W. (1974) *Hedges*. Published Collins, London

Rackham, O (1986) *History of the Countryside*. Dent

Solomon, M.G (1981) Windbreaks as a source of orchard pests and predators. In: *Pests, Pathogens and vegetation*, J.M.Thresh (ed), Pitman

Solomon, M.G. (1987) Fruit and Hops, in *Integrated Pest Management*, A.J.Burn, T.H.Croaker, and P.C.Jepson (eds), Academic Press pages 329-360

Solomon, M.G.; Cranham, J.E.; Easterbrook, M.A.; Fitzgerald (1989) Control of pear psyllid, *Cacopsylla pyricola*, in South East England by predators and pesticides. *Crop Protection*, **8**, 197-205

Sotherton N.W.; Rands M.R.W. (1987) The environmental interest of field margins to game and other wildlife: a Game Conservancy view. In *Field Margins*, J.M.Way and P.W.Greig-Smith (eds) BCPC monograph **35** pages 67-75

Sturrock, F.; Cathie,J. (1980) Farm Modernization and the Countryside. University of Cambridge, Department of Land Economy.

FIELD MARGINS: A MICROCOSM OF THE WIDER POLITICAL DEBATE

DR M BELL MA, FRAgS, MRTPI, MIEnvSc, AIAgrM
Ward Hadaway, Solicitors, Alliance House, Hood Street, Newcastle upon Tyne NE1 6LJ

ABSTRACT

Some years ago CPRE published a report asking for conservation elements of rural policy to come "out of the field corners". To focus conservation benefits on field margins follows much excellent landscape ecological work on corridors to link wildlife 'islands'. This is a practical, sensible and well advised approach. Inevitably, however, it can be attacked as having only 'marginal' impact when all countryside is said to be 'environmentally sensitive'. This paper reviews the context of EC agro-policy post GATT. It is naive to see conservation outside of the need for output - restriction. Factors of production must be reduced but it is politically inexpedient to create wasteland or be seen to destroy jobs. The range of technological, regulatory, advisory, voluntary and contracted options for field margins echoes the wider EEC, and particularly British, debate. Useful gains are made but the impact is necessarily limited.

INTRODUCTION

This is a practitioner's view from the field. What were the organisers doing wasting a space for a learned academic on a day to day pragmatic rural planner you may ask? A planner based in a northern solicitors' office at that. I have often asked it myself. It seems that somewhere in the quotidian dealings with farm planning, taxes, estate inheritance, neighbour disputes, prosecutions, easements, quotas, buying and selling land, which are the lot of a specialist practice, we may gain a feel for how policy impacts on those who actually own and farm the margins. Only you can judge.

THE END OF THE GREAT SCOTT AND ABERCROMBIE CONSENSUS

You are hardly likely to know where you are going if you forget where you came from. There seems something near consensus amongst the best social science commentators on countryside affairs that one has to begin analysis with the breakdown of the enormously potent post-war belief that protecting agriculture equals protecting the countryside. I characterise that view as embodied in the approach arising for example from the Scott Report's seminal views on land use after World War Two; and alongside it the type of "town" planning which saw countryside effectively as that which was not town. This I see in dominant planners such as Abercrombie contrasting cities with the countryside as a 'Ceres', a bountiful, farmed land. In Britain we look to field margins because our low land is fields. What we do with non-developed land is farm it.

I acknowledge immediately debts to people such as Howard Newby, Philip Lowe and Mark Felton for helping confirm my feelings. The collapse of that ideology remains arguably the most potent force in the rural political debate. There is a void to fill.

I would like to make three points about that breakdown in terms of this specific audience:

i) Was it ever correct?

It was always a more potent force than it should have been because of a lack of competing ideas with any base of political strength. There is no pure Hegelian idealism here; it was only in part "the right idea at the right time". It was at least equally important that it suited so many of the key parties. If you remain naive enough to think it was only to do with farming and food production go and talk to some poor devil trapped in an inner city tower block built supposedly to save farm land. (I say 'supposedly' because good, empirical work showed quite early that (a) tower blocks probably used no less land than giving people decent houses, and (b) if you really wanted home food production the best thing to do was to give people big gardens. These remain true but we still force people to live in boxes without enough space).

ii) Is there a new dominant ideology amongst farmers?

Thus, whilst I have followed the popular approach in heading this section so as to suggest the era was led by ideas I am not really such a Platonist. But nor is it a crudely Thatcherite matter of everyone simply pursuing their own interest. Put in academic terms one would say a number of political actors seem to have accepted the hegemony of perceived superiors to act to their own detriment. In English, I mean that most days I hear clients say things which are the "accepted thing to say" in clear disregard of the facts in front of them. Significant sections of the rural working class, farm females, family farmers' children (particularly younger siblings) have done this in the past. I perceive that a considerable number of the more "production-minded" farmers are doing it now. From the time of Durkheim social scientists have noted that suicide is one of the most dramatic indicators of social conditions. As I now observe a number of farmers I like and respect help sharpen the knife for their own execution I find that even more intriguing. Again, I am conscious this is an audience including horny handed practical folk so let me put it for disbelievers in less florid language. In my early years at National Farmers' Union (NFU) HQ the standard answer to the question "why is the farm going broke?" was "because I believed the Agricultural White Papers". That is, people carried on investing whilst quotas and price reductions were being negotiated. Now, the Milk Marketing Board (MMB) is on its way to the block and the

Potato Marketing Board being prepared to follow. Recently my senior partner, the North's leading agricultural solicitor, was given a torrid time by milk producers for daring to suggest that, purely as a lawyer, he would not advise people to rush out and sign the Milk Marque contracts. This was seen as disloyal to a good friend of farming. Yet, there was no social uprising about MMB abolition in the first place which was what really mattered. There were no anti-GATT riots in Britain. In the broad choice between highly supported conservation-farming and keeping such support as you can in what is otherwise free trade British farmers largely support the latter. In that context field margin gains are a sop by the hard men and a cop out option for the softies.

iii) <u>Is farming the enemy of conservation?</u>

As dialectical arguments tend to, things swung too far in the early and mid-1980s toward the argument that farming was a positive enemy of conservation. The most sure-footed political work over the last decade has been in re-balancing that debate. Field headlands have been a key part of it. In reality many habitats and treasured landscapes in the UK and Europe are farm-made and require people on the land.

I will return to these three themes as the essence of my title but first I should explore how far the wider debate goes.

FIELD HEADLANDS AS A MICROCOSM OF SOCIAL DISLOCATION IN MODERN BRITAIN

How is that for the best overstated heading you have seen in quite a long time? At least it is a counterbalance to the frightfully specific ecological ones. However, I am going to try and make the argument stick, so here goes.

That era of farming certainty I dealt with above was also an era of social certainty for the elite. A farmer knew what he was (and he almost certainly was "he"). Farmers were an elite with corporate access for the NFU into government. It was a different certainty for farm workers; most were going to be replaced by a machine. But let us concentrate on those who mattered.

When invited to close this illustrious conference it was suggested I had some repute (or notoriety) for rebutting unsupportable nostrums. So I will say for the most heretical thing of the day: rural sociology is wonderful, useful and important.

Out of the superb work supported by the Economic and Social Research Council over recent years I want to emphasise certain factors.

Being a rural dweller is now part of the pattern of consumption. It is a lifestyle choice for those able (usually rich enough) to escape the city or suburbia. A large part of my workload derives from the jarring of new countryside dwellers with those making a living in the area or trying also to live there. Within a generation farmers have become strangers in their own villages. Consuming the countryside involves demands for more space. Incomers have a £150,000 house but not the land. The epitome of the new villagers' desire is an unsprayed, uncropped corridor to walk with weekend visitors, act like they own and allow the dogs to defecate on. They like headland set aside.

Across much of developed lowland UK farmers have moved to a wider view of their role. The Country Landowners Association (CLA) and National Farmers Union astutely promote this as the wider rural business involving stewardship. Equally, it is seeing land as a means more than an end. Laying out golf courses and looking for development opportunities land owners are more "instrumentalist" in being willing to view field margin management as one part of their choice of lifestyle. Food production is only one element.

The defining fact of much traditional rural sociology had been that the simple Marxists and other 19th and early 20th Century commentators had been wrong in seeing food production moving from family to big farm control. Food manufacture and distribution were dominated by large companies but the farm family survived as a disciplined production unit. Compare this heresy from the CLA Journal at the end of 1993:

"What determines the future of the farm should be its own viability and integrity as a business and not our ability to brainwash or emotionally blackmail our offspring. This is an unpopular view, because it hits hard at an old and well-tested assumption that the young will provide for the old, rather than the old provide a realistic pension for themselves in their own working life."

The brave writer of that described herself threefold in the article "First as a farmer' wife ... As landowners .... as a CLA committee member ...". This is the very stuff of dislocated Britain. People have to self-ascribe because it is not obvious who you are.

Simply producing food is no longer next to godliness. A well fed, consumerist, lifestyle - choosing West (ie forgetting a few million "poor" who do not exercise choice) wants perfect vegetables without sprays, organic production without muck and a grazed countryside filled by vegetarians. This is not mass society as we knew it in the 50s and 60s.

In sum, the old certainties based on birthplace, class, job and gender have gone. Britain is socially dislocated. The field margin certainties have gone likewise (rip it out, keep it sheep-proof or shoot it). Now one chooses the field margin to suit the lifestyle.

MULTI-SPEED AGRICULTURE, THE AGENDA APPROACH AND NEW OWNERSHIP

Thus if farming and style of farming/ownership is nowadays a social choice it is important for policy to reflect that new world. So let us start with the world agenda.

Internationally we have GATT.

The pressures for a world trade deal reinforced concerns at the EEC about over production which would exist anyway. When the "Protect Agriculture = Protect the Rural Environment" nexus broke down the Shoard-Bowers - Cheshire attack found its public strength from the argument "why are we destroying so much to produce food we do not need?". All the field margin elements, length of hedgerows lost, destruction of macro and micro habitats, use of chemicals in its broadest sense became judged in that context.

So, we have a fundamental EEC need to reduce the productiveness of agriculture. This debate focuses on what to do with the 1 - 2 million hectares of land we do not need. Gillian Shepherd's policy of "reaping the green dividend" from set aside and related instruments starts there. If we needed the food this conference agenda would be very different.

Classical economists, of course, remind one that the factors of production are not just land but labour and capital also. In post war agriculture I find it helpful to take technology as a separate resource factor. In the high tide of the past the same land with less labour and some capital investment produced a great deal more not just because it could be ploughed by tractor rather than horse but because of the application of nitrogen, improved seeds and increasingly effective sprays. Now the tide has turned. Arable farmers have superbly productive land which may have its capacity vastly diminished because it is in a Nitrate Sensitive Area. Livestock farmers have stock with top class genetics but can no longer maximise growth by the use of hormones. I choose deliberately two examples of political restraints which I understand from my reading as an interested layman are scientifically irrational decisions. Politically I note, however, that they both serve to reduce productive capacity and one would be pretty naive to think that may not have been in the Council of Ministers' Minds.

Moving down to the National Policy Level

So, we find it having to pull two ways at once (and probably more). There is to be a reduced EC cake and Britain wants its best share. Agriculture Ministers therefore announce specific forms of capital aid and especially support marketing initiatives. Marketing is the vogue word in agriculture not just because it is a Thatcherite vogue word anyway but because it carries the implication that we can ignore the problematic production element and simply concentrate on selling more. To call this a half truth would be giving it too much weight. Perhaps a sixteenth or a thirty second of a truth is closer.

I am focusing all the time on the particular field margin as a microcosm of this wider debate. The individual field margin exists in the specific field on the particular farm. To any such specific business struggling to survive then "margin" in its other sense is vital. If there is no greater demand for wheat than last year (or indeed less) then my farm has to capture more of the marginal market than my neighbour's. At a regional level we rejoiced in 1992 because the North East had better harvests than elsewhere in England so it helped our clients settle their accounts (always welcome) but it was on the basis of beggaring their neighbours in the south. Nationally, the Government must push British producers to get their proper share of goat cheese/sheep cheese/garlic production; or whatever is the identified area for reducing imports. But that must be at the cost of impoverishing some part of rural France, Spain, Greece or "at best" somewhere outside the EEC entirely.

Few cases are as extreme as sugar. The continued prosperity of beet growers on Nottinghamshire sands (and their call for crop protection products) may be at the price of economically dislocating a Caribbean island. Nonetheless, you do not have to talk long to sheep producers in France or Spain to see that Britain's continued success in sheep breeding and export is destroying their livelihoods.

It may be possible to persuade people to consume more computer games but one cannot readily persuade fat Europeans to consume more food.

Thus, I submit the field margin debate is a particularly potent sub set of the environment/agriculture balance which itself exists within the context of EC surplus reduction set against the framework of the World Trade talks. In the British context this expresses itself as being seen to do our bit for Europe-wide production restraint whilst gaining the maximum from the home electorate by politically sexy environmental schemes which tend to keep people on the land, or at least the land tended. Some aspects of field margins, particularly miles of hedgerow and stone wall, are politically salient; others, such as Carabid Beetle populations, appeal to a more specialist audience.

PIC'N'MIX POLICY

At the broadest level the political culture of British countryside politics is individualist and voluntarist.

It hardly needs saying that some of the "voluntary" schemes have an iron fist inside the velvet glove. SSSI management, nitrate sensitive areas and pollution control on livestock farms for example all work within the context of the idea that all the interests solve the problem together or something worse is around the corner. I am not knocking the way it is presented. Presentation is a great part of politics and I am unstinting in my praise for how far, fast and sensitively MAFF, DoE and the other national agencies have come.

The present policy approach which I characterise as tailoring particular solutions to particular problems, introducing a regional or habitat type dimension and offering the opportunity for those farmers who want to get off the treadmill so to do is quite simply correct. I had the honour of working for some years to Sir Richard Butler as President of the NFU. His old dad had some very sensible words about "politics being the art of the possible". Present policy achieves a great deal of what is possible and probably more than might have been thought only a few years ago. The question in countryside management, including such things as stewardship and the Hedgerow Incentive Scheme is how far, how fast and how much more of the same can we achieve or afford.

This broad success in the area of countryside management is all the more striking when contrasted with the same Government's failure in statutory land use planning. There, the last decade and a quarter has been characterised by lurches of policy, unclear statements on the diversification of the rural economy and in more recent times has come down, in my view, to Ministers misleading people by saying one thing to farming audiences and quite another in policy pronouncements and the actual decisions of their regional offices and inspectorate. Above all there has been a failure to bring countryside management and statutory land use planning into a common framework.

As things stand, however, we can, and frequently do, sit down with clients and plan their withdrawal from mainstream productive farming.

Depending on where they are one would consider:-

- Environmentally Sensitive Area scheme
- Countryside Stewardship
- SSSI management
- Nitrate Sensitive Area scheme
- Hedgerow Incentive scheme
- Heather restoration or other upland management
- Farm Woodland Premium scheme/Woodland Grant Scheme
- Woodland but with specific community forest or National Forest objectives
- the former Countryside Premium Scheme
- Various local or regional initiatives

These things do not exist in a vacuum, of course. Such approaches are predicated upon the leasing away or sale of the milk quota, disposal of other dairy capital assets, sale or leasing away of sheep quota, or the cow quota. The scheme will usually involve some form of non-permanent tenancy on parts of the land to allow a more productive minded neighbour to use them. The so-called "Gladstone v Bower" arrangements remain favoured; under which land is let without security for between one and two years. Land which looks as if it can be "got away" for minerals, opencasting, town or village edge development, barn conversions or whatever is kept.

What remains so good about this broad policy approach is that those selecting it do so because they are committed. You can always plane far more wood with the grain than against it. There remain a large number of farmers and other landowners who wish to do something other than produce the maximum rape output each year. There seems little point in trying to force universal policies on young men committed to being the best arable farmers in South Yorkshire (or wherever) when there is a willing population to work on instead.

Almost all the schemes above have field margin elements. Either explicitly or implicitly they reduce or eliminate spraying, they often involve specific wildlife friendly management of marginal habitats.

In this broad area we need technical and particular improvements and I identify the key ones (at the time of writing) as being:-

i A simple option within all types of set aside or agri-environmental packages to allow woodland planting to "count" as a farm's contribution to the set aside requirement.

ii Acceptance sooner rather than later that the complete removal of forestry from taxation was not correct. At least in targeted areas there need to be tax based incentives to woodland planting.

iii The "bring us your own ideas" paragraph in Countryside Stewardship to be brought up front and made the dominant theme. Even "Jack and Jill do Welfare Economics" would tell you that you will get far more for your money if landholders put in bids to win the Stewardship funds rather than read down the menu to see what is the best deal they can get away with.

iv Each and every countryside scheme which can do it should give attention to or have special discretionary elements for the most important habitat links. A mixed broad leaved woodland planted to replace an open, hard-sprayed, arable field and linking two areas of ancient semi-natural woodland counts for more than the same acreage somewhere else.

v Continue to cut down the paperwork.

CHANGE BY AGE, STAGE OR LAW

Field margin management also falls into the wider debate where it involves or requires some type of change in management style. Over recent years a number of socio-economic researchers have identified stages, normal in the family farming cycle, when environmental damage was most likely to occur. Typically, these have been such times as a son taking over with new aims and ideas.

In a rural solicitors' practice one sees that the family cycle is a far more complicated animal involving partnerships, calculations of taxation efficiency and attempts to balance equitable distribution of the property with the maintenance of the core business.

Some basic changes in land holding are usually bad for field margins. A farmer buying next door's fields thus rendering the hedge between them functionless is the most obvious. For the maximum public good field margins generally benefit not only from sympathetic management but from consistent management.

Practical experience over recent years, and particularly the directions of change in the last few months, lead me to conclude land use change is, on balance, malevolent. Principally I see:

i In milk, uncertainty over the world post-MMB is continuing to concentrate milk on those farms which seriously want to produce it. Such farms tend, anyway, to be carefully costed and keen managers of reseeded grass land. Buying or leasing quota puts them further on the treadmill and means they crop their grass tightly hedgerow to hedgerow. (Indeed, I wrote this section of the paper whilst working on the Staffordshire/Cheshire borders and one might better describe it as post and wire fence to post and wire fence).

ii I suspect many land agents or fellow specialist practices would echo those to whom I have spoken in the preparation of this paper; they have more instructions to acquire land than to sell it. I say "acquire" not "buy" specifically because many of those predatory, expansion minded farmers will take tenancies, share farming or the "Gladstone v Bower" arrangements described above. None of these are designed to encourage caring, long term farming.

iii In livestock again the more specialist farmers are concentrating production. They too may have to buy in sheep or beef quota rights. They have an eye on the increasing extensification requirements in EC livestock regimes. Along with reduced labour availability this means there is less reason to keep up internal hedgerows. Hedgerows can be lost as a feature by simply growing out into scrubby trees as well as by actual removal.

Against these trends in the main commodity areas one sets people opting for environmental schemes and a counter balancing weight of planting/management from new smaller occupiers of land. These purchasers, often near town or village edges, take land less wanted by farming. Whether for their own small scale agricultural business, equestrian enterprise or simply recreational use such new occupiers sub-divide land and should create a greater ratio of boundary per hectare.

In my experience such enterprises tend to run into development control problems (not least since changes in the General Development Order on units of less than 5 hectares). Thus they often live in a limbo world of low investment and limited environmental care. In most Planning Authorities benefits to wildlife, the semi-natural landscape of hedgerows, sympathetic rural land management and tree planting will count for very little if the new farm or holding needs a stable, shed or lock-up. If there is to be planting then screening leylandii still seem to outweigh slower growing indigenous species. It is to that statutory planning framework that I turn finally.

STATUTORY TOWN AND COUNTRY PLANNING IN ITS OWN VACUUM PACK

It has been unfortunate, sad but essential that the boom in countryside management from around 1981 has existed outside the statutory Town and Country Planning framework. We could certainly not have moved so far, so fast and with such understanding between the actors within a model based upon the statutory system. I would take a good deal of persuading that ESAs would be a success if the early ones had been made a National Park function.

To the extent then that field margin initiatives have stood almost entirely outside statutory planning they have **not** been part of the wider debate.

I emphasise "system" because the individuals comprising it have more sense than the system as an institution. I am involved in successful examples where officers or members have taken the wider view. I cannot say where they are because they have normally involved gaining a lot of good at the cost of the rules.

I have referred before to an example astonishing even in terms of some of the unthinking applications of supposed green belt policy that are my bread and butter.

This was a Lancashire mossland Authority who would not let a young couple born and bred out on the moss to rebuild, after fire, one of the typical wooden structures of the area. They told them they would have to justify an agricultural dwelling and therefore they should drain and improve their bit of moss land (neither of them being farmers) and make a viable holding with carrots, celery etc.

It was only upon visiting the Authority to check their files before enquiry that I saw proudly for sale on the front desk, the Mossland Strategy identifying my clients' holding as one of the last pieces of natural moss land undamaged by agricultural activity. Sponsored and supported and adopted by the self same Authority this called for the protection of the cultural heritage of the mosslands, the maintenance of the traditional structures and the non-improvement of the remaining most precious parts.

We won the enquiry but primarily on technical legal grounds as to whether the use had been abandoned following the fire. The inspector gave no obvious weight to the ecological aspects.

I can contrast that however with a much pressured South Eastern Authority where the Planning Officer, looking to the greater good, accepted from a Local Agent an agricultural appraisal showing potential viability from a farm system that retained around 50 acres as organic, completely unsprayed hay meadow with superb old hedgerows.

Broadly, however, one starts from a position wherein the statutory framework has not yet found an approach to replace the early consensus. The unhappier side of the new emphasis from Section 54(a) of the Town and Country Planning Act is the reinforcement of a view that everything can be achieved via the Statutory Plan. I suspect and fear this will reinforce a prevalent mode of thought that there are "right" and "wrong" places for particular uses to occur. In this world view it is some kind of prostitution of the planning process to allow something which can be made acceptable. Indeed, it is a regular part of my work to try to walk a tightrope whereby if we offered too much at the outset it would be argued that it must have been a fundamentally bad site because of all the landscaping work which we are offering to do.

The reciprocal is a fear by Authorities of asking for too much by way of landscape or environmental gain which may be open to attack on appeal or in the Courts.

All that needs to be done in that broad area is a paper for another time and venue but one welcomes the Countryside Commission's increasing espousal of "green planning" deals.

The bottom line remains that you are still more likely to get consent on a farm for a holiday caravan park to be located in a wood (where it must inevitably mean the loss or destruction of some of that wood) rather than in mediocre arable land as part of a scheme for replenishing, reinvigorating and replanting its margins and copses. Whilst that remains the case thousands of planning decisions take place with little contribution to the gains which could be made and conservation remains too often in the margins.

AN EXPERIMENT WITH PAYING FOR CONSERVATION RESULTTS CONCERNING DITCH BANK VEGETATIONS IN THE WESTERN PEAT DISTRICT OF THE NETHERLANDS

M. KRUK, W. TWISK, H.J. DE GRAAF & W.J. TER KEURS

Environmental Biology, Institute for Evolutionary and Ecological Sciences, University of Leiden, P.O. Box 9516, 2300 RA Leiden, The Netherlands

ABSTRACT

A system of paying farmers for conservation results is put forward as an additional system to the Dutch Environmentally Sensitive Areas scheme. This new scheme is called "nature production payments". The aim is to pay farmers for conservation results without prescribing specific rules for the management. Such a system has a strong appeal to the farmer as an entrepreneur, using his craftmanship and creativity. For nature conservation bodies, the system is considered "good value for less money". In order to make the scheme operational, 1) it must have a clear definition of the "product", 2) the measurement of the product (by the farmer) must be easy, 3) the returns must be easy to check and 4) it should provide fair prices to encourage the farmer to take part. To develop such a system an experiment has been carried out, focussing on the way the "nature poduction" was monitored. The results show that such a system is feasible and has much support from both farmers and nature conservationists.

INTRODUCTION

The typical Dutch polder landscape can be found mainly in the peat areas in the western and northern part of the Netherlands. Because of the relatively wet soils, the agricultural use in these areas is mainly restricted to dairy farming. The dairy farming practice in the Netherlands is, although getting more extensive in recent years, still very intensive (Clausman & Melman, 1991), causing among others a severe decline in the plant species diversity of the grasslands and ditch banks (Van Strien, 1991). Nevertheless, botanical conservation and modern dairy farming could be combined (see Twisk *et al.* elsewhere in this issue), in particular if farmers are left free to take their own conservation measures and rewarded for their conservation results.

NATURE CONSERVATION IN RURAL AREAS

Until recently, nature conservation measures in the rural areas usually consist of reducing the agricultural intensity. The approach is based on the assumption that only traditional, more extensive farming practices will benefit wildlife. In other words: modern agriculture and nature conservation are fundamentally conflicting activities (Reyrink, 1988) and therefore have to be separated from each other. In this conservation strategy, small nature reserves are set aside and, due to the high costs, only in a limited (designated) area farmers are financially compensated for loss of income due to *restrictions* on their farm management (see Melman elsewhere in this issue). This kind of measure has been implemented in other European countries as well (Mathers & Woods, 1989).

Although this system of *'payments for means to farmers'* has its advantages (like "everybody knows what he is in for"), it has important disadvantages too (Van Strien *et al.*, 1988; Van Paassen *et al.*, 1991):

- the relation between means and results is not always clear;
- the compensations vary in a limited range, i.e. not all possible extra efforts of farmers are rewarded. Knowledge and craftmanship of the farmer concerning conservation are not used;
- the prescribed means often do not fit into the farm management. This goes for both technical and psychological aspects of farming;
- not all means can be described unambigiously and/or checked.

Because of these (and other) disadvantages, many farmers have a resistance towards management agreements prescribing restrictions in their farm management. Therefore less farmers implement conservation measures on their farms than may be desired. This is a reason to develop more stimulating instruments (also see Melman elsewhere in this issue).

NATURE PRODUCTION PAYMENTS SCHEME

The general idea for a new instrument for nature conservation on modern farms is to pay for the amount of nature the farmer "produces", e.g. rare plant- or bird species. Regulations of this type were first proposed in the Netherlands by De Meijere (1979), and more recently new attention has been paid to this idea by Van Strien *et al.* (1988) and Van Paassen *et al.* (1991). The main advantages of such an approach are that the farmer is considered as a producer and is paid for positive results and not for omitting things, the payment is only for concrete results and not for creating conditions and there are no fixed prescriptions, so that the farmer is able to bring in his own skill and creativity.

Since the introduction of the idea it has been applied in the Netherlands on a small scale for a few threatened species (such as Barn owl (*Tyto alba*) and Swallow (*Hirundo rustica*). In the English Peak District such a system has been used for threatened plant species (Van Paassen *et al.*, 1991). These applications appear to be reasonably successful, as farmers often become more motivated for and active in nature conservation on their farm. To stimulate agricultural nature management it seems useful to investigate if the system can be applied to other conservation values as well. To achieve nature conservation goals like preservation of species, the system should be applied on a large scale. Research is needed to find out if the system is feasible on such a scale. For the system to work, it is necessary that:

- there is sufficient knowledge about the possibilities for nature conservation on modern farms and there are sufficient facilities for farmers to obtain this information;
- there is a method for measuring the nature production results (in order to be able to reward the conservation results). If applied on a larger scale, the measurements should ideally be done by the farmers themselves to lower the costs. In that case it also should be possible to inspect the famers' returns, in order to be sure that his measurements are in accordance with the real situation. Furthermore there has to be a way to see of no fraud is being committed;
- there are fair prices for the nature products. Too low product prices will not be stimulating, too high prices will lead to high costs or - with a limited budget - to application on a small area only.

All these points will determine the feasibility of the system as well as the enthusiasm of farmers for the system, the conservation measures they will take and thus the results which may be reached. It is necessary to investigate the feasibility for every nature aspect one wishes to apply the system to. If there are severe an insuperable problems beforehand, one should abandon the idea. However, if there are only technical probelems to be expected which cannot be considered insuperable beforehand, then these should be the object of research.

FEASIBILITY FOR DITCH BANK VEGEATION

An experiment

To investigate if the nature prodcution payments scheme is feasible for ditch bank vegetation an experiment was set up to test the system. The experiment was started in 1992 and will continue to 1995. A (more or less) similar experiment for meadow birds was started in 1993. It is a cooperative experiment between the Dutch Farmers Organization, the South Hollandish Environmental Federation and the Department of Environmental Biology at Leiden. The experiment was financed by the provincial and national government bodies. The Bureau for Land Management of the Dutch Ministry of Agriculture in 1993 also started an experiment with the management of field margins, in which certain aspects of a nature production scheme are studied as well (see Melman elsewhere in this issue).

Design of the experiment

The most important questions of the experiment are:

1. How can a farmer produce a species-rich ditch bank vegetation?
2. How can the nature production easily be measured and checked?
3. Which rewards should be given per unit of product?
4. To which conservation results will the application of a system of nature production payments eventually lead?

The emphasis in the first two years of the experiment was laid on adequate measurement and inspection techniques for the nature production. This was because most doubts on the system are centred on these issues (Van Paassen *et al.*, 1991). For answering questions 3 and 4 we plan to use the results of the nature management of the farmers during the experiment. Because the ditch bank vegetation changes only slowly when the management is changed, this means that we cannot yet answer these questions. We will therefore restrict the discussion on the experiment in this paper to questions 1 and 2.

The experiment is carried out in the western peat district of the Netherlands. Ten farmers joined the experiment, with about 65 kilometers of ditch banks. They received a small, fixed sum of money for joining the experiment. No rewards according to the produced amount of nature were given yet, for one of the very aims of the experiment is to determine the right product prices. Another ten 'control' farmers without any nature management were invited to join the experiment to compare the nature results. Previous to the experiment a measurement and inspection technique was developed. These techniques were tested in practice and adapted during the course of the experiment.

Nature production

Much information is already available in the Netherlands on how to produce species-rich ditch banks combined with modern farming (Van

Strien, 1991). However, every farm is different and therefore this general information has to be translated to specific situations. On-site advice is necessary in order to meet the specific requirements and conditions and possibilities of the farm(er). In the experiment farmers and investigators together have tried and (if necessary) adapted different nature production methods. We measured the quantity of manure and fertilizer applied in the ditch banks for example and lowered this (if required) by changing the distance of the tractor to the ditch. In order to help the famers in their nature production, the participating farmers also had the possibility to buy new equipment with a 50% subsidy.

Nature production measurements

An average dairy farm in the peat district has about 10 kilometers of ditch banks, containing 40-70 plant species on average. A survey of all these plants on the whole length of dicht banks would require a lot of knowledge and a tremendous amount of time. Therefore, we developed a system which requires only a sample of the banks per field to be surveyed. In that part, the farmer had to score the presence (or absence) of about fifteen conspicious and easily recognizable plant species. These are indicator species, the presence of which indicates a species rich vegetation and the absence for a species poor vegetation (Jansen *et al.*, 1990 and in prep.). The more indicator species present, the higher the botanical value (expressed as the "nature-value index") of the vegetation figure 1). The nature-value index is a measure for the floristic richness which is based on the regional, national and world rarity of the plant species as well as the rate of decline of the species, taking into account the cover percentages (Clausman & Van Wijngaarden, 1984). Examples of the used indicator species are Ragged Robin (*Lychnis flos-cuculi*), Marsh-marigold (*Caltha palustris*) and Yellow flag (*Iris pseudacorus*).

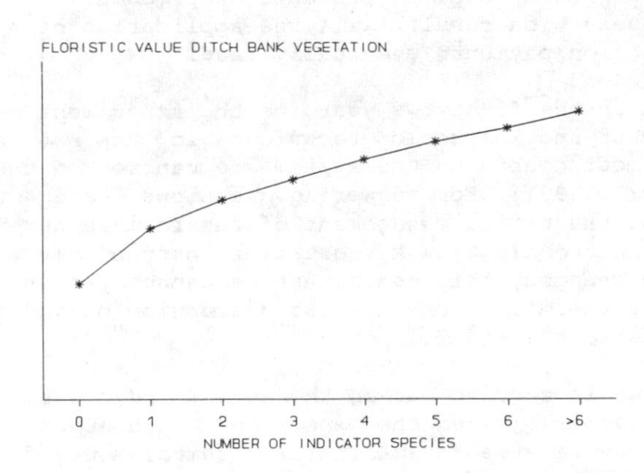

FIGURE 1. The relationship between the number of indicator plant species and the floristic value (see text) of the vegetation of the whole ditch bank.

The scores of the farmers were checked by the investigators. The total difference between the scores was 7% in the first year, which is quite low (table 1). However, farmers relatively often thought that some species were absent while in fact they were present (according to the investigators). For this reason, we added some indicator species and omitted other ones, because they appeared not to be conspicious or recognizable enough for the famers. The knowledge of the farmers was enlarged by organizing a field excursion at the start of the second year, during which we familiarized them with most of the plant species. When we look at results of the second year (table 1), we can see that the farmers missed the indicator species less often.

TABLE 1. The agreement and disagreement (italic) of the presence and absence of indicator species according to the farmers and the investigators.

Investigators → Farmers ↓	Year	Species absent	Species present	Total	Total level of disagreements
Species absent	1992	2026 (96%)	*67 (26%)*	2093	
	1993	1791 (96%)	*48 (11%)*	1839	
Species present	1992	*89 (4%)*	195 (74%)	284	
	1993	*81 (4%)*	395 (89%)	476	
Total	1992	2115 (100%)	262 (100%)	2377	*89+67/2377 = 7%*
	1993	1872 (100%)	443 (100%)	2315	*81+48/2315 = 6%*

When we evaluated the measuring method with the farmers, they all were confident about it. All participating farmers found the monitoring "nice to do". Most of them got also more interested in their "product" and wanted to know more about other plant species as well. All participants found the time they needed for the survey (2-8 hours, depending on the farm size) acceptable, and even were prepared to spend more time if necessary.

<u>Inspection method</u>

As public money will be spent in the system, a certain degree of control of the farmers' returns will be required. To check the returns of the nature production, a minimum number of "agreements" is required by which the farmers should stand. These "agreements" are meant to minimise the difference between the farmer's returns and the inspection and to carry out this inspection is fast and simple. The agreements we made were:

- the returns had to be sent to the inspection agency at the last day of the monitoring and at the same date the farmer had to phone to say that he had sent in his returns;
- every part of the ditch bank he had surveyed should be marked with two garden canes we supplied;
- the vegetation should not be mown or grazed for one week, or until the day of inspection.

In the first year some mistakes were made by the farmers: not all surveyed parts of the ditches were marked, hampering the inspection. In the second year, however, the marking of the plots was omitted only twice in 162 cases. Not one of the farmers had any problems with the agreement that the vegetation should not be mown or grazed during a week.

In consultation with the farmers we studied whether fraud, i.e. the planting of indicator species, was feasible and effective. To prevent fraud, the part of the ditch banks which should be surveyed by the farmer changes every year and is known to the farmer only shortly before the survey is due to start. Our own plantation experiments showed that the plants do not survive if the ditch bank management is not changed to the benefit of the plants at the same time. Moreover: in about 75% of the cases the fraud could be noticed easily by the inspector. The farmers themselves state that fraud should be punished severely.

CONCLUSIONS

The results of the experiment for ditch bank vegetation offer good perspectives for a system of nature production payments, both for farmers and nature conservationists. On the one hand the study of the feasability of the system does not show any major problem, both the tested measurement of the nature product (by means of indicator species) and the inspection method are adequate. Moreover, the participating farmers are enthusiastic, illustrated by remarks as "this system is absolutely perfect". By this system nature becomes a product of the farmer himself, not just something others "demand" him to protect. As a farmer remarked "The power of the system is that nature becomes your own product".

REFERENCES

Clausman, P.H.M.A. & Th.C.P. Melman (1991). Instruments for combining intensive dairy farming and nature conservation in the Netherlands. *Landscape and Urban Planning*, **20**, 205-210.

Clausman, P.H.M.A. & W. van Wijngaarden m.m.v. A.J. de Held (1984). Het vegetatie-onderzoek van de provincie Zuid-Holland; deelrapport I: verspreiding en ecologie van wilde planten in Zuid-Holland. Deel A: waarderingsparameters. Provinciale Planologische Dienst Zuid-Holland, 's-Gravenhage.

Jansen. F.A., M. Kruk & W.J. ter Keurs, 1989. Natuurproduktie-betaling in het veenweidengebied - een eenvoudige bepaling van de natuurwaarde van slootkantvegetaties. Landbouwkundig Tijdschrift **101**(10), 5-9.

Mathers, M. & A. Woods (1989). Making the most of Environmentally Sensitive Areas. *RSPB Conservation Review*, **3**, 50-55.

Meijere, J.C. de (1979). De boer als producent van natuur en landschap. *NRC Handelsblad* jan. 1979.

Paassen, A.G. van, P. Terwan & J.M. Stoop (1991). Resultaatbeloning in het agrarisch natuurbeheer. Centrum Landbouw en Milieu, Utrecht.

Reyrink, L.A.F. (1988). Bird protection in grassland in the Netherlands. In: Park, J.R. (ed.). Environmental management in agriculture; European perspectives. *Proceedings of a workshop held from 14 to 17 July at Bristol, United Kingdom*, pp. 159-169. Belhaven Press, London/New York.

Strien, A.J. van (1991). Maintenance of plant species divesity on dairy farms. PhD Thesis, University of Leiden.

Strien, A.J. van, M.M. Dorenbosch, M. Kruk & W.J. ter Keurs, 1988. Natuurproduktie-betaling - betalingen aan boeren voor geproduceerde natuur. *Landschap* **2**, 131-138.

EUROPEAN RESEARCH NETWORK ON FIELD MARGIN ECOLOGY

E. J. P. MARSHALL
Department of Agricultural Sciences, University of Bristol, Institute of Arable Crops Research, Long Ashton Research Station, Bristol, BS18 9AF, UK

W. JOENJE
Section of Weed Science, Agricultural University, Bornsesteeg 69, 6708 PD Wageningen, The Netherlands

F. BUREL
Laboratoire d'Évolution des Systèmes Naturels et Modifiés, Université de Rennes, Campus de Beaulieu I, 35042 Rennes Cedex, France

ABSTRACT

A European Research Network on Field Margin Ecology was established in 1993 as part of a Commission of the European Community-funded AIR3 project on aspects of field boundary ecotones. The objectives of the Network are to foster collaboration between researchers, to promote information flow and to produce a Code of Good Practice in regard to boundary management for use by extension services. A Newsletter, titled Field Margins, is circulated to researchers via the national coordinator for each European Union member state and via observers from other European countries.

THE RESEARCH NETWORK

Field margins constitute a network of semi-natural habitat in farm landscapes which interact with adjacent agriculture and which have a number of potential roles in more sustainable farming systems (Marshall, 1993). Considerable research effort has been made on aspects of the ecology and management of such areas across Europe. Both similar and different approaches have been taken in effecting policy on the management of margins in different countries. The Network seeks to improve coordination of research effort and information flow between interested researchers and policy makers across Europe.

The Network was established within the European Commission's AIR3 Research Programme under contract AIR3 CT-920476/477 with the Directorate General for Agriculture. The objectives for the Network are to:

- to establish links between and a register of researchers and projects on aspects of field margin ecology
- to produce a Community code of good practice for field boundary management
- to produce and circulate a Newsletter to enhance collaboration and information flow

National coordinators were appointed to the Network by the European Commission in 1993 (see on). Further additions will be made to cover most of Europe. The coordinators are gathering relevant information on researchers and projects that will be fed into AGREP, the European database on agricultural research. A bibliographic database is being created by the French national coordinator based on the EndNote program. The Newsletter, which is produced on an occasional basis and circulated by coordinators, is a simple means of passing information on projects, national schemes, for example the UK Hedgerow Incentive Scheme, and other data to a wide but select audience. Potential contributors are invited to contact their national coordinator listed below:

Belgium:
Dr Jean-Pierre Maelfait
Ministrie van de Vlaamse Gemeenschap
Institut voor Natuurbehoud
Kiewitdreef 3
B 3500 Hasselt
Belgium
Tel: +32 11 210110; FAX: +32 11 242262

Denmark:
Dr Anna Bodil Hald
DMU National Environmental Research Institute
P.O. Box 358
DK-4000 Roskilde
Denmark
Tel: +45 46 30 12 00; FAX: +45 46 30 11 14

France:
Dr Françoise Burel
Lab. d'Évolution des Systèmes Naturels et Modifiés
Université de Rennes I
Avenue du Général Leclerc
35042 RENNES Cedex
France
Tel: +33 99 28 61 45; FAX: +33 99 38 15 71

Germany:
Dr Bärbel Gerowitt
Forschungs- und Studienzentrum Landwirtschaft und Umwelt
Am Vogelsang 6
D-37075 Göttingen
Germany
Tel: +49 551 395537; FAX: +49 551 396034

Greece:
Dr Sotiris Tsiouris
Aristotelian University of Thessaloniki
Faculty of Agriculture
Lab. of Ecology & Environmental Protection
P.O. Box 251
54006 Thessaloniki
Greece
Tel: +30 31 99 2562; FAX: +30 31 998655

Italy:
Dr Maurizio G. Paoletti
Dipartimento di Biologia
Università Degli Studi di Padova
Via Trieste, 75
35121 Padova
Italy
Tel: +39 49 828 6308/6309;
FAX: +39 49 828 6300

Ireland:
Dr Gordon Purvis
Department of Environmental Resource Management
Faculty of Agriculture
University College Dublin
Belfield
Dublin 4
Ireland
Tel: +353 1 7067741; FAX: +353 1 2837328

Netherlands:
Dr Wouter Joenje
Section of Weed Science
Agricultural University
Bornsesteeg 69
6708 PD Wageningen
The Netherlands
Tel: +31 8370 82678; FAX: +31 8370 84845

UK: **Project Coordinator**
Dr Jon Marshall
Dept. Agricultural Sciences
University of Bristol
Institute of Arable Crops Research
Long Ashton Research Station
Bristol BS18 9AF
UK
Tel: +44 275 392181; FAX: +44 275 394007

Coordinators will be appointed for Spain and Portugal. Once these are in place, observers from Norway (Dr G.L.A. Fry), Sweden (Dr J. Lagerlöf), Austria (Dr B. Kromp) and Switzerland (Dr J. Lys) will be invited as observers to join the Network.

REFERENCE

Marshall, E.J.P. (1993) Exploiting semi-natural habitats as part of good agricultural practice. In: *Scientific basis for codes of good agricultural practice.* V.W.L. Jordan (Ed.) *EUR 14957*. Commission of the European Communities, Luxembourg. pp. 95-100.